Superconductors: Theoretical Aspects

Volume III

Superconductors: Theoretical Aspects

Volume III

Edited by **Jared Jones**

New York

Published by NY Research Press,
23 West, 55th Street, Suite 816,
New York, NY 10019, USA
www.nyresearchpress.com

Superconductors: Theoretical Aspects
Volume III
Edited by Jared Jones

International Standard Book Number: 978-1-63238-431-7 (Hardback)

Printed in the United States of America.

Contents

             Permissions

             List of Contributors

Preface

The purpose of the book is to provide a glimpse into the dynamics and to present opinions and studies of some of the scientists engaged in the development of new ideas in the field from very different standpoints. This book will prove useful to students and researchers owing to its high content quality.

The science of superconductivity is expanding speedily and new inventions are possible in the near future. This book is a compilation of researches accomplished by experts dealing with a variety of theoretical aspects of superconductivity such as eilenberger approach, cooper pairing, theory of ferromagnetic, etc. The authors have attempted to express their unique vision and give an insight into the examined area of research. We intend to help our readers in understanding the above stated topic in a better way.

At the end, I would like to appreciate all the efforts made by the authors in completing their chapters professionally. I express my deepest gratitude to all of them for contributing to this book by sharing their valuable works. A special thanks to my family and friends for their constant support in this journey.

Editor

Theory

Eilenberger Approach to the Vortex State in Iron Pnictide Superconductors

I. Zakharchuk, P. Belova, K. B. Traito and E. Lähderanta

Additional information is available at the end of the chapter

1. Introduction

The SC gap, which characterizes the energy cost for breaking a Cooper pair, is an important quantity when clarifying the SC mechanism. The gap size and its momentum dependence reflect the strength and anisotropy of the pairing interactions, respectively. Some experiment executed by Li *et al.* [1] in response to a suggestion by Klemm [2] tested the phase of the wave function in $Bi_2Sr_2CaCu_2O_8$ and revived the *s*-wave viewpoint [3, 4]], which, although championed by Dynes's group [4], had been out of favor even for $Bi_2Sr_2CaCu_2O_8$, although not disproven. This experiment once more created uncertainty over whether the superconducting pairs are consistent with *s*-wave or *d*-wave superconductivity (Van Harlingen [5], Ginsberg [6], Tsuei and Kirtley [7]).

The discovery of Fe-based superconductors [8] generated intensive debate on the superconducting (SC) mechanism. Motivated by high-T_c values up to 56 K [9], the possibility of unconventional superconductivity has been intensively discussed. A plausible candidate is the SC pairing mediated by antiferromagnetic (AFM) interactions. Two different approaches, based on the itinerant spin fluctuations promoted by Fermi-surface (FS) nesting [10, 11], and the local AFM exchange couplings [12], predict the so-called $s^{\pm}$-wave pairing state, in which the gap shows a *s*-wave symmetry that changes sign between different FSs. Owing to the multiorbital nature and the characteristic crystal symmetry of Fe-based superconductors, s_{++}-wave pairing without sign reversal originating from novel orbital fluctuations has also been proposed [13, 14]. The unconventional nature of the superconductivity is supported by experimental observations such as strongly FS-dependent anomalously large SC gaps [15–17] and the possible sign change in the gap function [18, 19] on moderately doped $BaFe_2As_2$, NdFeAsO and $FeTe_{1-x}Se_x$. However, a resonance like peak structure, observed by neutron scattering measurements [18], is reproduced by considering the strong correlation effect via quasiparticle damping, without the necessity of sign reversal in the SC gap [20]. Although the $s^{\pm}$-wave state is expected to be very fragile as regards impurities due to the interband scattering [21], the superconducting state is remarkably robust regarding impurities and α-particle irradiation [22].

There is growing evidence that the superconducting gap structure is not universal in the iron-based superconductors [23, 24]. In certain materials, such as optimally doped $BaKFe_2As_2$ and $BaFeCo_2As_2$, strong evidence for a fully gapped superconducting state has been observed from several low-energy quasiparticle excitation probes, including magnetic penetration depth [25, 26], and thermal conductivity measurements [27]. In contrast, significant excitations at low temperatures due to nodes in the energy gap have been detected in several Fe-pnictide superconductors. These include $LaFePO$ ($T_c = 6$ K) [28, 29], $BaFe_2AsP_2$ ($T_c = 31$ K) [30–32], and KFe_2As_2 ($T_c = 4$ K) [33, 34].

At a very early stage, it was realized that electron and hole doping can have qualitatively different effects in the pnictides [35]. Hole doping should increase the propensity to a nodeless ($s^{\pm}$) SC phase. The qualitative picture applies to both the "122" as the "1111" compounds: As the Fermi level is lowered, the M h pocket becomes more relevant and the $M \leftrightarrow X$ scattering adds to the $(\pi, 0)/(0, \pi)$ scattering from Γ to X. As such, the anisotropy-driving scattering, such as interelectron pocket scattering, becomes less relevant and yields a nodeless, less anisotropic, and more stable $s^{\pm}$ [36]. This picture is qualitatively confirmed by experiments. While thermoelectric, transport, and specific heat measurements have been performed for $K_xBa_{1-x}Fe_2As_2$ from $x = 0$ to the strongly hole-doped case $x = 1$ [37, 38], more detailed studies have previously focused on the optimally doped case $x = 0.4$ with $T_c = 37$ K, where all measurements such as penetration depth and thermal conductivity find indication for a moderately anisotropic nodeless gap [39, 40]. Similarly, angle-resolved photoemission spectroscopy (ARPES) on doped $BaFe_2As_2$ reveals a nodeless SC gap [16, 41].

The experimental findings for the SC phase in KFe_2As_2 were surprising. Thermal conductivity [33], penetration depth [34], and NMR [42] provide a clear indication of nodal SC. The critical temperature for KFe_2As_2 is ~ 3 K, an order of magnitude less than the optimally doped samples. ARPES measurements [43] show that the e pockets have nearly disappeared, while the h pockets at the folded Γ point are large and have a linear dimension close to π/a. A detailed picture of how the SC phase evolves under hole doping in $K_xBa_{1-x}Fe_2As_2$ was found and that the nodal phase observed for $x = 1$ is of the (extended) d-wave type [44]. The functional renormalization group was used to investigate how the SC form factor evolves under doping from the nodeless anisotropic $s^{\pm}$ in the moderately hole-doped regime to a d-wave in the strongly hole-doped regime, where the e pockets are assumed to be gapped out. The d-wave SC minimizes the on-pocket hole interaction energy. It was found that the critical divergence scale to be of an order of magnitude lower than for the optimally doped $s^{\pm}$ scenario, which is consistent with experimental evidence [44].

The synthesis of another iron superconductor immediately attracted much attention for several reasons [9, 45]. LiFeAs is one of the few superconductors which does not require additional charge carriers and is characterized by T_c approaching the boiling point of hydrogen. Similar to $AeFe_2As_2$ (Ae = Ba, Sr, Ca "122") and LnOFeAs ("1111") parent compounds, LiFeAs ($T_c = 18$ K) consists of nearly identical $(Fe_2As_2)^{2-}$ structural units and all three are isoelectronic, though the former do not superconduct. The band structure calculations unanimously yield the same shapes for the FS, as well as very similar densities of states, and low energy electronic dispersions [46, 47]. Moreover the calculations even find in LiFeAs an energetically favorable magnetic solution which exactly corresponds to the famous stripelike antiferromagnetic order in "122" and "1111" systems [46, 48]. The experiments, however, show a rather different situation. The structural transition peculiar to "122" and "1111" families is remarkably absent in LiFeAs and is not observed under an applied pressure of up to 20 GPa [49]. Resistivity and susceptibility as well as μ-spin rotation experiments show no evidence of magnetic transition [50, 51]. Only a weak magnetic background [51] and field induced magnetism in the doped compound have been detected [50]. What was identified was a notable absence of the Fermi surface nesting, a strong renormalization of the conduction bands by a factor of 3, a high density of states at the Fermi level caused by a van Hove singularity, and no evidence of either a static or a fluctuating order; although superconductivity with in-plane isotropic energy gaps have been found implying the s_{++} pairing state [52]. However, a gap anisotropy along the Fermi surface up to $\sim 30\%$ was observed in Ref. [53]. Thus, the type of the superconducting gap symmetry in LiFeAs is still an open question.

The aim of our paper is to apply quasiclassical Eilenberger approach to the vortex state considering $s^{\pm}$, s_{++} and $d_{x^2-y^2}$-wave pairing symmetries as presumable states for the different levels of impurity scattering rates Γ^*, to calculate the cutoff parameter ξ_h [54, 55] and to compare results with experimental data for iron pnictides. As described in Ref. [56], ξ_h is important for the description of the muon spin rotation (μSR) experiments and can be directly measured.

The London model used for the analysis of the experimental data does not account for the spatial dependence of the superconducting order parameter and it fails down at distances of the order of coherence length from the vortex core center, i.e., $B(r)$ logarithmically diverges as $r \to 0$. To correct this, the **G** sum in the expression for the vortex lattice free energy can be truncated by multiplying each term by a cutoff function $F(G)$. Here, **G** is a reciprocal vortex lattice vector. In this method the sum is cut off at high $G_{max} \approx 2\pi/\xi_h$, where ξ_h is the cutoff parameter. The characteristic length ξ_h accommodates a number of inherent uncertainties of the London approach; the question was discussed originally by de Gennes group [57] and discussed in some detail in Ref. [58]. It is important to stress that the appropriate form of $F(G)$ depends on the precise spatial dependence of the order parameter in the the vortex core region, and this, in general, depends on the temperature and the magnetic field.

A smooth Gaussian cutoff factor $F(G) = exp(-\alpha G^2 \xi^2)$ was phenomenologically suggested. Here, ξ is the Gizburg-Landau coherence length. If there is no dependence of the superconducting coherence length on temperature and magnetic field, then changes in the spatial dependence of the order parameter around a vortex correspond to changes in α. By solving the Ginzburg-Landau (GL) equations, Brandt determined that $\alpha = 1/2$ at fields near B_{c2} [59], and arbitrarily determined it to be $\alpha \approx 2$ at fields immediately above B_{c1} [60]. For an isolated vortex in an isotropic extreme (the GL parameter $\kappa_{GL} \gg 1$) s-wave superconductor, α was obtained by numerical calculation of GL equations. It was found that α decreases smoothly from $\alpha = 1$ at B_{c1} to $\alpha \approx 0.2$ at B_{c2} [61]. The analytical GL expression was obtained by [62] for isotropic superconductors at low inductions $B \ll B_{c2}$. Using a Lorentzian trial function for the order parameter of an isolated vortex, Clem found for large $\kappa_{GL} \gg 1$ that $F(G)$ is proportional to the modified Bessel function. In Ref. [63], the Clem model [62] was extended to larger magnetic fields up to B_{c2} through the linear superposition of the field profiles of individual vortices. In this model, the Clem trial function [62] is multiplied by a second variational parameter f_∞ to account for the suppression of the order parameter due to the overlapping of vortex cores. This model gave the method for calculating the magnetization of type-II superconductors in the full range $B_{c1} < B < B_{c2}$. Their analytical formula is in a good agreement with the well-known Abrikosov high-field result and considerably corrects the results obtained with an exponential cutoff function at low fields [64]. This approximation was widely used for the analysis of the experimental data on magnetization of type-II superconductors (see references 27-29 in Ref. [65]). The improved approximate Ginzburg-Landau solution for the regular flux-line lattice using circular cell method was obtained in Ref. [65]. This solution gives better correlation with the numerical solution of GL equations.

The Ginzburg-Landau theory, strictly speaking, is only valid near T_c but it is often used in the whole temperature range taking the cutoff parameter ξ_h and penetration depth λ as a fitting parameters. Recently, an effective London model with the effective cutoff parameter $\xi_h(B)$ as a fitting parameter was obtained for clean [54] and dirty [55] superconductors, using self-consistent solution of quasiclassical nonlinear Eilenberger equations. In this approach, λ is not a fitting parameter but calculated from the microscopical theory of the Meissner state. As was shown in Ref. [66], the reduction of the amount of the fitting parameters to one, considerably simplifies the fitting procedure. In this method, the cutoff parameter obtained from the Ginzburg-Landau model was extended over the whole field and temperature ranges. In this case, the effects of the bound states in the vortex cores lead to the Kramer-Pesch effect [67], i.e. delocalization between the vortices [68, 69], nonlocal electrodynamic [58] and nonlinear effects [70] being self-consistently included.

Following the microscopical Eilenberger theory, ξ_h can be found from the fitting of the calculated magnetic field distribution $h_E(\mathbf{r})$ to the Eilenberger - Hao-Clem (EHC) field distribution $h_{EHC}(\mathbf{r})$ [54, 55]

$$h_{EHC}(\mathbf{r}) = \frac{\Phi_0}{S} \sum_{\mathbf{G}} \frac{F(G)e^{i\mathbf{G}\mathbf{r}}}{1 + \lambda^2 G^2},\tag{1}$$

where

$$F(\mathbf{G}) = uK_1(u),\tag{2}$$

where $K_1(u)$ is modified Bessel function, $u = \xi_h G$ and S is the area of the vortex lattice unit cell. It is important to note that ξ_h in Eq. (1) is obtained from solving the Eilenberger equations and does not coincide with the variational parameter ξ_v of the analytical Ginzburg-Landau (AGL) model.

In **chapter 2** and **3** we solve the Eilenberger equations for $s^{\pm}$, s_{++} and $d_{x^2-y^2}$-wave pairing symmetries, fit the solution to Eq. (1) and find the cutoff parameter ξ_h. In this approach all nonlinear and nonlocal effects connected with vortex core and extended quasiclassical states are described by one effective cutoff parameter ξ_h. The nonlocal generalized London equation with separated quasiclassical states was also developed as regards the description of the mixed state in high-T_c superconductors such as YBa$_2$Cu$_3$O$_{7-\delta}$ compounds (the Amin-Franz-Affleck (AFA) model) [70, 71]. In this case, fourfold anisotropy arises from d-wave pairing. This theory was applied to the investigation of the flux line lattice (FLL) structures [72] and effective penetration depth measured by μSR experiments [73]. This approach will be considered in **chapter 4**.

2. The cutoff parameter for the field distribution in the mixed states of $s^{\pm}$- and s_{++}-wave pairing symmetries

In this chapter, we consider the model of the iron pnictides, where the Fermi surface is approximated by two cylindrical pockets centered at Γ (hole) and M (electron) points of the

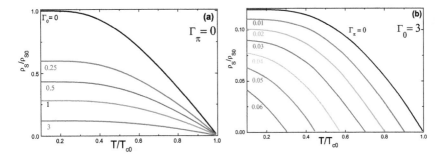

Figure 1. (Color online) The temperature dependence of superfluid density $\rho_S(T)/\rho_{S0}$ at (a) interband scattering rate $\Gamma_\pi = 0$ with different values of intraband scattering Γ_0 and (b) intraband scattering rate $\Gamma_0 = 3$ with different values of interband scattering Γ_π.

Fermi surface, i.e. a two dimensional limit of the five-band model [74]. In Eq. (1) $\lambda(T)$ is the penetration depth in the Meissner state. In this model $\lambda(T)$ is given as

$$\frac{\lambda_{L0}^2}{\lambda^2(T)} = 2\pi T \sum_{\omega_n > 0} \frac{\bar{\Delta}_n^2}{\eta_n(\bar{\Delta}_n^2 + \omega_n^2)^{3/2}}, \tag{3}$$

where $\lambda_{L0} = (c^2/4\pi e^2 v_F^2 N_0)^{1/2}$ is the London penetration depth at $T = 0$ including the Fermi velocity v_F and the density of states N_0 at the Fermi surface and $\eta_n = 1 + 2\pi(\Gamma_0 + \Gamma_\pi)/(\sqrt{\bar{\Delta}_n^2 + \omega_n^2})$. Here, $\Gamma_0 = \pi n_i N_F |u_0|^2$ and $\Gamma_\pi = \pi n_i N_F |u_\pi|^2$ are the intra- and interband impurity scattering rates, respectively ($u_{0,\pi}$ are impurity scattering amplitudes with correspondingly small, or close to $\pi = (\pi, \pi)$, momentum transfer). In this work, we investigate the field distribution in the vortex lattice by systematically changing the impurity concentration in the Born approximation, and analyzing the field dependence of the cutoff parameter. In particular, we consider two limits: small $\Gamma^* \ll 1$ (referred to as the "stoichiometric" case) and relatively high $\Gamma^* \geq 1$ ("nonstoichiometric" case). Here, Γ^* is measured in the units of $2\pi T_{c0}$. We consider Γ^* as intraband scattering Γ_0 with constant interband scattering $\Gamma_\pi = 0$.

In Eq. (3), $\bar{\Delta}_n = \Delta(T) - 4\pi\Gamma_\pi \bar{\Delta}_n/\sqrt{\bar{\Delta}_n^2 + \omega_n^2}$ for the $s^\pm$ pairing and $\bar{\Delta}_n = \Delta(T)$ for the s_{++} pairing symmetry. The order parameter $\Delta(T)$ in Meissner state is determined by the self-consistent equation

$$\Delta(T) = 2\pi T \sum_{0 < \omega_n < \omega_c} \frac{V^{SC}\bar{\Delta}_n}{\sqrt{\bar{\Delta}_n^2 + \omega_n^2}}. \tag{4}$$

Experimentally, $\lambda(T)$ can be obtained by radio-frequency measurements [75] and magnetization measurements of nanoparticles [76]. Fig. 1 shows the calculated temperature dependence of the superfluid density $\rho_S(T)/\rho_{S0} = \lambda_{L0}^2/\lambda^2(T)$, with different values of impurity scattering Γ for $s^\pm$-wave pairing symmetry. With the Riccati transformation of the Eilenberger equations, quasiclassical Green functions f and g can be parameterized via functions a and b [77]

$$\bar{f} = \frac{2a}{1 + ab}, \quad f^\dagger = \frac{2b}{1 + ab}, \quad g = \frac{1 - ab}{1 + ab}, \tag{5}$$

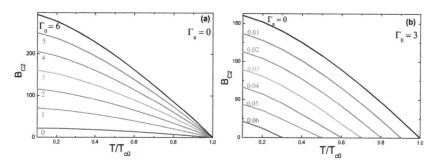

Figure 2. (Color online) (a) The temperature dependence of the upper critical field B_{c2} at interband scattering $\Gamma_\pi = 0$ with different values of intraband scattering values Γ_0. (b) The calculated temperature dependence of B_{c2} at intraband scattering rate $\Gamma_0 = 3$ with different values of interband scattering Γ_π.

satisfying the nonlinear Riccati equations. In Born approximation for impurity scattering we have

$$\mathbf{u} \cdot \nabla a = -a \left[2(\omega_n + G) + i\mathbf{u} \cdot \mathbf{A}_s \right] + (\Delta + F) - a^2 (\Delta^* + F^*), \tag{6}$$

$$\mathbf{u} \cdot \nabla b = b \left[2(\omega_n + G) + i\mathbf{u} \cdot \mathbf{A}_s \right] - (\Delta^* + F^*) + b^2 (\Delta + F), \tag{7}$$

where $\omega_n = \pi T(2n + 1)$, $G = 2\pi \langle g \rangle (\Gamma_0 + \Gamma_\pi) \equiv 2\pi \langle g \rangle \Gamma^*$, $F = 2\pi \langle f \rangle (\Gamma_0 - \Gamma_\pi)$ for $s^\pm$ pairing symmetry and $F = 2\pi \langle f \rangle \Gamma^*$ for the s_{++} pairing symmetry. Here, $\mathbf{u}$ is a unit vector of the Fermi velocity. In the new gauge vector-potential $\mathbf{A}_s = \mathbf{A} - \nabla \phi$ is proportional to the superfluid velocity. It diverges as $1/r$ at the vortex center (index s is put to denote its singular nature). The FLL creates the anisotropy of the electron spectrum. Therefore, the impurity renormalization correction in Eqs. (6) and (7), averaged over the Fermi surface, can be reduced to averages over the polar angle θ, i.e. $\langle \ldots \rangle = (1/2\pi) \int \ldots d\theta$.

To take into account the influence of screening the vector potential $\mathbf{A}(\mathbf{r})$ in Eqs. (6) and (7) is obtained from the equation

$$\nabla \times \nabla \times \mathbf{A_E} = \frac{4}{\kappa^2} \mathbf{J}, \tag{8}$$

where the supercurrent $\mathbf{J}(\mathbf{r})$ is given in terms of $g(\omega_n, \theta, \mathbf{r})$ by

$$\mathbf{J}(\mathbf{r}) = 2\pi T \sum_{\omega_n > 0} \int_0^{2\pi} \frac{d\theta}{2\pi} \frac{\hat{\mathbf{k}}}{i} g(\omega_n, \theta, \mathbf{r}). \tag{9}$$

Here $\mathbf{A}$ and $\mathbf{J}$ are measured in units of $\Phi_0/2\pi\xi_0$ and $2ev_F N_0 T_c$, respectively. The spatial variation of the internal field $h(\mathbf{r})$ is determined through

$$\nabla \times \mathbf{A} = \mathbf{h}(\mathbf{r}), \tag{10}$$

where $\mathbf{h}$ is measured in units of $\Phi_0/2\pi\xi_0^2$.

The self-consistent condition for the pairing potential $\Delta(\mathbf{r})$ in the vortex state is given by

$$\Delta(\mathbf{r}) = V^{SC} 2\pi T \sum_{\omega_n > 0}^{\omega_c} \int_0^{2\pi} \frac{d\theta}{2\pi} f(\omega_n, \theta, \mathbf{r}), \tag{11}$$

where V^{SC} is the coupling constant and ω_c is the ultraviolet cutoff determining T_{c0} [55]. Consistently throughout our paper energy, temperature, and length are measured in units of T_{c0} and the coherence length $\xi_0 = v_F/T_{c0}$, where v_F is the Fermi velocity. The magnetic field $\mathbf{h}$ is given in units of $\Phi_0/2\pi\xi_0^2$. The impurity scattering rates are in units of $2\pi T_{c0}$. In calculations the ratio $\kappa = \lambda_{L0}/\xi_0 = 10$ is used. It corresponds to $\kappa_{GL} = 43.3$ [77].

To obtain the quasiclassical Green function, the Riccati equations [Eq. (6, 7)] are solved by the Fast Fourier Transform (FFT) method for triangular FLL [55]. This method is reasonable for the dense FLL, discussed in this paper. In the high field the pinning effects are weak and they are not considered in our paper. To study the high field regime we needed to calculate the upper critical field $B_{c2}(T)$. This was found from using the similarity of the considered model to the model of spin-flip superconductors from the equations [78]

$$\ln\left(\frac{T_{c0}}{T}\right) = 2\pi T \sum_{n \geq 0} [\omega_n^{-1} - 2D_1(\omega_n, B_{c2})], \tag{12}$$

where

$$D_1(\omega_n, B_{c2}) = J(\omega_n, B_{c2}) \times [1 - 2(\Gamma_0 - \Gamma_\pi)J(\omega_n, B_{c2})]^{-1}, \tag{13}$$

$$J(\omega_n, B_{c2}) = \left(\frac{4}{\pi B_{c2}}\right)^{1/2} \times \int_0^\infty dy \exp(-y) \arctan\left[\frac{(B_{c2}y)^{1/2}}{\alpha}\right], \tag{14}$$

where $\alpha = 2(\omega_n + \Gamma_0 + \Gamma_\pi)$.

Fig. 2 shows $B_{c2}(T)$ dependences at (a) $\Gamma_\pi = 0$, $\Gamma_0 = 0, 1, 2, 3, 4, 5, 6$ and (b) $\Gamma_0 = 3$, $\Gamma_\pi = 0.01, 0.02, 0.03, 0.04, 0.05, 0.06$ calculated from Eqs. (12-14). In Fig. 2 the different influence of the intraband and interband scattering on $B_{c2}(T)$ dependence can be seen. The $B_{c2}(T)$ curve increases with Γ_0 (ξ_{c2} decreases with Γ_0), but Γ_π results in decreasing $B_{c2}(T)$ (increasing of ξ_{c2}).

Fig. 3 (a) shows magnetic field dependence $\xi_h(B)$ in reduced units at $T/T_{c0} = 0.5$ for the $s^\pm$ pairing with $\Gamma_0 = 3$, $\Gamma_\pi = 0.02$ and $\Gamma_0 = 0.5$, $\Gamma_\pi = 0.03$ and "clean" case (solid lines) and for the s_{++} pairing with $\Gamma^* = 0.5$ and $\Gamma^* = 3$ (dotted lines). The dashed line shows the analytical solution of the AGL theory [63]

$$\xi_v = \xi_{c2}(\sqrt{2} - \frac{0.75}{\kappa_{GL}})(1 + b^4)^{1/2}[1 - 2b(1 - b)^2]^{1/2}. \tag{15}$$

This dependence with ξ_{c2} as a fitting parameter is often used for the description of the experimental μSR results [56, 79]. As can be seen from Fig. 3 (a), the magnetic field dependence of ξ_h/ξ_{c2} is nonuniversal because it depends not only on B/B_{c2} (as in the AGL theory, dashed line in Fig. 3 (a)), but also on interband and intraband impurity scattering parameters. In the cases where $\Gamma_0 = \Gamma_\pi = 0$, the results are the same for $s^\pm$ and s_{++} pairing symmetries. We indicated that this curve is "clean" one. In this figure, the case $\Gamma_0 \gg \Gamma_\pi$ is considered

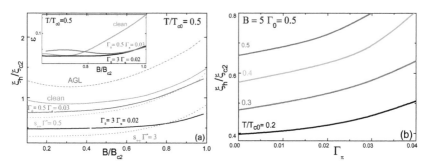

Figure 3. (Color online) (a) The magnetic field dependence of ξ_h/ξ_{c2} for superconductors with impurity scattering. The solid lines represent our solution of Eilenberger equations at $T/T_{c0} = 0.5$ for "clean" case ($\Gamma_0 = \Gamma_\pi = 0$) and $s^\pm$ model ($\Gamma_0 = 0.5$, $\Gamma_\pi = 0.03$ and $\Gamma_0 = 3$, $\Gamma_\pi = 0.02$). The dotted lines show result for s_{++} model ($\Gamma^* = 0.5$ and $\Gamma^* = 3$). Dashed line demonstrates the result of the AGL theory for ξ_v from Eq. 15. The inset shows the magnetic field dependence of mean square deviation of the h_{EHC} distribution from the Eilenberger distribution normalized by the variance of the Eilenberger distribution, ε, for $T/T_{c0} = 0.5$ at $\Gamma_0 = \Gamma_\pi = 0$ ("clean"); $\Gamma_0 = 3$, $\Gamma_\pi = 0.02$ and $\Gamma_0 = 0.5$, $\Gamma_\pi = 0.03$. (b) The interband scattering Γ_π dependence of ξ_h/ξ_{c2} at different temperatures T/T_{c0} (intraband scattering $\Gamma_0 = 0.5$ and $B = 5$) for the $s^\pm$ pairing.

and the value of ξ_h is reduced considerably in comparison with the clean case. One can compare the observed behavior with that in s_{++} pairing model. In s_{++} pairing symmetry the intraband and interband scattering rates act in a similar way and ξ_h/ξ_{c2} decreases always with impurity scattering. In contrast, in $s^\pm$ model $\xi_h/\xi_{c2}(B/B_{c2})$ dependences show different forms of behavior with Γ_π. Here, ξ_h/ξ_{c2} increases with Γ_π at $B/B_{c2} < 0.8$ and decreases at higher fields, i.e. the curves become more flattened. A crossing point appears in the dependences $\xi_h/\xi_{c2}(B/B_{c2})$ for $s^\pm$ and s_{++} pairing. We also calculated the magnetic field dependence of mean square deviation of h_{EHC} distribution of the magnetic field from the Eilenberger distribution normalized by the variance of the Eilenberger distribution $\varepsilon = \sqrt{\overline{(h_E - h_{EHC})^2}/\overline{(h_E - B)^2}}$, where $\overline{\cdots}$ is the average over a unit vortex cell. The inset to Fig. 3 (a) demonstrates $\varepsilon(B)$ dependence for $T/T_{c0} = 0.5$ at $\Gamma_0 = 0$, $\Gamma_\pi = 0$; $\Gamma_0 = 3$, $\Gamma_\pi = 0.02$ and $\Gamma_0 = 0.5$, $\Gamma_\pi = 0.03$. From this figure, it can be seen that the accuracy of effective London model is deteriorating as the magnetic field increases; however, in superconductors with impurity scattering the accuracy is below 6% even when it is close to the second critical field (the inset to Fig. 3 (a)).

In Fig. 3 (b), the interband scattering Γ_π dependences of ξ_h are presented in low fields for the $s^\pm$ pairing at different temperatures T. As can be seen ξ_h/ξ_{c2} increases with the interband scattering rate Γ_π. Strong decreasing of ξ_h/ξ_{c2} with a decrease in the temperature can be explained by the Kramer-Pesch effect [67]. It should be noted that the normalization constant ξ_{c2} increases with Γ_π because Γ_π suppress T_c similar to superconductors with spin-flip scattering (violation of the Anderson theorem). Thus, the rising ξ_h/ξ_{c2} implies more strong growth of ξ_h than ξ_{c2} (from GL theory one can expect $\xi_h/\xi_{c2} = Const$). Qualitatively, it can be explained by the strong temperature dependence of $\xi_h(B, T/T_c)$, which is connected to the Kramer-Pesch effect [67]. Increasing Γ_π results in suppression of T_c, i.e. effective increasing of T and $\xi_h(T/T_c)$. $\xi_{c2}(T/T_c)$ has not such a strong T_c dependence, thus leading to the increasing of the ratio ξ_h/ξ_{c2} with Γ_π.

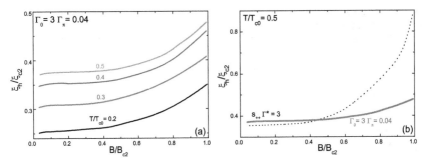

Figure 4. (Color online) (a) The magnetic field dependence of cutoff parameter ξ_h/ξ_{c2} at different temperatures ($T/T_{c0} = 0.2, 0.3, 0.4, 0.5$) for $s^\pm$ pairing with $\Gamma_0 = 3$, $\Gamma_\pi = 0.04$. (b) The magnetic field dependence of ξ_h/ξ_{c2} for $s^\pm$ model ($\Gamma_0 = 3$, $\Gamma_\pi = 0.04$, solid line) and s_{++} model ($\Gamma^* = 3$, dotted line) at $T/T_{c0} = 0.5$.

The superfluid density in iron pnictides often shows a power law dependence with theexponent, which is approximately equal to two at low temperatures [39, 74]. This law was explained by $s^\pm$ model with parameters $\Gamma_0 = 3$ and $\Gamma_\pi = 0.04 - 0.06$. Fig. 4 (a) shows $\xi_h/\xi_{c2}(B/B_{c2})$ dependence with $\Gamma_0 = 3$ and $\Gamma_\pi = 0.04$ at different temperatures. All curves demonstrate rising behavior with values much less than one in the whole field range, i.e. they are under the AGL curve of ξ_v. The small value of the cutoff parameter was observed in iron pnictide $BaFe_{1.82}Co_{0.18}As$, where $\xi_h/\xi_{c2}(\sim 0.4) < 1$ [80]. Fig. 4 (b) shows $\xi_h/\xi_{c2}(B/B_{c2})$ for $\Gamma_0 = 3$, $\Gamma_\pi = 0.04$ ($s^\pm$ pairing) and $\Gamma^* = 3$ (s_{++} pairing). It can be seen from the graph that ξ_h/ξ_{c2} is strongly suppressed in $s^\pm$ pairing with comparison to the s_{++} pairing. This can be explained by the fact that in superconductors, without interband pair breaking, the increase in high field is connected with the field-dependent pair breaking, as the upper critical field is approached. The physics of unconventional superconductors depends on impurity pair breaking and introducing characteristic field B^* in the field dependence by the substitution $B/B_{c2} \rightarrow (B + B^*(\Gamma_\pi))/B_{c2}(\Gamma_\pi)$. The crossing point between $s^\pm$ and s_{++} curves depends on Γ_π and it shifts to the lower field in comparison with case $\Gamma_\pi = 0.02$ shown in Fig. 3 (a).

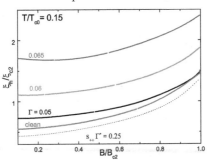

Figure 5. (Color online) The magnetic field dependence of the cutoff parameter at $T/T_{c0} = 0.15$ with the same values of intraband Γ_0 and interband Γ_π scattering rate Γ ($\Gamma = 0$ for "clean" case and $\Gamma = 0.05, 0.06, 0.065$ for the $s^\pm$ pairing). Dotted line shows result for s_{++} model ($\Gamma^* = 0.25$).

The case of weak intraband scattering was also studied. This case can be realized in stoichiometrical pnictides such as LiFeAs. Fig. 5 presents the magnetic field dependence of ξ_h/ξ_{c2} with scattering parameters $\Gamma_0 = \Gamma_\pi = \Gamma$ equal to 0, 0.05, 0.06 and 0.065 at $T/T_{c0} = 0.15$. The dotted line shows the result for s_{++} model ($\Gamma^* = 0.25$). The $\xi_h(B)$ dependence shifts upward from the "clean" curve and has a higher values in $s^\pm$ model. In contrast, the ξ_h/ξ_{c2} curve shifts downward with impurity scattering in s_{++} model. The high values of ξ_h observed in μSR measurements in LiFeAs [81] supports the $s^\pm$ pairing.

3. The cutoff parameter in the mixed state of $d_{x^2-y^2}$-wave pairing symmetry

A nontrivial orbital structure of the order parameter, in particular the presence of the gap nodes, leads to an effect in which the disorder is much richer in $d_{x^2-y^2}$-wave superconductors than in conventional materials. For instance, in contrast to the s-wave case, the Anderson theorem does not work, and nonmagnetic impurities exhibit a strong pair-breaking effect. In addition, a finite concentration of disorder produces a nonzero density of quasiparticle states at zero energy, which results in a considerable modification of the thermodynamic and transport properties at low temperatures. For a pure superconductor in a d-wave-like state at temperatures T well below the critical temperature T_c, the deviation $\Delta\lambda$ of the penetration depth from its zero-temperature value $\lambda(0)$ is proportional to T. When the concentration n_i of strongly scattering impurities is nonzero, $\Delta\lambda \propto T^n$, where $n = 2$ for $T < T^* \ll T_c$ and $n = 1$ for $T^* < T \ll T_c$ [24]. Unlike s-wave superconductor, impurity scattering suppresses both the transition temperature T_c and the upper critical field $H_{c2}(T)$ [82].

The presence of the nodes in the superconducting gap can also result in unusual properties of the vortex state in $d_{x^2-y^2}$-wave superconductors. At intermediate fields $H_{c1} < H \ll H_{c2}$, properties of the flux lattice are determined primarily by the superfluid response of the condensate, i.e., by the relation between the supercurrent $\vec{j}$ and the superfluid velocity $\vec{v}_s$. In conventional isotropic strong type-II superconductors, this relation is to a good approximation that of simple proportionality,

$$\vec{j} = -e\rho_s\vec{v}_s, \tag{16}$$

where ρ_s is a superfluid density. More generally, however, this relation can be both nonlocal and nonlinear. The concept of nonlocal response dates is a return to the ideas of Pippard [83] and is related to the fact that the current response must be averaged over the finite size of the Cooper pair given by the coherence length ξ_0. In strongly type-II materials the magnetic field varies on a length scale given by the London penetration depth λ_0, which is much larger than ξ_0 and, therefore, nonlocality is typically unimportant unless there exist strong anisotropies in the electronic band structure [84]. Nonlinear corrections arise from the change of quasiparticle population due to superflow which, to the leading order, modifies the excitation spectrum by a quasiclassical Doppler shift [85]

$$\varepsilon_k = E_k + \vec{v}_f\vec{v}_s, \tag{17}$$

where $E_k = \sqrt{\epsilon_k^2 + \Delta_k^2}$ is the BCS energy. Once again, in clean, fully gapped conventional superconductors, this effect is typically negligible except when the current approaches the pair breaking value. In the mixed state, this happens only in the close vicinity of the vortex cores that occupy a small fraction of the total sample volume at fields well below H_{c2}. The situation changes dramatically when the order parameter has nodes, such as in $d_{x^2-y^2}$ superconductors.

Nonlocal corrections to Eq. (16) become important for the response of electrons with momenta on the Fermi surface close to the gap nodes, even in strongly type-II materials. This can be understood by realizing that the coherence length, being inversely proportional to the gap [85], becomes very large close to the node and formally diverges at the nodal point. Thus, quite generally, there exists a locus of points on the Fermi surface where $\xi \gg \lambda_0$ and the response becomes highly nonlocal. This effect was first discussed in Refs. [72, 86] in the mixed state. Similarly, the nonlinear corrections become important in a d-wave superconductors. Eq. (17) indicates that finite areas of gapless excitations appear near the node for arbitrarily small v_s.

Low temperature physics of the vortex state in s-wave superconductors is connected with the nature of the current-carrying quantum states of the quasiparticles in the vortex core (formed due to particle-hole coherence and Andreev reflection [87]). The current distribution can be decomposed in terms of bound states and extended states contributions [88]. Close to the vortex core, the current density arises mainly from the occupation of the bound states. The effect of extended states becomes important only at distances larger than the coherence length. The bound states and the extended states contributions to the current density have opposite signs. The current density originating from the bound states is paramagnetic, whereas extended states contribute a diamagnetic term. At distances larger than the penetration depth, the paramagnetic and diamagnetic parts essentially cancel out each other, resulting in exponential decay of the total current density. The vortex core structure in the d-wave superconductors can be more complicated because there are important contributions coming from core states, which extend far from the vortex core into the nodal directions and significantly effect the density of states at low energy [89]. The possibility of the bound states forming in the vortex core of d-wave superconductors was widely discussed in terms of the Bogoliubov-de Gennes equation. For example, Franz and Tešanović claimed that there should be no bound states [90]. However, a considerable number of bound states were found in Ref.[91] which were localized around the vortex core. Extended states, which are rather uniform, for $|E| < \Delta$ where E is the quasiparticle energy and Δ is the asymptotic value of the order parameter, were also found far away from the vortex. In the problem of the bound states, the conservation of the angular momentum around the vortex is important. In spite of the strict conservation of the angular momentum it is broken due to the fourfold symmetry of $\Delta(k)$, however, the angular momentum is still conserved by modulo 4, and this is adequate to guarantee the presence of bound states.

Taking into account all these effects, the applicability of EHC theory regarding the description of the vortex state in $d_{x^2-y^2}$-wave superconductors is not evident *apriori*. In this chapter, we numerically solve the quasiclassical Eilenberger equations for the mixed state of a $d_{x^2-y^2}$-wave superconductor for the pairing potential $\Delta(\theta, \mathbf{r}) = \Delta(\mathbf{r}) \cos(2\theta)$, where θ is the angle between the $\mathbf{k}$ vector and the a axis (or x axis). We check the applicability of Eq. (1) and find the cutoff parameter ξ_h. The anisotropic extension of Eq. (1) to Amin-Franz-Affleck will be discussed in chapter 4.

To consider the mixed state of a d-wave superconductor we take the center of the vortex as the origin and assume that the Fermi surface is isotropic and cylindrical. The Riccatti equations for $d_{x^2-y^2}$-wave superconductivity are [92]

$$\mathbf{u} \cdot \nabla a = -a \left[2(\omega_n + G) + i\mathbf{u} \cdot \mathbf{A}_s \right] + \Delta - a^2 \Delta^*, \tag{18}$$

$$\mathbf{u} \cdot \nabla b = b \left[2(\omega_n + G) + i\mathbf{u} \cdot \mathbf{A}_s \right] - \Delta^* + b^2 \Delta, \tag{19}$$

where $G = 2\pi \langle g \rangle \Gamma$ with d-wave pairing potential $\Delta(\mathbf{r})$

$$\Delta(\theta, \mathbf{r}) = V_{d_{x^2-y^2}}^{SC} 2\pi T \cos(2\theta) \sum_{\omega_n > 0}^{\omega_c} \int_0^{2\pi} \frac{d\bar{\theta}}{2\pi} f(\omega_n, \bar{\theta}, \mathbf{r}) \cos(2\bar{\theta}), \tag{20}$$

where $V_{d_{x^2-y^2}}^{SC}$ is a coupling constant in the $d_{x^2-y^2}$ pairing channel. The obtained solution is fitted to Eq. (1) giving the value of cutoff parameter ξ_h for $d_{x^2-y^2}$-wave pairing symmetry.

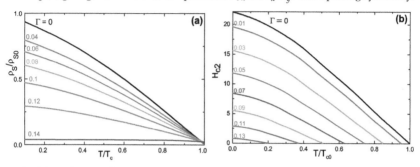

Figure 6. (Color online) (a) The temperature dependence of superfluid density $\rho_S(T)/\rho_{S0}$ with different values of impurity scattering Γ. (b) The temperature dependence of the upper critical field B_{c2} with different values of impurity scattering Γ.

In $d_{x^2-y^2}$-wave superconductor $\lambda(T)$ in Eq. (1) is given as [85]

$$\frac{\lambda_{L0}^2}{\lambda^2(T)} = 2\pi T \oint \frac{d\theta}{2\pi} \sum_{\omega_n > 0} \frac{|\tilde{\Delta}(\theta)|^2}{(\tilde{\omega}_n^2 + |\tilde{\Delta}(\theta)|^2)^{3/2}}, \tag{21}$$

where

$$\tilde{\omega}_n = \omega_n + \Gamma \langle \frac{\tilde{\omega}_n}{\sqrt{\tilde{\omega}_n^2 + |\tilde{\Delta}(\vec{p}_f'; \omega_n)|^2}} \rangle_{\vec{p}_f'}, \tag{22}$$

$$\tilde{\Delta}(\vec{p}_f; \omega_n) = \Delta(\vec{p}_f) + \Gamma \langle \frac{\tilde{\Delta}(\vec{p}_f'; \omega_n)}{\sqrt{\tilde{\omega}_n^2 + |\tilde{\Delta}(\vec{p}_f'; \omega_n)|^2}} \rangle_{\vec{p}_f'}, \tag{23}$$

$$\Delta(\vec{p}_f) = \int d\vec{p}_f' V(\vec{p}_f, \vec{p}_f') \pi T \sum_{\omega_n}^{|\omega_n| < \omega_c} \frac{\tilde{\Delta}(\vec{p}_f')}{\sqrt{\tilde{\omega}_n^2 + |\tilde{\Delta}(\vec{p}_f')|^2}}. \tag{24}$$

Because of the symmetry of $d_{x^2-y^2}$-wave pairing the impurity induced corrections for the pairing potential in Eq. (23) are zero and $\tilde{\Delta} = \Delta$. This is different from the $s^{\pm}$- and s_{++} cases, where the corrections are not zero. Fig. 6 (a) shows the calculated temperature dependence of the superfluid density $\rho_S(T)/\rho_{S0} = \lambda_{L0}^2/\lambda^2(T)$ with different values of impurity scattering Γ for $d_{x^2-y^2}$-wave pairing symmetry.

To study high the field regime we need to calculate the upper critical field $B_{c2}(T)$. For $d_{x^2-y^2}$-wave $B_{c2}(T)$ is given as [82]

$$\ln(\frac{T}{T_c}) - \Psi(\frac{1}{2} + \frac{v}{2t_c}) + \Psi(\frac{1}{2} + \frac{v}{2t}) = \frac{3}{2} \int_0^\infty \frac{du}{shu} \int_0^1 dz(1-z^2)[e^{-x}(1-2xc)^{-1}]e^{-\frac{v}{t}u}, \tag{25}$$

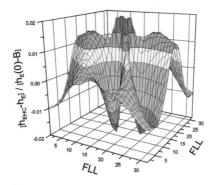

Figure 7. (Color online) Normalized differences between the fields calculated with the London model and the Eilenberger equation for $d_{x^2-y^2}$-wave pairing with $\Gamma = 0.03$, $B/B_{c2} = 0.1$ and $T/T_{c0} = 0.3$.

$$c[\ln(\frac{T}{T_c}) - \Psi(\frac{1}{2} + \frac{v}{2t_c}) + \Psi(\frac{1}{2} + \frac{v}{2t})] =$$

$$= \frac{3}{2}\int_0^\infty \frac{du}{shu}x\int_0^1 dz|(1-z^2)[e^{-x}(-x+c(1-4x+2x^2))-c]e^{-\frac{v}{t}u}, \qquad (26)$$

where $v = 2\Gamma$, $t = T/T_{c0}$, $t_c = T_c/T_{c0}$ and $x = \rho u^2(1-z^2)$, $\rho = B/(4\pi t)^2$. Fig. 6 (b) depicts the temperature dependence of the upper critical field B_{c2} with different values of impurity scattering Γ. Figs. 6 (a) and (b) are similar to those in $s^\pm$-wave superconductors. T_c is suppressed by impurity scattering resulting in the same expressions for $s^\pm$ and d-wave superconductors with replacing $\Gamma_\pi \to \Gamma/2$.

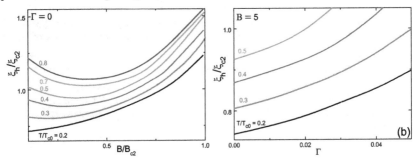

Figure 8. (Color online) (a) The magnetic field dependence of the cutoff parameter ξ_h/ξ_{c2} with different temperatures ($T/T_{c0} = 0.2, 0.3, 0.4, 0.5, 0.7, 0.8$) for $d_{x^2-y^2}$ pairing with $\Gamma = 0$. (b) The impurity scattering Γ dependence of ξ_h/ξ_{c2} at different temperatures for $d_{x^2-y^2}$ pairing with $B = 5$.

Fig. 7 shows the normalized differences between the fields calculated with the London model and the Eilenberger equations for $d_{x^2-y^2}$-wave pairing symmetry for the values of $\Gamma = 0.03$, $B/B_{c2} = 0.1$ and $T/T_{c0} = 0.3$. The accuracy of the fitting is better than 2%.

Fig. 8 (a) demonstrates the magnetic field dependence of cutoff parameter ξ_h/ξ_{c2} at different temperatures ($T/T_{c0} = 0.2, 0.3, 0.4, 0.5, 0.7, 0.8$) for $d_{x^2-y^2}$ pairing with $\Gamma = 0$. Fig. 8 (b) shows the impurity scattering Γ dependence of ξ_h/ξ_{c2} at different temperatures for $d_{x^2-y^2}$

pairing with $B = 5$. For clean superconductors (Fig. 8 (a)) ζ_h/ζ_{c2} has a minimum in its field dependence similar to usual s-wave superconductors [93]. However, this ratio decreases with temperature due to Kramer-Pesch effect. It was demonstrated theoretically and experimentally that the low energy density of states $N(E)$ is described by the same singular V-shape form $N(E) = N_0(H) + \alpha|E| + O(E^2)$ for all clean superconductors in a vortex state, irrespective of the underlying gap structure [94]. This explains the similarity in the behavior between s- and d-wave pairing symmetries.

The difference between pairing symmetries reveals itself in impurity scattering dependence ζ_h/ζ_{c2}. In s_{++} symmetry ζ_h/ζ_{c2} always decreases with impurity scattering rate Γ (Fig. 3 (a)), in $s^{\pm}$ symmetry its behavior depends on the field range and relative values of intraband and interband impurity scattering rates: it can be a decreasing function of Γ_π (Fig. 4 (b)) or an increasing function of Γ_π (Fig. 3 (b)). In d-wave superconductors ζ_h/ζ_{c2} always increases with Γ (Fig. 8 (b)) similar to the case of $s^{\pm}$ symmetry with $\Gamma_0 = \Gamma_\pi$ (Fig. 5). This can be understood from the comparison of the Ricatti equations of the $s^{\pm}$ and d-wave pairing. In both cases the renormalization factor $F = 0$ due to a cancelation of the intraband and interband impurity scattering rates in $s^{\pm}$ pairing or symmetry reason $\langle f \rangle = 0$ for d-wave pairing.

4. The quasiclassical approach to the Amin-Franz-Affleck model and the effective penetration depth in the mixed state in $d_{x^2-y^2}$-wave pairing symmetry

In this chapter, we construct a model where the nonlinear corrections arising from the Doppler energy shift of the quasiparticle states by the supercurrent [85] and effects of the vortex core states are described by an effective cutoff function. Nonlocal effects of the extended quasiparticle states are included in our model explicitly, i. e. instead of $\lambda(T)$ in Eq. (1) we use an analytically obtained anisotropic electromagnetic response tensor [70, 72, 73]. Because the nonlocal effects are assumed to be effective in clean superconductors we limit our consideration to the case $\Gamma = 0$.

For a better comparison with the nonlocal generalized London equation (NGLE) and the AGL theory we used another normalization of the cutoff parameter in Eq. (1), $u = k_1\sqrt{2}\zeta_{BCS}G$. This form of $F(G)$ correctly describes the high temperature regime. We compare our results with those obtained from the NGLE theory in a wide field and temperature range considering k_1 as the fitting parameter.

The magnetic field distribution in the mixed state in the NGLE approximation is given by [72]

$$h_{NGLE}(\mathbf{r}) = \frac{\Phi_0}{S} \sum_{\mathbf{G}} \frac{F(G)e^{i\mathbf{G}\mathbf{r}}}{1 + L_{ij}(\mathbf{G})G_iG_j}, \tag{27}$$

where

$$L_{ij}(\mathbf{G}) = \frac{Q_{ij}(\mathbf{G})}{\det\hat{Q}(\mathbf{G})}. \tag{28}$$

The anisotropic electromagnetic response tensor is defined by

$$Q_{ij}(\mathbf{G}) = \frac{4\pi T}{\lambda_{L0}^2} \sum_{\omega_n>0} \int_0^{2\pi} \frac{d\theta}{2\pi} \frac{\Delta(\theta)^2 \hat{v}_{Fi}\hat{v}_{Fj}}{\sqrt{\omega_n^2 + |\Delta(\theta)|^2}(\omega_n^2 + |\Delta(\theta)|^2 + \gamma_{\mathbf{G}}^2)}, \tag{29}$$

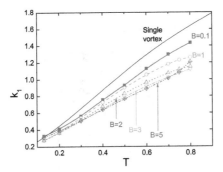

Figure 9. (Color online) The temperature dependence of the coefficient k_1 in the NGLE model obtained at $\kappa = 10$ and $B = 0.1, 1, 2, 3, 5$ from a fitting made with the solution of the Eilenberger equations.

where $\gamma_G = \mathbf{v}_F \cdot \mathbf{G}/2$. In Eq. (29) the term with γ_G describes the nonlocal correction to the London equation. Putting $\gamma_G = 0$ we obtain the London result $L_{ij}(\mathbf{G}) = \lambda(T)^2 \delta_{ij}$. We use the same shape of the cutoff function as in Eq. (1) but the values of the cutoff parameters are different because of fitting them to the various field distributions. In presentation of h_{NGLE} the anisotropy effects of the Eilenberger theory remain.

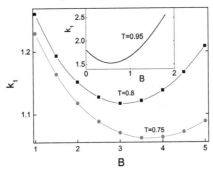

Figure 10. (Color online) Field dependence of k_1 at $T = 0.75$ and 0.8 obtained from the fitting to the Eilenberger equations. The inset shows $k_1(B)$ calculated from the Hao-Clem theory at $T = 0.95$.

Fig. 9 shows the $k_1(T)$ dependence in the NGLE model obtained at $\kappa = 10$ and $B = 0.1, 1, 2, 3, 5$ from the fitting to the solution of the Eilenberger equations. As can be seen from Fig. 9 the coefficient k_1 is strongly reduced at low temperatures. This is a reminiscent of the Kramer-Pesch result for s-wave superconductors (shrinking of the vortex core with decreasing temperature) [95]. It is also found that k_1 is a decreasing function of B. This can be explained by reduction of the vortex core size by the field [68]. The topmost curve in Fig. 9 gives the values of k_1 calculated for a single vortex [96]. At high temperatures the Ginzburg-Landau theory can be applied. Using the values of the parameters of this theory for d-wave superconductors [97] $\xi_{GL} = \xi_{BCS}\pi/\sqrt{3}$ is obtained. A variational approach of the Ginzburg-Landau equations for the single vortex [62] gives $k_1 = \pi/\sqrt{3} \approx 1.81$ is in reasonable agreement with the high temperature limit of k_1 for a single vortex in Fig. 9. Another interesting observation is the nonmonotonic behavior of $k_1(B)$ in low fields at high

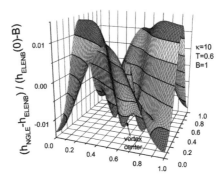

Figure 11. (Color online) Normalized differences between the fields calculated with the London model (NGLE) and the Eilenberger equation (ELENB) for $B = 1$ and $T = 0.6$. The scales of lengths are those of the flux line lattice unit vectors.

temperatures. Fig. 10 depicts the field dependence of k_1 at $T = 0.75$ and 0.8 showing a minimum which moves to lower fields with increasing of the temperature. This result agrees qualitatively with the Hao-Clem theory [63] which also predicts a minimum in the $k_1(B)$ dependence. This is demonstrated in the inset to Fig. 10, where $k_1(B)$ is shown at $T = 0.95$.

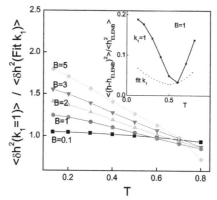

Figure 12. (Color online) Temperature dependence of the ratio of the second moment of the magnetic field distributions obtained from the NGLE model with the fixed and fitted parameter k_1 (see the text below). The inset shows the mean-square deviation of the magnetic field distribution from the origin for parameter k_1 set to unity (solid line) and fitted (dotted line).

The quality of the fitting can be seen from Fig. 11 where the normalized difference between the fields calculated in the NGLE model and the Eilenberger equations at $B = 1$, $T = 0.6$ and $\kappa = 10$ is shown. The accuracy of the fitting is about 1 percent. Thus, there is only a little improvement in the Eilenberger equations fitting to NGLE theory in comparison with local London theory (Eq. (1)). The similarity of the field and temperature dependences of the cutoff parameter in these theories are shown in Fig. 9 and Fig. 10.

To show the influence of the magnetic field and temperature on k_1 dependence, we calculate the values of $\langle \delta h^2_{NGLE} \rangle$ using the field distribution obtained in the Eq. 27. Fig. 12 shows the temperature dependence of the ratio $\langle \delta h^2_{NGLE} \rangle$ with the cutoff parameter obtained from the solution of the Eilenberger equations to that with $k_1 = 1$. From the data presented in Fig. 12, it can be sen that this ratio deviates considerably from unity when the temperature is lowered, which points to the importance of the proper determination of the value for the cutoff parameter. For the magnetic field distribution, obtained from solving the NGLE, we also calculate the mean-square deviation of this distribution from the origin (the Eilenberger equations solution). The inset demonstrates this deviation for fixed and fitted parameter k_1.

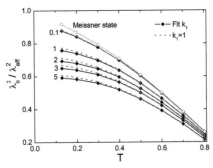

Figure 13. (Color online) The ratio of λ_0 to λ_{eff} calculated from the NGLE equation with $k_1 = 1$ and k_1 from Fig. 9.

This consideration proves that the nonlocal generalized London model with $h_{NGLE}(\mathbf{r})$ distribution also needs the properly determined cutoff parameter k_1, *i.e.* introducing only nonlocal extended electronic states does not allow the avoidance of the problem of vortex core solving.

In the analysis of the experimental μSR and SANS data the field dependent penetration depth $\lambda_{eff}(B)$ is often introduced [56]. It has physical sense even if it is not dependent on the core effects, *i.e.* it should be an invariant of the cutoff parameter. One such way of doing this was suggested in the AFA model [70, 73]:

$$\frac{\lambda_{eff}}{\lambda} = \left(\frac{|\delta h^2_0|}{|\delta h^2_{NGLE}|} \right)^{1/4}. \tag{30}$$

Here, $|\delta h^2_0|$ is the variance of the magnetic field $h_0(\mathbf{r})$ obtained by applying the ordinary London model with the same average field B and λ and with the same cutoff parameter as in the field distribution $h_{NGLE}(\mathbf{r})$.

In Fig. 13 establishes the temperature dependence of the ratio $\lambda^2_0 / \lambda^2_{eff}$ calculated from the h_{NGLE} distribution with $k_1 = 1$ and with Fit k_1 from the solution of Eilenberger equations for the different field value. The obtained $\lambda_{eff}(B)$ dependences are quite similar in these cases. The low-field result $(B/B_0 = 0.1)$ for λ_{eff} is close to $\lambda(T)$ in the Meissner state. This demonstrates that λ_{eff} is determined by a large scale of the order of FLL period and is not very sensitive to details of the microscopical core structure and the cutoff parameter [98]. The AFA model was originally developed in order to explain the structural transition in FLL in d-wave superconductors where anisotropy and nonlocal effects arise from nodes in the gap at the Fermi surface and the appearance there of the long extending electronic states [72].

The obtained anisotropy of superconducting current around the single vortex in AFA theory agrees reasonably with that found from the Eilenberger equations [96]. Extending electronic states also results in the observed field dependent flattening of $\lambda_{eff}(B)$ at low temperatures [73]. Thus, our microscopical consideration justifies the phenomenological AFA model and the separation between localized and extended states appears to be quite reasonable.

5. Conclusions

The core structure of the vortices is studied for $s^{\pm}$, $d_{x^2-y^2}$ symmetries (connected with interband and intraband antiferromagnetic spin fluctuation mechanism, respectively) and s_{++} symmetry (mediated by moderate electron-phonon interaction due to Fe-ion oscillation and the critical orbital fluctuation) using Eilenberger approach and compared with the experimental data for iron pnictides. It is assumed [99] that the nodeless $s^{\pm}$ pairing state is realized in all optimally-doped iron pnictides, while nodes in the gap are observed in the over-doped KFe_2As_2 compound, implying a $d_{x^2-y^2}$-wave pairing state, there are also other points of view [10, 13]. The stoichiometrical LiFeAs, without antifferromagmetic ordering, is considered as a candidate for the implementation of the s_{++} symmetry. Different impurity scattering rate dependences of cutoff parameter ξ_h are found for $s^{\pm}$ and s_{++} cases. In the nonstoichiometric case, when intraband impurity scattering (Γ_0) is much larger than the interband impurity scattering rate (Γ_π) the ξ_h/ξ_{c2} ratio is less in $s^{\pm}$ symmetry. When $\Gamma_0 \approx \Gamma_\pi$ (stoichiometric case) opposite tendencies are found, in $s^{\pm}$ symmetry the ξ_h/ξ_{c2} rises above the "clean" case curve ($\Gamma_0 = \Gamma_\pi = 0$) while it decreases below the curve in the s_{++} case. In d-wave superconductors ξ_h/ξ_{c2} always increases with Γ. For $d_{x^2-y^2}$ pairing the nonlocal generalized London equation and its connection with the Eilenberger theory are also considered. The problem of the effective penetration depth in the vortex state for d-wave superconductors is discussed. In this case, the field dependence of λ_{eff} is connected with the extended quasiclassical state near the nodes of the superconducting gap.

Author details

I. Zakharchuk, P. Belova, K. B. Traito and E. Lähderanta
Lappeenranta University of Technology, Finland

6. References

[1] Q. Li, Y. N. Tsay, M. Suenaga, R. A. Klemm, G. D. Gu, and N. Koshizuka. *Phys. Rev. Lett.*, 83:4160, 1999.

[2] R. A. Klemm. *Int. J. Mod. Phys. B*, 12:2920, 1983.

[3] D. R. Harshman, W. J. Kossler, X. Wan, A. T. Fiory, A. J. Greer, D. R. Noakes, C. E. Stronach, E. Koster, and J. D. Dow. *Phys. Rev. B*, 69:174505, 2004.

[4] A. G. Sun, D. A. Cajewski, M. B. Maple, and R. C. Dynes. *Phys. Rev. Lett.*, 72:2267, 1994.

[5] D. A. Wollmann, D. J. Van Harlingen, J. Giapintzakis, and D. M. Ginsberg. *Phys. Rev. Lett.*, 74:797, 1995.

[6] J. P. Rice, N. Rigakis, D. M. Ginsberg, and J. M. Mochel. *Phys. Rev. B*, 46:11050, 1992.

[7] C. C. Tsuei, J. R. Kirtley, C. C. Chi, L. S. Yu-Jahnes, A. Gupta, T. Shaw, J. Z. Sun, and M. B. Ketchen. *Phys. Rev. Lett.*, 73:593, 1994.

[8] Y. Kamihara, T. Watanabe, M. Hirano, and H. Hosono. *J. Am. Chem. Soc.*, 130:3296, 2008.

[9] X.C. Wang, Q.Q. Liu, Y.X. Lv, W.B. Gao, L.X. Yang, R.C. Yu, F.Y. Li, and C.Q. Jin. *Solid State Commun.*, 148:538, 2008.

[10] I. I. Mazin, D. J. Singh, M. D. Johannes, and M. H. Du. *Phys. Rev. Lett.*, 101:057003, 2008.

[11] K. Kuroki, S. Onari, R. Arita, H. Usui, Y. Tanaka, H. Kontani, and H. Aoki. *Phys. Rev. Lett.*, 101:087004, 2008.

[12] K. Seo, B. A. Bernevig, and J. Hu. *Phys. Rev. Lett.*, 101:206404, 2008.

[13] H. Kontani and S. Onari. *Phys. Rev. Lett.*, 104:157001, 2010.

[14] Y. Yanagi, Y. Yamakawa, and Y. Ōno. *Phys. Rev. B*, 81:054518, 2010.

[15] H. Ding, K. Nakayama P. Richard and, K. Sugawara, T. Arakane, Y. Sekiba, A. Takayama, S. Souma, T. Sato, T. Takahashi, Z. Wang, X. Dai, Z. Fang, G. F. Chen, J. L. Luo, and N. L. Wang. *Europhys. Lett.*, 83:47001, 2008.

[16] L. Wray, D. Qian, D. Hsieh, Y. Xia, L. Li, J. G. Checkelsky, A. Pasupathy, K.K. Gomes, C.V. Parker, A.V. Fedorov, G. F. Chen, J. L. Luo, A. Yazdani, N. P. Ong, N. L. Wang, and M. Z. Hasan. *Phys. Rev. B*, 78:184508, 2008.

[17] Z. H. Liu, P. Richard, K. Nakayama, G.-F. Chen, S. Dong, J. B. He, D. M. Wang, T.-L. Xia, K. Umezawa, T. Kawahara, S. Souma, T. Sato, T. Takahashi, T. Qian, Y. Huang, N. Xu, Y. Shi, H. Ding, and S. C. Wang. *Phys. Rev. B*, 84:064519, 2011.

[18] A. D. Christianson, E. A. Goremychkin, R. Osborn, S. Rosenkranz, M. D. Lumsden, C. D. Malliakas, I. S. Todorov, H. Claus, D. Y. Chung, M. G. Kanatzidis, R. I. Bewley, and T. Guidi. *Nature (London)*, 456:930, 2008.

[19] T. Hanaguri, S. Niitaka, K. Kuroki, and H. Takagi. *Science*, 328:474, 2010.

[20] S. Onari, H. Kontani, and M. Sato. *Phys. Rev. B*, 81:060504(R), 2010.

[21] S. Onari and H. Kontani. *Phys. Rev. Lett.*, 103:177001, 2009.

[22] C. Tarantini, M. Putti, A. Gurevich, Y. Shen, R. K. Singh, J. M. Rowell, N. Newman, D. C. Larbalestier, P. Cheng, Y.Jia, and H.-H. Wen. *Phys. Rev. Lett.*, 104:087002, 2010.

[23] G. R. Stewart. *Rep. Prog. Phys.*, 83:1589, 2011.

[24] P. J. Hirschfeld and N. Goldenfeld. *Phys. Rev. B*, 48:4219, 1993.

[25] K. Hashimoto, T. Shibauchi, S. Kasahara, K. Ikada, S. Tonegawa, T. Kato, R. Okazaki, C. J. van der Beek, M. Konczykowski, H. Takeya, K. Hirata, T. Terashima, and Y. Matsuda. *Phys. Rev. Lett.*, 102:207001, 2009.

[26] L. Luan, T. M. Lippman, C. W. Hicks, J. A. Bert, O. M. Auslaender, J.-H. Chu, J. G. Analytis, I. R. Fisher, and K. A. Moler. *Phys. Rev. Lett.*, 106:067001, 2011.

[27] M. A. Tanatar, J.-Ph. Reid, H. Shakeripour, X. G. Luo, N. Doiron-Leyraud, N. Ni, S. L. Bud'ko, P. C. Canfield, R. Prozorov, and L. Taillefer. *Phys. Rev. Lett.*, 104:067002, 2010.

[28] J. D. Fletcher, A. Serafin, L. Malone, J. G. Analytis, J.-H. Chu, A. S. Erickson, I. R. Fisher, and A. Carrington. *Phys. Rev. Lett.*, 102:147001, 2009.

[29] C. W. Hicks, T. M. Lippman, M. E. Huber, J. G. Analytis, J.-H. Chu, A. S. Erickson, I. R. Fisher, and K. A. Moler. *Phys. Rev. Lett.*, 103:127003, 2009.

[30] K. Hashimoto, M. Yamashita, S. Kasahara, Y. Senshu, N. Nakata, S. Tonegawa, K. Ikada, A. Serafin, A. Carrington, T. Terashima, H. Ikeda, T. Shibauchi, and Y. Matsuda. *Phys. Rev. B*, 81:220501(R), 2010.

[31] Y. Nakai, T. Iye, S. Kitagawa, K. Ishida, S. Kasahara, T. Shibauchi, Y. Matsuda, and T. Terashima. *Phys. Rev. B*, 81:020503(R), 2010.

[32] M. Yamashita, Y. Senshu, T. Shibauchi, S. Kasahara, K. Hashimoto, D. Watanabe, H. Ikeda, T. Terashima, I. Vekhter, A. B. Vorontsov, and Y. Matsuda. *Phys. Rev. B*, 84:060507(R), 2011.

[33] J. K. Dong, S. Y. Zhou, T. Y. Guan, H. Zhang, Y. F. Dai, X. Qiu, X. F. Wang, Y. He, X. H. Chen, and S. Y. Li. *Phys. Rev. Lett.*, 104:087005, 2010.

[34] K. Hashimoto, A. Serafin, S. Tonegawa, R. Katsumata, R. Okazaki, T. Saito, H. Fukazawa, Y. Kohori, K. Kihou, C. H. Lee, A. Iyo, H. Eisaki, H. Ikeda, Y. Matsuda, A. Carrington, and T. Shibauchi. *Phys. Rev. B*, 82:014526, 2010.

[35] G. Xu, H. Zhang, X. Dai, and Z. Fang. *Europhys. Lett.*, 84:67015, 2008.

[36] R. Thomale, C. Platt, W. Hanke, and B. A. Bernevig. *Phys. Rev. Lett.*, 106:187003, 2011.

[37] Y. J. Yan, X. F. Wang, R. H. Liu, H. Chen, Y. L. Xie, J. J. Ying, and X. H. Chen. *Phys. Rev. B*, 81:235107, 2010.

[38] N. Ni, S. L. Bud'ko, A. Kreyssig, S. Nandi, G. E. Rustan, A. I. Goldman, S. Gupta, J. D. Corbett, A. Kracher, and P. C. Canfield. *Phys. Rev. B*, 78:014507, 2008.

[39] C. Martin, R. T. Gordon, M. A. Tanatar, H. Kim, N. Ni, S. L. Bud'ko, P. C. Canfield, H. Luo, H. H. Wen, Z. Wang, A. B. Vorontsov, V. G. Kogan, and R. Prozorov. *Phys. Rev. B*, 80:020501(R), 2009.

[40] M. Rotter, M. Tegel, and D. Johrendt. *Phys. Rev. Lett.*, 101:107006, 2008.

[41] Y. Zhang, L. X. Yang, F. Chen, B. Zhou, X. F. Wang, X. H. Chen, M. Arita, K. Shimada, H. Namatame, M. Taniguchi, J. P. Hu, B. P. Xie, and D. L. Feng. *Phys. Rev. Lett.*, 105:117003, 2010.

[42] S.W. Zhang, L. Ma, Y. D. Hou, J. Zhang, T.-L. Xia, G. F. Chen, J. P. Hu, G. M. Luke, and W. Yu. *Phys. Rev. B*, 81:012503, 2010.

[43] T. Sato, K. Nakayama, Y. Sekiba, P. Richard, Y.-M. Xu, S. Souma, T. Takahashi, G. F. Chen, J. L. Luo, N. L. Wang, and H. Ding. *Phys. Rev. Lett.*, 103:047002, 2009.

[44] R. Thomale, C. Platt, W. Hanke, J. Hu, and B. A. Bernevig. *Phys. Rev. Lett.*, 107:117001, 2011.

[45] J. H. Tapp, Z. J. Tang, B. Lv, K. Sasmal, B. Lorenz, P. C. W. Chu, and A. M. Guloy. *Phys. Rev. B*, 78:060505(R), 2008.

[46] D. J. Singh. *Phys. Rev. B*, 78:094511, 2008.

[47] I. A. Nekrasov, Z.V. Pchelkina, and M.V. Sadovskii. *JETP Lett.*, 88:543, 2008.

[48] R. A. Jishi and H. M. Alyahyaei. *Adv. Condens. Matter Phys.*, 2010:1, 2010.

[49] S. J. Zhang, X. C. Wang, R. Sammynaiken, J. S. Tse, L. X. Yang, Z. Li, Q. Q. Liu, S. Desgreniers, Y. Yao, H. Z. Liu, and C. Q. Jin. *Phys. Rev. B*, 80:014506, 2009.

[50] F. L. Pratt, P. J. Baker, S. J. Blundell, T. Lancaster, H. J. Lewtas, P. Adamson, M. J. Pitcher, D. R. Parker, and S. J. Clarke. *Phys. Rev. B*, 79:052508, 2009.

[51] C.W. Chu, F. Chen, M. Gooch, A.M. Guloy, B. Lorenz, B. Lv, K. Sasmal, Z.J. Tang, J.H. Tapp, and Y.Y. Xue. *Physica C*, 469:326, 2009.

[52] S. V. Borisenko, V. B. Zabolotnyy, D. V. Evtushinsky, T. K. Kim, I. V. Morozov, A. N. Yaresko, A. A. Kordyuk, G. Behr, A. Vasiliev, R. Follath, and B. Büchner. *Phys. Rev. Lett.*, 105:067002, 2010.

[53] K. Umezawa, Y. Li, H. Miao, K. Nakayama, Z.-H. Liu, P. Richard, T. Sato, J. B. He, D.-M. Wang, G. F. Chen, H. Ding, T. Takahashi, and S.-C. Wang. *Phys. Rev. Lett.*, 108:037002, 2012.

[54] R. Laiho, M. Safonchik, and K. B. Traito. *Phys. Rev. B*, 76:140501(R), 2007.

[55] R. Laiho, M. Safonchik, and K. B. Traito. *Phys. Rev. B*, 78:064521, 2008.

[56] J. E. Sonier. *Rep. Prog. Phys.*, 70:1717, 2007.

[57] P. deGennes. *Superconductivity of Metals and Alloys*, pages Addison–Wesley, New York, 1989.

[58] V. G. Kogan, A. Gurevich, J. H. Cho, D. C. Johnston, M. Xu, J. R. Thompson, and A. Martynovich. *Phys. Rev. B*, 54:12386, 1996.

[59] E. H. Brandt. *Phys. Rev. B*, 37:2349(R), 1988.

[60] E. H. Brandt. *Physica C*, 195:1, 1992.

[61] I. G. de Oliveira and A. M. Thompson. *Phys. Rev. B*, 57:7477, 1998.

[62] J. R. Clem. *J. Low Temp. Phys*, 18:427, 1975.

[63] Z. Hao, J. R. Clem, M. W. McElfresh, L. Civale, A. P. Malozemoff, and F. Holtzberg. *Phys. Rev. B*, 43:2844, 1991.

[64] A. Yaouanc, P. Dalmas de Reotier, and E. H. Brandt. *Phys. Rev. B*, 55:11107, 1997.

[65] W. V. Pogosov, K. I. Kugel, A. L. Rakhmanov, and E. H. Brandt. *Phys. Rev. B*, 64:064517, 2001.

[66] A. Maisuradze, R. Khasanov, A. Shengelaya, and H. Keller. *J. Phys.: Condens. Matter*, 21:S075701, 2009.

[67] L. Kramer and W. Pesch. *Z. Phys.*, 269:59, 1974.

[68] M. Ichioka, A. Hasegawa, and K. Machida. *Phys. Rev. B*, 59:8902, 1999.

[69] M. Ichioka, A. Hasegawa, and K. Machida. *Phys. Rev. B*, 59:184, 1999.

[70] M. H. S. Amin, I. Affleck, and M. Franz. *Phys. Rev. B*, 58:5848, 1998.

[71] I. Affleck, M. Franz, and M. H. Sharifzadeh Amin. *Phys. Rev. B*, 55, 1997.

[72] M. Franz, I. Affleck, and M. H. S. Amin. *Phys. Rev. Lett.*, 79:1555, 1997.

[73] M. H. S. Amin, M. Franz, and I. Affleck. *Phys. Rev. Lett.*, 84:5864, 2000.

[74] A. B. Vorontsov, M. G. Vavilov, and A. V. Chubukov. *Phys. Rev. B*, 79:140507(R), 2009.

[75] W. A. Huttema, J. S. Bobowski, P. J. Turner, R.Liang, W. N. Hardy, D. A. Bonn, and D. M. Broun. *Phys. Rev. B*, 80:104509, 2009.

[76] V. G. Fleisher, Yu. P. Stepanov, K. B. Traito, E. Lähderanta, and R. Laiho. *Physica C*, 264:295, 1996.

[77] P. Miranović, M. Ichioka, and K. Machida. *Phys. Rev. B*, 70:104510, 2004.

[78] Y. N. Ovchinnikov and V. Z. Kresin. *Phys. Rev. B*, 52:3075, 1995.

[79] J. E. Sonier. *J. Phys.: Condens. Matter*, 16:S4499, 2004.

[80] J. E. Sonier, W. Huang, C. V. Kaiser, C. Cochrane, V. Pacradouni, S. A. Sabok-Sayr, M. D. Lumsden, B. C. Sales, M. A. McGuire, A. S. Sefat, and D. Mandrus. *Phys. Rev. Lett.*, 106:127002, 2011.

[81] D. S. Inosov, J. S. White, D. V. Evtushinsky, I. V. Morozov, A. Cameron, U. Stockert, V. B. Zabolotnyy, T. K. Kim, A. A. Kordyuk, S. V. Borisenko, E. M. Forgan, R. Klingeler, J. T. Park, S. Wurmehl, A. N. Vasiliev, G. Behr, C. D. Dewhurst, and V. Hinkov. *Phys. Rev. Lett.*, 104:187001, 2010.

[82] G. Yin and K. Maki. *Physica B*, 194-196:2025, 1994.

[83] A. B. Pippard. *Proc. R. Soc. London A*, 216:547, 1953.

[84] V. G. Kogan, M. Bullock, B. Harmon, P. Miranovič, Lj. Dobrosavljevič-Grujič, P. L. Gammel, and D. J. Bishop. *Phys. Rev. B*, 55:8693 (R), 1997.

[85] D. Xu, S. K. Yip, and J. A. Sauls. *Phys. Rev. B*, 51:16233, 1995.

[86] I. Kosztin and A. J. Leggett. *Phys. Rev. Lett.*, 79:135, 1997.

[87] D. Rainer, J. A. Sauls, and D. Waxman. *Phys. Rev. B*, 54:10094, 1996.

[88] F. Gygi and M. Schluter. *Phys. Rev. B*, 41:822, 1990.

[89] T. Dahm, S. Graser, C. Iniotakis, and N. Schopohl. *Phys. Rev. B*, 66:144515, 2002.

[90] M. Franz and Z. Tešanovič. *Phys. Rev. Lett.*, 80:4763, 1998.

[91] M. Kato and K. Maki. *Physica B*, 284:739, 2000.

[92] A. Zare, A. Markowsky, T. Dahm, and N. Schopohl. *Phys. Rev. B*, 78:104524, 2008.

[93] P. Belova, K. B. Traito, and E. Lähderanta. *J. Appl. Phys.*, 110:033911, 2011.

[94] N. Nakai, P. Miranovic, M. Ichioka, H. F. Hess, K. Uchiyama, H. Nishimori, S. Kaneko, N. Nishida, and K. Machida. *Phys. Rev. Lett.*, 97:147001, 2006.

[95] L. Kramer and W. Pesch. *J. Low Temp. Phys.*, 15:367, 1974.

[96] R. Laiho, E. Lähderanta, M. Safonchik, and K. B. Traito. *Phys. Rev. B*, 71:024521, 2005.

[97] Y. Ren, J.-H. Xu, and C. S. Ting. *Phys. Rev. Lett.*, 74:3680, 1995.

[98] R. Laiho, M. Safonchik, and K. B. Traito. *Phys. Rev. B*, 73:024507, 2006.

[99] D. C. Johnston. *Advances in Physics*, 59:803, 2010.

Chapter 2

Microwave Absorption by Vortices in Superconductors with a Washboard Pinning Potential

Valerij A. Shklovskij and Oleksandr V. Dobrovolskiy

Additional information is available at the end of the chapter

1. Introduction

1.1. The essential physical background

It is well-known that a type-II superconductor, while exposed to a magnetic field **B** whose magnitude is between the lower and upper critical field, is penetrated by a flux-line array of Abrikosov vortices, or *fluxons* [1–3]. Each vortex contains one magnetic flux quantum, $\Phi_0 = 2.07 \times 10^{-15}$ Wb, and the repulsive interaction between vortices makes them to arrange in a triangular lattice, with the vortex lattice parameter $a_L \simeq \sqrt{\Phi_0/B}$ where $B = |\mathbf{B}|$. A vortex is often simplified by the hard-core model [4], where the core is a cylinder of normal material with a diameter of the order of the coherence length. In this model, the magnetic field is constant in the core but decays exponentially outside the core over a distance of the order of the effective magnetic penetration depth.

In an ideal material, the vortex array would move with average velocity **v** under the action of the Lorentz force $\mathbf{F}_L$ essentially perpendicular to the transport current. Due to the nonzero viscosity experienced by the vortices when moving through a superconductor, a faster vortex motion corresponds to a larger dissipation. In experiments, inhomogeneities are usually present or can intentionally be introduced in a sample [5] which may give rise to local variations of the superconducting order parameter. This may cause the vortices to be pinned. By this way, the resistive properties of a type-II superconductor are determined by the vortex dynamics, which due to the presence of pinning centers can be described as the motion of vortices in some *pinning potential* (PP) [6]. In particular, randomly arranged and chaotically distributed point-like pinning sites give rise to an ubiquitous, *isotropic* (*i*) pinning contribution, as said of the "background nature". Depending on the relative strength between the Lorentz and pinning forces, the vortex lattice can be either pinned or on move, with a nonlinear transition between these regimes. Thus, the current-voltage characteristics (CVC) of such a sample is strongly nonlinear.

The importance of flux-line pinning in preserving superconductivity in a magnetic field and the reduction of dissipation via control of the vortex motion has been in general recognized since the discovery of type-II superconductivity [1, 2, 7, 8]. Later on, it has been found that the dissipation by vortices can be suppressed to a large degree if the intervortex spacing a_L geometrically matches the period length of the PP [9]. Moreover, artificially created linearly-extended pinning sites are known to be very effective for the reduction of the dissipation by vortices in one [10, 11] or several particular directions. Indeed, if the PP ensued in a superconductor is *anisotropic* (*a*), the direction of vortex motion can be deflected away from the direction of the Lorentz force. In this case, the nonlinear vortex dynamics becomes two-dimensional (*2D*) so that $\mathbf{v} \nparallel \mathbf{F}_L$. The non-collinearity between $\mathbf{v}$ and $\mathbf{F}_L$ is evidently more drastic the weaker the background i pinning is [12], which can otherwise mask this effect [13]. The most important manifestation of the pinning anisotropy is known as guided vortex motion, or the *guiding* effect [14], meaning that vortices tend to move along the PP channels rather than to overcome the PP barriers. As a consequence of the guided vortex motion, an *even-in-field* reversal transverse resistivity component appears, unlike the ordinary Hall resistivity which is *odd* regarding the field reversal. A guiding of vortices can be achieved with different sorts of PP landscapes [15, 16] though it is more strongly enhanced and can be more easily treated theoretically when using PPs of the washboard type (WPP).

One more intriguing effect appears when the PP profile is asymmetric. In this case the reflection symmetry of the pinning force is broken and thus, the critical currents measured under current reversal are not equal. As a result, while subjected to an ac current drive of zero mean a net rectified motion of vortices occurs. This is known as a rocking *ratchet* effect and has been widely used for studying the basics of mixed-state physics, e.g., by removing the vortices from conventional superconductors [17], as well as to verify ideas of a number of nanoscale systems, both solid state and biological [18, 19].

1.2. Experimental systems with a washboard pinning potential

The first experimental realization of a WPP used a periodic modulation of the thickness of cold-rolled sheets of a Nb-Ta alloy [14]. In this work the influence of isotropic pointlike disorder on the guiding of vortices was discussed for the first time. Later on, lithographic techniques have been routinely employed to create periodic pinning arrays consisting of practically identical nanostructures in the form of, e.g, microholes [20, 21], magnetic dots [22], and stripes [23]. The main idea all these works share is to suppress periodically the superconducting order parameter.

With regard to a theoretical description it has to be stated that a full and exact account of the nonlinear vortex dynamics in superconducting devices proposed in these works [20–23] in a wide range of external parameters is not available due to the complexity of the periodic PP used in these references. Due to this reason it has been proposed by the authors in a number of articles to study a simpler case, such as a WPP periodic in one direction [12, 24–27] or bianisotropic [28]. The main advantage of these approaches implies the possibility to describe the phenomenon of guided vortex motion along the WPP channels, i.e., the directional anisotropy of the vortex velocity, if the transport current is applied under various in-plane angles. For instance, self-organization has been used [10, 29] to provide semi-periodic, linearly extended pinning "sites" by spontaneous facetting of m-plane sapphire substrate surfaces on which Nb films have been grown. It has been demonstrated that pronounced guiding of vortices occurs. Experimental data [10] were in good agreement with theoretical results [12] that allowed to estimate both, the i and a PP parameters.

From the viewpoint of theoretical modeling, saw-tooth and harmonic PPs represent the most simple forms of the WPP. On the one hand, these simple forms of the WPP allow one to explicitly calculate the dc magneto-resistivity and the ac impedance tensor as the physical quantities of interest in the problem. On the other hand, these WPP's forms are highly realistic in the sense of appropriate experimental realizations which range from naturally occurring pinning sites in high temperature superconductors (HTSCs) to artificially created linearly-extended pinning sites in high-T_c and conventional superconductors, more often in thin films. Some experimental systems exhibiting a WPP are exemplified in Fig. 1. The experimental geometry of the model discussed below implies the standard four-point bridge of a thin-film superconductor with a WPP placed into a small perpendicular magnetic field with a magnitude $B \ll B_{c2}$ such that our theoretical treatment can be performed in the *single-vortex approximation*.

Summarizing what has been said so far, by tuning the intensity, form, and asymmetry of the WPP in a superconductor, one can manipulate the fluxons via dynamical, directional, and orientational control of their motion. Evidently, a number of more sophisticated phenomena arise due to the variety of the dynamical regimes which the vortex ensemble passes through. In the next sections we will consider a particular problem in the vortex dynamics when the vortices are subjected to superimposed subcritical dc $j_0 < j_c$ and small ac $j_1 \to 0$ current drives at frequencies ω in the microwave range. What is discussed below can be directly employed to the wide class of thin superconductors with a WPP, including but not limited to those examples shown in Fig. 1. In particular, by looking for the dc magneto-resistivity and the ac impedance responses, we will elucidate: a) how to derive the absorbed power by vortices in such a superconductor as function of all the driving parameters of the problem and b) how to solve the inverse problem, i.e., to reconstruct the coordinate dependence of a PP from the dc current-induced shift in the *depinning* frequency ω_p [30], deduced from the curves $P(\omega|j_0)$. The importance and further aspects of this issue are detailed next.

1.3. Which information can be deduced from microwave measurements?

The measurement of the complex impedance response accompanied by its power absorption $P(\omega)$ in the radiofrequency and microwave range represents a powerful approach to investigate pinning mechanisms and the vortex dynamics in type-II superconductors. The reason for this is that at frequencies $\omega \ll \omega_B$, substantially smaller than those invoking the breakdown of the zero-temperature energy gap ($\omega_B = 2\Delta(0)/\hbar \approx$ 100 GHz for a superconductor with a critical temperature T_c of 10 K), high-frequency and microwave impedance measurements of the mixed state yield information about flux pinning mechanisms, peculiarities in the vortex dynamics, and dissipative processes in a superconductor. It should be stressed that this information can not be extracted from the dc resistivity data obtained in the steady state regime when pinning is strong in the sample. This is due to the fact that in the last case when the critical current density j_c is rather large, the realization of the dissipative mode, in which the flux-flow resistivity ρ_f can be measured, requires $j_0 \gtrsim j_c$. This is commonly accompanied by a non-negligible electron overheating in

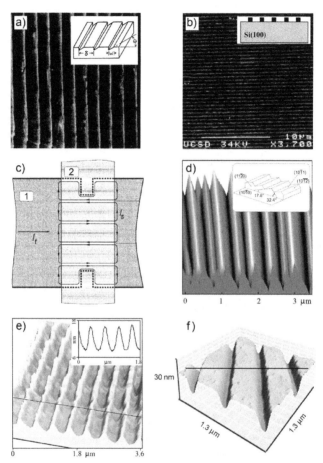

Figure 1. Examples of selected experimental systems exhibiting a washboard pinning potential re-printed after original research papers: a) In-2% Bi foil imprinted with diffraction grating [31]. b) Parallel lines of Ni prepared by electron-beam lithography on a Si substrate onto which a Nb film was sputtered [32]. c) Superconducting microbridge (1) with an overlaying magnetic tape (2) containing a pre-recorded magnetization distribution [33]. d) Nb film deposited onto faceted $\alpha - Al_2O_3$ substrate surface [10]. e) Nb film surface with an array of ferromagnetic Co stripes fabricated by focused electron beam-induced deposition [11]. f) Nb film surface with an array of grooves etched by focused electron beam milling [34]. In addition, there is a large number of HTSC-based experimental systems with a WPP, ranging from uniaxially twinned films to the usage of the intrinsic layers within a HTSC [35–39].

the sample [40, 41] which changes the value of the sought ρ_f. At the same time, measurements of the absorbed power by vortices from an ac current with amplitude $j_1 \ll j_c$ allow one to determine ρ_f at a dissipative power level of $P_1 \sim \rho_f j_1^2$ which can be many orders of magnitude smaller than $P_0 \sim \rho_f j_0^2$. Consequently, measurements of the complex ac response versus frequency ω probe the pinning forces virtually in the absence of overheating effects which are otherwise unavoidable at overcritical steady-state dc current densities.

The multitude of experimental works published recently utilizing the usual four-point scheme [42], strip-line coplanar waveguides (CPWs) [43], the Corbino geometry [44, 45], or the cavity method [46] to investigate the microwave vortex response in as-grown thin-film superconductors or in those containing some nano-tailored PP landscape reflects the explosively growing interest in the subject. In connection with this, from the microwave power absorption further insight into the pinning mechanisms can be gained. In particular, artificially fabricated pinning nanostructures provide a PP of unknown shape that requires certain assumptions concerning its coordinate dependence in order to fit the measured data. At the same time, in a real sample a certain amount of disorder is always presented, acting as pinning sites for a vortex as well. By this way, an approach how to reconstruct the form of the PP experimentally realized in a sample is of self-evident importance for both, application-related and fundamental reasons. A scheme how to reconstruct the coordinate dependence of a PP has been recently proposed by the authors [47] and will be elucidated in Sec. 3.3.

1.4. Development of the theory in the field

A very early model to describe the absorbed power by vortices refers to the work of Gittleman and Rosenblum (GR) [30]. GR measured the power absorption by vortices in PbIn and NbTa films over a wide range of frequencies ω and successfully analyzed their data on the basis of a simple model for a 1D parabolic PP. In their pioneering work, a small ac excitation of vortices in the absence of a dc current was considered. Later on, GR have supplemented their equation of motion for a vortex with a dc current and have introduced a cosine PP [48]. The GR results have been obtained at $T = 0$ in a linear approximation for the pinning force and will be presented here not only for their historical importance but rather to provide the foundation for the subsequent generalization of the model.

Later on, the theory accounting also for vortex creep at non-zero temperature in a 1D cosine PP has been extended by Coffey and Clem (CC) [49]. In the following, the CC theory has been experimentally proved to be very successful [16] to describe the high-frequency electromagnetic properties of superconductors. However, it had been developed for a small microwave current and in the absence of a dc drive.

Recently, the CC results have been substantially generalized by the authors [25, 27] for a 2D cosine WPP. The washboard form of the PP allowed for an exact theoretical description of the 2D anisotropic nonlinear vortex dynamics for any arbitrary values of the ac and dc current amplitudes, temperature, and the angle between the transport current direction with respect to the guiding direction of the WPP. The influence of the Hall effect and anisotropy of the vortex viscosity on the absorbed power by vortices has also been analyzed [50, 51]. Among other nontrivial results obtained, an enhancement [25] and a sign change [27] in the power absorption for $j_0 \gtrsim j_c$ have been predicted.

Whereas the general *exact* solution of the problem [25, 27] has been obtained for non-zero temperature in terms of a matrix continued fraction [52], here we treat the problem analytically in terms of only elementary functions which allow a more intuitive description of the main effects. Solving the equation of motion for a vortex at $T = 0$, $j_0 < j_c$, and $j_1 \to 0$ in the general case, we also consider some important limiting cases of isotropic vortex viscosity and zero Hall constant provided it substantially helps us to elucidate the physical picture. The theoretical treatment of the problem is provided next.

2. General formulation of the problem to be solved

Let the x axis with the unit vector x (see Fig. 2) be directed perpendicular to the washboard channels, while the y axis with the unit vector y is along these channels. The equation of motion for a vortex moving with velocity v in a magnetic field $\mathbf{B} = B\mathbf{n}$, where $B \equiv |\mathbf{B}|$, $\mathbf{n} = n\mathbf{z}$, $\mathbf{z}$ is the unit vector in the z direction, and $n \pm 1$, has the form

$$\hat{\eta}\mathbf{v} + \alpha_H \mathbf{v} \times \mathbf{n} = \mathbf{F} + \mathbf{F}_p, \tag{1}$$

where $\mathbf{F} = (\Phi_0/c)\mathbf{j} \times \mathbf{n}$ is the Lorentz force, $\mathbf{j} = \mathbf{j}_0 + \mathbf{j}_1(t)$, and $\mathbf{j}_1(t) = \mathbf{j}_1 \exp iwt$, where $\mathbf{j}_0$ and $\mathbf{j}_1$ are the densities of dc and small ac currents, respectively, and ω is the ac frequency. Φ_0 is the magnetic flux quantum and c is the speed of light. $\hat{\eta}$ is the vortex viscosity tensor and α_H is the Hall coefficient. In Eq. (1) $\mathbf{F}_p = -\nabla U_p(x)$ is the anisotropic pinning force, where $U_p(x)$ is some periodic pinning potential (PP).

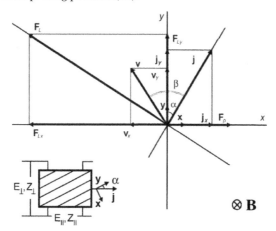

Figure 2. The system of coordinates xy with the unit vectors x and y is associated with the WPP channels which are parallel to the vector y. The transport current density vector $\mathbf{j} = \mathbf{j}_0 + \mathbf{j}_1 \exp iwt$ is directed at an angle α with respect to y. β is the angle between the average velocity vector v and j. $\mathbf{F}_p$ is the average pinning force provided by the WPP and $\mathbf{F}_L$ is the Lorenz force for a vortex. Inset: a schematic sample configuration in the general case. A thin type-II superconductor (foil, thin film, or thin layer of crystal) is placed into a small perpendicular magnetic field $\mathbf{B}$. A WPP is formed in the sample and the direction of the WPP channels is shown by hatching. Experimentally deducible values are the dc voltages $E_\parallel$ and $E_\perp$ as well as the ac impedances $Z_\parallel$ and $Z_\perp$.

If x and y are the coordinates along and across the anisotropy axis, respectively, tensor $\hat{\eta}$ is diagonal in the xy representation, and it is convenient to define η_0 and γ by the formulas

$$\eta_0 = \sqrt{\eta_{xx}\eta_{yy}}, \qquad \gamma = \sqrt{\eta_{xx}/\eta_{yy}}, \qquad \eta_{xx} = \gamma\eta_0, \qquad \eta_{yy} = \eta_0/\gamma, \tag{2}$$

where η_0 is the averaged viscous friction coefficient, and γ is the anisotropy parameter.

Since $U_p(x)$ depends only on the x coordinate and is periodic, i.e., $U_p(x) = U_p(x+a)$, where a is the period of the PP, the pinning force $\mathbf{F}_p$ is directed always along the anisotropy axis x and has no component along the y axis, i.e., $F_{py} = 0$. As usually [25, 27, 48, 49, 53, 54], we use a WPP of the cosine form

$$U_p(x) = (U_p/2)(1 - \cos kx), \tag{3}$$

where $k = 2\pi/a$, $\mathbf{F}_p = -(dU_p/dx)\mathbf{x} = F_{px}\mathbf{x}$, and $F_{px} = -F_c \sin kx$, where $F_c = U_p k/2$ is the maximum value of the pinning force. Because $\mathbf{F} \equiv \mathbf{F}(t) = \mathbf{F}_0 + \mathbf{F}_1(t)$, where $\mathbf{F}_0 = (\Phi_0/c)\mathbf{j}_0 \times \mathbf{n}$ is the Lorentz force invoked by the dc current and $\mathbf{F}_1 = (\Phi_0/c)\mathbf{j}_1(t) \times \mathbf{n}$ is the Lorentz force invoked by the small ac current, we assume that $\mathbf{v}(t) = \mathbf{v}_0 + \mathbf{v}_1(t)$, where $\mathbf{v}_0$ is time-independent, while $\mathbf{v}_1(t) = \mathbf{v}_1 \exp i\omega t$.

Our goal is to determine $\mathbf{v}$ from Eq. (1) and to substitute it then in the expression for the electric field. To accomplish this, Eq. (1) can be rewritten in projections on the coordinate axes

$$\begin{cases} \gamma[v_{0x} + v_{1x}(t)] + \delta[v_{0y} + v_{1y}(t)] = [F_{0x} + F_{1x}(t) + F_{px}]/\eta_0, \\ (1/\gamma)[v_{0y} + v_{1y}(t)] - \delta[v_{0x} + v_{1x}(t)] = [F_{0y} + F_{1y}(t)]/\eta_0, \end{cases} \tag{4}$$

where $\epsilon = \alpha_H/\eta_0$ and $\delta = n\epsilon$.

However, instead of to straightforwardly proceed with the solution of Eqs. (4), we first consider some physically important limiting cases in which Eq. (1) is substantially simplified. Later on, in Sec. 4.3, Eq. (1) will be dealt with in its general form.

3. The Gittleman-Rosenblum model

3.1. Dynamics of pinned vortices on a small microwave current

Let us consider the case of an isotropic vortex viscosity, i.e., $\gamma = 1$ while $\eta_0 = \eta$ in the absence of Hall effect, i.e., $\epsilon = 0$. We restrict our analysis to the consideration of the vortex motion with velocity $v(t)$ only along the x-axis. This case corresponds to $\alpha = 0$ when the vortices move across the PP barriers. Furthermore, we first assume that $j_0 = 0$. Then Eq. (1) can be rewritten in the form originally used [30] for a parabolic pinning potential

$$\eta\dot{x} + k_p x = f_L, \tag{5}$$

where x is the vortex displacement, η is the vortex viscosity, k_p is the constant which characterizes the restoring force f_p in the PP well $U_p(x) = (1/2)k_p x^2$ and $f_p = -dU_p/dx = -k_p x$. In Eq. (5) $f_L = (\Phi_0/c)j_1(t)$ is the Lorentz force acting on a vortex, and $j_1(t) = j_1 \exp i\omega t$ is the density of a small microwave current with the amplitude j_1. Looking for the solution

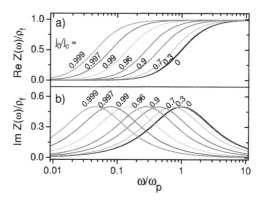

Figure 3. The frequency dependences of a) real and b) imaginary parts of the ac impedance calculated for a cosine pinning potential $U_p(x) = (U_p/2)(1 - \cos kx)$ at a series of dc current densities, as indicated. In the absence of a dc current, the GR results are revealed in accordance with Eqs. (9).

of Eq. (5) in the form $x(t) = x \exp i\omega t$, where x is the complex amplitude of the vortex displacement, one immediately gets $\dot{x}(t) = i\omega x(t)$ and

$$x = \frac{(\Phi_0/\eta c)j_1}{i\omega + \omega_p},$$ (6)

where $\omega_p \equiv k_p/\eta$ is the depinning frequency. This frequency ω_p determines the transition from the non-dissipative to dissipative regimes in the vortex dynamics in response to a small microwave signal, as will be elucidated in the text later. To calculate the magnitude of the complex electric field arising due to the vortex on move, one takes $E = B\dot{x}/c$. Then

$$E(\omega) = \frac{\rho_f j_1}{1 - i\omega_p/\omega} \equiv Z(\omega)j_1.$$ (7)

Here $\rho_f = B\Phi_0/\eta c^2$ is the flux-flow resistivity and $Z(\omega) \equiv \rho_f/(1 - i\omega_p/\omega)$ is the microwave impedance of the sample.

In order to calculate the power P absorbed per unit volume and averaged over the period of an ac cycle, the standard relation $P = (1/2)\text{Re}(E \cdot J^*)$ is used, where E and J are the complex amplitudes of the ac electric field and current density, respectively. The asterisk denotes the complex conjugate. Then, from Eq. (7) it follows

$$P(\omega) = (1/2)\text{Re}Z(\omega)j_1^2 = (1/2)\rho_f j_1^2/[1 + (\omega_p/\omega)^2].$$ (8)

For the subsequent analysis, it is convenient to write out real and imaginary parts of the impedance $Z = \text{Re}Z + i\text{Im}Z$, namely

$$\text{Re}Z(\omega) = \rho_f/[1 + (\omega_p/\omega)^2], \qquad \text{Im}Z(\omega) = \rho_f(\omega/\omega_p)/[1 + (\omega/\omega_p)^2].$$ (9)

The frequency dependences (9) are plotted in dimensionless units Z/ρ_f and ω/ω_p in Fig. 3 (see the curve for $j_0/j_c = 0$). From Eqs. (5), (6), and (8) it follows that *pinning forces dominate*

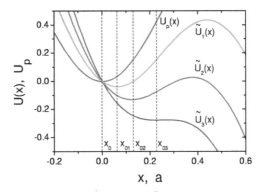

Figure 4. Modification of the effective pinning potential $\tilde{U}_i(x) \equiv U_p(x) - f_{0i}x$, where $U_p(x) = (U_p/2)(1 - \cos kx)$ is the WPP, with the gradual increase of f_0 such as $0 = f_0 < f_{01} < f_{02} \lesssim f_{03} = f_c$, i.e., a vortex is oscillating in the gradually tilting pinning potential well in the vicinity of the rest coordinate x_{0i}.

at low frequencies, i.e., when $\omega \ll \omega_p$ and $Z(\omega)$ is mainly nondissipative with $\mathrm{Re}Z(\omega) \approx (\omega/\omega_p)^2 \ll 1$, whereas *frictional forces dominate at higher frequencies*, i.e., when $\omega \gg \omega_p$ and $Z(\omega)$ is dissipative with $\mathrm{Re}Z(\omega) \approx \rho_f[1 - (\omega_p/\omega)^2]$. In other words, due to the reduction of the amplitude of the vortex displacement with the increase of the ac frequency, a vortex is getting not influenced by the pinning force. This can be seen from Eq. (6) where $x \sim 1/\omega$ for $\omega \gg \omega_p$; this is accompanied, however, with the independence of the vortex velocity of ω in this regime in accordance with Eq. (7).

3.2. Influence of a dc current on the depinning frequency

The GR model can be generalized for an arbitrary PP and can also account for an arbitrary dc current superimposed on a small microwave signal. For determinacy, let us consider a subcritical dc current with the density $j_0 < j_c$, where j_c is the critical current density in the absence of a microwave current. Our aim now is to determine to which changes in the effective PP parameters the superimposition of the dc current leads, because $\tilde{U}(x) \equiv U_p(x) - xf_0$ in the presence of $j_0 \neq 0$. Here $U_p(x)$ is the x-coordinate dependence of the PP when $j_0 = 0$. The modification of the effective PP with the gradual increase of f_0 is illustrated in Fig. 4 for the WPP by Eq. (3). Note also that $f_0 < f_c$, where f_0 and f_c are the Lorentz forces which correspond to the current densities j_0 and j_c, respectively.

In the presence of an arbitrary dc current, the equation of motion for a vortex (5) has the form

$$\eta v(t) = f(t) + f_p, \tag{10}$$

where $f(t) = (\Phi_0/c)j(t)$ is the Lorentz force with $j(t) = j_0 + j_1(t)$, where $j_1(t) = j_1 \exp i\omega t$, and j_1 is the amplitude of a small microwave current. Due to the fact that $f(t) = f_0 + f_1(t)$, where $f_0 = (\Phi_0/c)j_0$ and $f_1(t) = (\Phi_0/c)j_1(t)$ are the Lorentz forces for the subcritical dc and microwave currents, respectively, one can naturally assume that $v(t) = v_0 + v_1(t)$, where v_0 is time-independent, whereas $v_1(t) = v_1 \exp i\omega t$. In Eq. (10) the pinning force is $f_p =$

$-dU_p(x)/dx$, where $U_p(x)$ is some PP. Our aim is to determine $v(t)$ from Eq. (10) which, taking into account the considerations above, acquires the next form

$$\eta[v_0 + v_1(t)] = f_0 + f_p + f_1(t), \tag{11}$$

Let us consider the case when $j_1 = 0$. If $j_0 < j_c$, i.e., $f_0 < f_c$, where f_c is the maximal value of the pinning force, then $v_0 = 0$, i.e., the vortex is in rest. As it is seen from Fig. 4 the rest coordinate x_0 of the vortex in this case depends on f_0 and is determined from the condition of equality to zero of the effective pinning force $\tilde{f}(x) = -d\tilde{U}(x)/dx = f_p(x) + f_0$, which reduces to the equation $f_p(x_0) + f_0 = 0$, or

$$f_0 = \frac{dU_p(x)}{dx}\Big|_{x=x_0}, \tag{12}$$

the solution of which is the function $x_0(f_0)$.

Let us now add a small oscillation of the vortex in the vicinity of x_0 under the action of the small external alternating force $f_1(t)$ with the frequency ω. For this we expand the effective pinning force $\tilde{f}(x)$ in the vicinity of $x = x_0$ into a series in terms of small displacements $u \equiv x - x_0$ which gives

$$\tilde{f}(x - x_0) \simeq \tilde{f}(x_0) + \tilde{f}'(x_0)u + \ldots \tag{13}$$

Then, taking into account that $\tilde{f}(x_0) = 0$ and $\tilde{f}'(x_0) = U_p''(x_0)$, Eq. (11) acquires the form

$$\eta \dot{u}_1 + \tilde{k}_p u = f_1, \tag{14}$$

where $\tilde{k}_p(x_0) = U_p''(x_0)$ is the effective constant characterizing the restoring force $\tilde{f}(u)$ at small oscillations of a vortex in the effective PP $\tilde{U}(x)$ close by $x_0(f_0)$, and $v_1 = \dot{u} = i\omega u$. Equation (14) for the determination of v_1 is physically equivalent to GR Eq. (5) with the only distinction that the vortex depinning frequency $\tilde{\omega}_p \equiv \tilde{k}_p/\eta$ now depends on f_0 through Eq. (12), i.e., on the dc transport current density j_0. Thereby, all the results of Sec. 3.1 [see Eqs. (6)-(9)] can be repeated here with the changes $x \to u$ and $\omega_p \to \tilde{\omega}_p$. It should be noted, that all the described till now did not require one to know the actual form of the PP. In order to discuss the changes in the dependences $\text{Re}Z(\omega)$ and $\text{Im}Z(\omega)$ caused by the dc current, the PP must be specified. We take the cosine WPP determined by Eq. (3); though any other non-periodic PP can also be used. For the curves $\text{Re}Z(\omega|j_0)$ and $\text{Im}Z(\omega|j_0)$ plotted in Fig. 3, the dependence $\tilde{\omega}_p(j_0/j_c) = \omega_p\sqrt{1 - (j_0/j_c)^2}$ for the cosine WPP is used [50]. Its derivation will also be outlined in Sec. 3.3.

Now we turn to the discussion of the figure data from which it is evident that with increase of j_0 the curves $\text{Re}Z(\omega|j_0)$ and $\text{Im}Z(\omega|j_0)$ shift to the left. The reason for this is that with increase of j_0 the PP well while tilted is broadening, as illustrated in Fig. 3. Thus, during the times shorter than $\tau_p = 1/\omega_p$, i.e., for $\omega > \omega_p$, a vortex can no longer non-dissipatively oscillate in the PP's well. As a consequence, the enhancement of $\text{Re}Z(\omega)$ occurs at lower frequencies. At the same time, the curves in Fig. 3 maintain their original shape. Thus, *the only universal parameter to be found experimentally is the depinning frequency ω_p.* For a fixed frequency and different j_0, real part of $Z(\omega)$ always acquires larger values for larger j_0, whereas the maximum in imaginary part of $Z(\omega)$ precisely corresponds to the middle point of the nonlinear transition in $\text{Re}Z(\omega)$. It should be noted that in the presence of $j_0 \neq 0$ *the dissipation remains non-zero* even at $T = 0$, though it is very small at very low frequencies.

3.3. Reconstruction of a pinning potential from the microwave absorption data

3.3.1. General scheme of the pinning potential reconstruction

We now turn to the detailed analytical description how to reconstruct the coordinate dependence of a PP experimentally ensued in the sample, on the basis of microwave power absorption data in the presence of a subcritical dc transport current. It will be shown that from the dependence of the depinning frequency $\tilde{\omega}_p(j_0)$ as a function of the dc transport current j_0 one can determine the coordinate dependence of the PP $U_p(x)$. The physical background for the possibility to solve such a problem is Eq. (12) which gives the correlation of the vortex rest coordinate x_0 with the value of the static force f_0 acting on the vortex and arising due to the dc current j_0.

From Eq. (12) it follows that while increasing f_0 from zero up to its critical value f_c one in fact "probes" all the points in the dependence $U_p(x)$. Taking the x_0-coordinate derivative in Eq. (12), one obtains

$$dx_0/df_0 = 1/U_p''(x_0) = 1/\tilde{k}_p(x_0), \tag{15}$$

where the relation $U''(x_0) = \tilde{k}_p(x_0)$ has been used [see Eq. (14) and the text below]. By substituting $x_0 = x_0(f_0)$, Eq. (15) can be rewritten as $dx_0/df_0 = 1/\tilde{k}_p[x_0(f_0)]$, and thus,

$$\frac{dx_0}{df_0} = \frac{1}{\eta \tilde{\omega}_p(f_0)}. \tag{16}$$

If the dependence $\tilde{\omega}(f_0)$ has been deduced from the experimental data, i.e., fitted by a known function, then Eq. (16) allows one to derive $x_0(f)$ by integrating

$$x_0(f_0) = \frac{1}{\eta} \int_0^{f_0} \frac{df}{\tilde{\omega}_p(f)}. \tag{17}$$

Then, having calculated the inverse function $f_0(x_0)$ to $x_0(f_0)$ and using the relation $f_0(x_0) = U_p'(x_0)$, i.e., Eq. (12), one finally obtains

$$U_p(x) = \int_0^x dx_0 f_0(x_0). \tag{18}$$

3.3.2. Example procedure to reconstruct a pinning potential

Here we would like to support the above-mentioned considerations by giving an example of the reconstruction procedure for a WPP. Let us suppose that a series of power absorption curves $P(\omega)$ has been measured for a set of subcritical dc currents j_0. Then for determinacy, let us imagine that each $i-$curve of $P(\omega|j_0)$ like those shown in Fig. 3 has been fitted with its fitting parameter $\tilde{\omega}_p$ so that one could map the points $[(\tilde{\omega}_p/\omega_p)_i, (j_0/j_c)_i]$, as shown by triangles in Fig. 5.

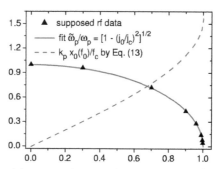

Figure 5. The pinning potential reconstruction procedure: step 1. A set of $[(\tilde{\omega}_p/\omega_p)_i, (j_0/j_c)_i]$ points (▲) has been deduced from the supposed measured data and fitted as $\tilde{\omega}_p/\omega_p = \sqrt{1-(j_0/j_c)^2}$ (solid line). Then by Eq. (17) $x_0(f_0) = (f_c/k_p)\arcsin(f_0/f_c)$ (dashed line).

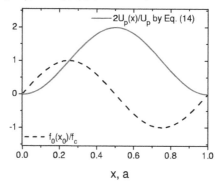

x, a

Figure 6. The pinning potential reconstruction procedure: step 2. The inverse function to $x_0(f_0)$ is $f_0(x_0) = f_c\sin(x_0 k_p/f_c)$ (dashed line). Then by Eq. (18) $U_p(x) = (U_p/2)(1-\cos kx)$ is the PP sought (solid line).

We fit the figure data in Fig. 5 by the function $\tilde{\omega}_p/\omega_p = \sqrt{1-(j_0/j_c)^2}$ and then substitute it into Eq. (17) from which one calculates $x_0(f_0)$. In the case, the function has a simple analytical form, namely $x_0(f_0) = (f_c/k_p)\arcsin(f_0/f_c)$. Evidently, the inverse to it function is $f_0(x_0) = f_c\sin(x_0 k_p/f_c)$ with the period $a = 2\pi f_c/k_p$ (see also Fig. 6). By taking the integral (18) one finally gets $U_p(x) = (U_p/2)(1-\cos kx)$, where $k = 2\pi/a$ and $U_p = 2f_c^2/k_p$.

3.4. Concluding remarks on the reconstruction scheme of a pinning potential

The problem to reconstruct the actual form of a potential subjected to superimposed constant and small alternation perturbations arises not only in the vortex physics but also in related fields such as charge-density-wave pinning [55] and Josephson junctions [56]. In the vortex physics, an early scheme how to reconstruct the coordinate dependence of the pinning force from measurements implying a small ripple magnetic field superposed on a larger dc magnetic field had been previously reported [57]. Also, due to the closest mathematical analogy should be also mentioned the Josephson junction problem wherein a plenty of non-sine current-phase relations is known to occur [56] and which could in turn benefit from the results reported here.

A respective experiment can be carried out at $T \ll T_c$ and implies a small microwave current density $j_1 \ll j_c$. Though the potential reconstruction scheme has been exemplified for a cosine WPP, i.e., for a periodic and symmetric PP, single PP wells [43] can also be proven in accordance with the provided approach. In the general case, *the elucidated here procedure does not require periodicity of the potential and can account also for asymmetric ones*. If this is the case, one has to perform the reconstruction procedure under the dc current reversal, i.e., two times: for $+j_0$ and $-j_0$.

Here, our consideration was limited to $T = 0$, $j_0 < j_c$, and $j_1 \to 0$ because this has allowed us to provide a clear reconstruction procedure in terms of elementary functions accompanying with a simple physical interpretation. Experimentally, adequate measurements can be performed, i.e., on conventional thin-film superconductors (e.g., Nb, NbN) at $T \ll T_c$. These are suitable due to substantially low temperatures of the superconducting state and that relatively strong pinning in these materials allows one to neglect thermal fluctuations of a vortex with regard to the PP's depth $U_p \simeq 1000 \div 5000$ K [10]. It should be stressed that due to the universal form of the dependences $P(\omega|j_0)$, the depinning frequency ω_p plays a role of the only fitting parameter for each of the curves $P(\omega|j_0)$, thus fitting of the measured data seems to be uncomplicated. However, one of most crucial issues for the experiment is to adequately superimpose the applied currents and then, to uncouple the picked-up dc and microwave signals maintaining the matching of the impedances of the line and the sample. Quantitatively, experimentally estimated values of the depinning frequency in the absence of a dc current at a temperature of about $0.6T_c$ are $\omega_p \approx 7$ GHz for a 20 nm-thick [45] and a 40 nm-thick [46] Nb films. This value is strongly suppressed with increase of both, the field magnitude and the film's thickness.

Concerning the general validity of the results obtained, one circumstance should be recalled. The theoretical consideration here has been performed in the single-vortex approximation, i.e., is valid only at small magnetic fields $B \ll B_{c2}$, when the distance between two neighboring vortices, i.e., the period of a PP is larger as compared with the effective magnetic field penetration depth, $a \gtrsim \lambda$.

4. Solution of the problem in the general case

4.1. General remarks on the Hall parameter and anisotropy of the vortex viscosity

Now, the extent to which the Hall term in the equation of motion of the vortex and a possible anisotropy of the viscous term affect the 2D dynamics and the resistive properties of the vortex ensemble both, at a direct (subcritical) current and at a small microwave ac current, will be investigated. It should be pointed out that, even though the Hall angle θ_H and, consequently, the dimensionless Hall coefficient ϵ is small for most superconductors, i.e., $\epsilon \ll 1$, anomalously large values of ϵ are observed in YBCO, NbSe$_2$, and Nb films in a number of cases at sufficiently low temperatures [58]; i.e., $\tan \theta_H \geq 1$. In the absence of pinning (Sec. 4.2), this means that the vortex velocity $\mathbf{v}$ in this case is directed predominantly along the direction of $\mathbf{j}_1(t)$, whereas, with a small Hall angle ($\tan \theta_H$), the directions of $\mathbf{v}$ and $\mathbf{j}_1(t)$ are virtually orthogonal. The influence of the Hall term on the vortex dynamics is taken into consideration because, as will be shown below, the absorption by vortices at an ac current substantially depends on both the magnitude of ϵ and the frequency ω, and angle α. Moreover, the analysis shows that the resistive characteristics of a sample at a subcritical dc is independent of the value of the Hall constant. In other words, it is impossible to extract the value of ϵ from experimental data at a dc, while ϵ *can be determined from an analysis of the power absorption at an ac* (Sec. 4.3). The physical cause of such a behavior of the transverse resistive response at a subcritical dc current is associated with the suppression of the Hall response as a consequence of vortex pinning in the transverse with respect to the WPP channels direction of their possible motion.

The viscosity anisotropy is taken into consideration because the anisotropy in the *ab* plane is fairly large in most HTSC crystals: for example, for YBCO crystals, the magneto-resistivity as the vortices move along the *a* or *b* axis can differ by more than a factor of 2 [35].

4.2. The case of zero pinning strength and arbitrary currents

4.2.1. Computing the dc resistivities

Let the Hall constant now be arbitrary ($\epsilon \neq 0$), while the vortex viscosity is arbitrary and anisotropic ($\gamma \neq 1$). We first consider the 2D vortex motion in the absence of pinning, i.e., when $\mathbf{F}_p = 0$. Note, that both the dc and the microwave ac currents are of arbitrary densities. The equation of motion for a vortex has the form

$$\hat{\eta}\mathbf{v} + \alpha_H \mathbf{v} \times \mathbf{n} = \mathbf{F}. \tag{19}$$

In this case, projections of the vortex velocity on the xy axes at constant current are $v_{0x} = \tilde{F}_{0x}/\tilde{\eta}_0$ and $v_{0y} = \tilde{F}_{0y}/\tilde{\eta}_0$, where $\tilde{\eta}_0 = \eta_0\gamma(1 + \delta^2)$, $\tilde{F}_{0x} = F_{0x} - \gamma\delta F_{0y}$, and $\tilde{F}_{0y} = \gamma^2 F_{0y} + \gamma\delta F_{0x}$. The main physical quantity that makes is possible to determine the resistive characteristics of the sample, i.e., its dc resistivity tensor $\hat{\rho}_0$ and the ac impedance tensor $\hat{Z}_1$ at frequency ω, is the electric field $\mathbf{E}(t)$ induced by the moving vortex system

$$\mathbf{E}(t) = [\mathbf{B} \times \mathbf{v}(t)]/c = (nB/c)[-v_y(t)\mathbf{x} + v_x(t)\mathbf{y}]. \tag{20}$$

We note that $\mathbf{E}(t) = \mathbf{E}_0 + \mathbf{E}_1(t)$, where $\mathbf{E}_0$ is the dc electric field, while $\mathbf{E}_1(t) = \mathbf{E}_1 \exp i\omega t$, where $\mathbf{E}_1$ is the complex amplitude of the ac electric field $\mathbf{E}_1(t)$. We next recall that the experimentally measurable resistive responses (longitudinal $E_\parallel$ and transverse $E_\perp$ with respect to the current direction) are associated with the responses E_x and E_y in the xy coordinate system by the relations $E_\parallel = E_x \sin\alpha + E_y \cos\alpha$ and $E_\perp = -E_x \cos\alpha + E_y \sin\alpha$, where $E_x = -n(B/c)v_y$ and $E_y = n(B/c)v_x$. At dc current, $E_{0\parallel}$ and $E_{0\perp}$ are respectively determined by

$$\begin{cases} E_{0\parallel} = [\rho_f j_0/\gamma\Delta](\gamma^2 \sin^2\alpha + \cos^2\alpha) \equiv j_0\rho_{0\parallel}, \\ E_{0\perp} = [\rho_f j_0/\gamma\Delta](\gamma\delta + (1 - \gamma^2)\sin\alpha\cos\alpha) \equiv j_0\rho_{0\perp}, \end{cases} \tag{21}$$

where $\rho_f \equiv B\Phi_0/\eta_0 c^2$ is the flux-flow resistivity and $\Delta = 1 + \delta^2$. Separating the even and odd in n components in Eqs. (21) one finally obtains

$$\begin{aligned} \rho_{0\parallel}^+ &= (\rho_f/\gamma\Delta)(\gamma^2 \sin^2\alpha + \cos^2\alpha), & \rho_{0,\parallel}^- &= 0, \\ \rho_{0\perp}^+ &= (\rho_f/\gamma\Delta)(1 - \gamma^2)\sin\alpha\cos\alpha, & \rho_{0\perp}^- &= \rho_f\delta/\Delta. \end{aligned} \tag{22}$$

It should be pointed out that Eqs. (22) are independent of j_0, i.e., the corresponding dc CVCs are *linear*. In other words, all three nonzero resistive responses correspond to the *flux-flow* regime of the vortex dynamics. The difference between them consists only in different dependences of the magnetoresistivities $\rho_{0\parallel}^+$, $\rho_{0\perp}^+$, and $\rho_{0\perp}^-$ on parameters α, γ, and ϵ. The presence of pinning (see Sec. 4.3.1) substantially changes these final conclusions.

4.2.2. Limiting cases of isotropic viscosity and/or zero Hall constant

Let us now consider several simple, physically different limiting cases which follow from Eq. (22). If there is no viscosity anisotropy ($\gamma = 1$) and no Hall effect ($\epsilon = 0$), the vortex dynamics is isotropic, i.e., is independent of the angle α. In this case, the vortex dynamics corresponds to the flux-flow mode which is *independent of the field reversal*. As expected, the only nonzero component is $\rho_{0\parallel}^+ = \rho_f$ which is even with respect to the change $\mathbf{B} \to -\mathbf{B}$.

However, if $\gamma = 1$ and only the Hall effect $\epsilon \neq 0$ is to be considered, the vortex dynamics becomes anisotropic in the sense that the directions of $\mathbf{F}_0$ and $\mathbf{v}_0$ no longer coincide. This *odd-in-field* anisotropy is of a Hall origin and can be quantitatively characterized by the Hall angle, θ_H, determined as

$$\tan \theta_H = \rho_{0\perp}^- / \rho_{0\parallel}^+ = \delta, \tag{23}$$

where $\rho_{0\perp}^-$ is the new transverse odd (Hall) component of the magneto-resistivity. If $|\delta| \ll 1$, the Hall anisotropy is weak, and the vortex velocity $\mathbf{v}_0$ is virtually perpendicular to the dc density $\mathbf{j}_0$, whereas, if $|\delta| \gg 1$, the directions of $\mathbf{v}_0$ and $\mathbf{j}_0$ virtually coincide.

The presence of a viscosity anisotropy $\gamma \neq 1$ even in the absence of the Hall effect ($\epsilon = 0$) results in the appearance of one more new magneto-resistivity $\rho_{0\perp}^+$. This component is even with respect to the inversion $\mathbf{B} \to -\mathbf{B}$ and, like the Hall effect, causes the vortex motion to be anisotropic, i.e., it causes the directions of $\mathbf{F}_0$ and $\mathbf{v}_0$ not to coincide. It is convenient to characterize the corresponding *even-in-field* anisotropy by the angle β defined as

$$\cot \beta = -\rho_{0\perp}^+ / \rho_{\parallel}^+ = (\gamma^2 - 1)/(\gamma^2 \tan \alpha + \cot \alpha). \tag{24}$$

By analogy with the appearance of directed motion of the vortices in the presence of a WPP, when there is no Hall effect and no viscosity anisotropy, the angle β defined by Eq. (24) can be treated similarly to that for the guiding effect in the problem with pinning.

4.2.3. Computing the ac impedance

Carrying out an analysis for ac current in the same way like for dc current, one has $v_{1x}(t) = \tilde{F}_{1x}(t)/\tilde{\eta}_0$ and $v_{1y}(t) = \tilde{F}_{1y}(t)/\tilde{\eta}_0$, where $\tilde{F}_{1x}(t) = F_{1x} - \gamma\delta F_{1y}$, and $\tilde{F}_{1y}(t) = \gamma^2 F_{1y} + \gamma\delta F_{1x}$. Then

$$\begin{cases} E_{1\parallel} = [\rho_f j_1(t)/\gamma\Delta](\gamma^2 \sin^2 \alpha + \cos^2 \alpha) \equiv j_1(t)Z_\parallel, \\ E_{1\perp} = [\rho_f j_1(t)/\gamma\Delta](\gamma\delta + (1 - \gamma^2) \sin \alpha \cos \alpha) \equiv j_1(t)Z_\perp. \end{cases} \tag{25}$$

Longitudinal ($\parallel$) and transverse ($\perp$) components are determined here with respect to the direction of $\mathbf{j}_1$. Separating the even and odd with respect to n components in Eqs. (25), one finally gets

$$Z_\parallel^+ = (\rho_f/\gamma\Delta)(\gamma^2 + \sin^2\alpha + \cos^2\alpha), \qquad Z_\parallel^- = 0,$$

$$Z_\perp = (\rho_f/\gamma\Delta)(1-\gamma^2)\sin\alpha\cos\alpha, \qquad Z_\perp^- = \rho_f\delta/\Delta. \tag{26}$$

It should be pointed out that, as it can be seen from Eqs. (26) and (22), the relationships for the transverse and longitudinal resistive responses at dc current formally coincide with those for the corresponding impedances in the absence of pinning, i.e., the impedance components are *real*. However, it should be recalled that $\rho_{\parallel,\perp} = \text{Re}[Z_{\parallel,\perp}\exp i\omega t]$.

4.2.4. Microwave absorption by vortices in the absence of pinning

To compute the absorbed power P in the ac response per unit volume and averaged over the period of an ac cycle, we use the standard expression $P = (1/2)\text{Re}(\mathbf{E}_1 \cdot \mathbf{j}_1^*)$, where $\mathbf{E}_1$ and $\mathbf{j}_1$ are the complex amplitudes of the ac electric field and the current density, respectively. Then, using Eqs. (25), it can be shown that

$$P = (j_1^2/2)\bar{\rho} \equiv (j_1^2/2)\text{Re}Z_\parallel = P_0(\gamma^2\sin^2\alpha + \cos^2\alpha)/\gamma\Delta, \tag{27}$$

where $P_0 = \rho_f(j_1^2/2)$. When $\gamma = 1$, the absorbed power becomes isotropic and depends only on the dimensionless Hall constant ϵ, i.e., $P = P_0/\Delta$ with P decreasing as ϵ increases. This is physically associated with the already established fact that, as ϵ increases, the direction of vector $\mathbf{v}_1$ approaches closer and closer to the direction of vector $\mathbf{j}_1$, so that the corresponding component of the longitudinal ac electric field $E_{1\parallel}$ decreases in amplitude as θ_H increases, while the absorbed power falls off. According to the physical picture and as it follows from Eq. (27), the power absorption is maximal and equals P_0 when $\gamma = 1$ and $\epsilon = 0$.

However, if $\gamma \neq 1$ and $\epsilon \neq 0$, one has $P = P(\alpha, \gamma, \epsilon)$, i.e., the absorbed power is anisotropic. Figure 7 shows the dependence of P/P_0 as a function of the anisotropy parameter γ at the Hall parameter $\delta = 0.1$ for various values of the angle α. It follows from Eq. (27) that the influence of parameter ϵ on $P(\alpha, \gamma, \epsilon)$ for any α and γ reduces to a reduction of the absorption with increasing ϵ, as well as the fact that the absorption anisotropy when $\gamma \neq 1$ is determined by the value of the nonlinear with respect to α and γ combination $\gamma\sin^2\alpha + (1/\gamma)\cos^2\alpha$. The latter implies that the term $(1/\gamma)\cos^2\alpha$ increases as $\alpha \to 0$ and $\gamma \to 0$, and that the term $\gamma\sin^2\alpha$ increases as $\alpha \to \pi/2$ and $\gamma \gg 1$. All these features are easy to see in Fig. (7).

4.3. The case of arbitrary pinning strength and subcritical currents

4.3.1. Computing the dc resistivities

Let us first consider the case in which there is no ac current, i.e., $j_1 = 0$. It then follows from Eq. (4) that

$$\begin{cases} \gamma v_{0x} + \delta v_{0y} = (F_{0x} + F_{px})/\eta_0, \\[2mm] (1/\gamma)v_{0y} - \delta v_{0x} = F_{0y}/\eta_0. \end{cases} \tag{28}$$

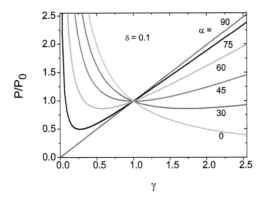

Figure 7. Dependence of the absorbed power P/P_0 on the anisotropy parameter γ at the Hall parameter $\delta = 0.1$ for a series of values of the angle α, as indicated.

The solution of this system of equation is

$$v_{0x} = (\tilde{F}_{0x} + F_{px})/\tilde{\eta}_0, \qquad v_{0y} = \gamma F_{0y}/\tilde{\eta}_0 + \gamma \delta(\tilde{F}_{0x} + F_{px})/\tilde{\eta}_0. \tag{29}$$

The motion of a vortex along the x axis will differ, depending on the values of the anisotropy parameter γ, the Hall coefficient ϵ, and the force $\tilde{F}_{0x}$. If $\tilde{F}_{0x} < F_c$, the vortex comes to rest in this direction, i.e., $V_{0x} = 0$. As follows from Fig. 4, the vortex's rest coordinate x_0 in this case depends on the value of $\tilde{F}_{0x}$. It then follows from Eq. (29) that, to determine the dependence $x_0(\tilde{F}_{0x})$, it is necessary to solve the equation $\tilde{F}_{0x} + F_{px} = 0$. For the WPP given by Eq. (3), the solution is

$$x_0 = (1/k) \arcsin(\tilde{F}_{0x}/F_c), \tag{30}$$

where $\tilde{F}_{0x}/F_c = \tilde{\jmath}_{0y}/j_c$, $\tilde{\jmath}_{0y} = n(j_{0y} + \gamma\delta j_{0x})$, and j_c is the critical current when $\alpha = 0$.

We now add a small ac signal $j_1(t)$ with frequency ω and consider how a small ac force $F_1(t)$ affects the vortex dynamics in the subcritical dc current regime, i.e., when $\tilde{\jmath}_{0y} < j_c$. It follows from Eq. (20) that, when $j_1 = 0$, the value of E_0 can be obtained by averaging $\mathbf{E}_1(t)$ over time, taking into account the periodicity of $\mathbf{F}_1(t)$. Then $\mathbf{E} = \langle\mathbf{E}(t)\rangle$, where $\langle\ldots\rangle = 1/T \int_{t_0}^{t_0+T} \ldots dt$, and $T = 2\pi/\omega$. We note that $v_{0x} = 0$ when $j_{0y} < j_c$. As a result, one gets

$$\mathbf{E}_0 = \frac{nB}{c} v_{0y}\mathbf{x} = \frac{nB}{c} \frac{\gamma F_{0y}}{\eta_0} \mathbf{x}. \tag{31}$$

Here $F_{0y} = -n(\Phi_0/c)j_{0x}$, and therefore $\mathbf{E}_0 = -\gamma\rho_f j_{0x}\mathbf{x}$ and $E_{0x} = -\gamma\rho_f j_{0x}$.

It follows from Eq. (20) that

$$\begin{cases} E_{0\|} = E_{0x}\sin\alpha = -\gamma\rho_f j_0 \sin^2\alpha \equiv j_0\rho_{0\|}, \\[2mm] E_{0\perp} = -E_{0x}\cos\alpha = \gamma\rho_f j_0 \sin\alpha\cos\alpha \equiv j_0\rho_{0\perp}. \end{cases} \tag{32}$$

Finally,

$$\rho_{0\|} = -\gamma\rho_f \sin^2\alpha, \qquad \rho_{0\perp} = \gamma\rho_f \sin\alpha\cos\alpha, \tag{33}$$

from which it immediately follows that these responses are independent of the **B**-reversal. The only information which can be extracted from Eq. (33) is concerning the angle α for the given sample, from the relation $\tan \alpha = -\rho_{0\parallel}/\rho_{0\perp}$ and the value of the product $\gamma \rho_f$. From Eq. (33) it also follows that the longitudinal and transverse responses are nondissipative only when $\alpha = 0$; this is caused by the subcritical nature of the transport current. A dissipation arises when $\alpha \neq 0$ due to the appearance of a component of the driving force F_{0y} that does not contain the Hall constant [see Eq. (29) for V_{0y}, taking into account that $V_{0x} = 0$] and is directed along the WPP channels. Thus, when $\tilde{F}_{0x} < F_c$, the vortex motion and the resistive response of the sample are independent of ϵ, i.e., *the Hall parameter can not be determined from experiment at a constant subcritical current*, unlike the case already described in Sec. 4.2.1.

4.3.2. Computing the ac impedance tensor

Let us now proceed to an analysis of the responses to an ac current, using the relationship $\mathbf{E}_1(t) = \mathbf{E}(t) - \mathbf{E}_0 = \mathbf{E}(t) - \langle \mathbf{E}(t) \rangle$. From this and from Eq. (4) one has that

$$
\begin{cases}
\gamma v_{1x}(t) + \delta v_{1y}(t) = [\tilde{F}_{0x} + F_{1x}(t) + F_{px}]/\eta_0, \\
v_{1y}(t)/\gamma - \delta v_{1x}(t) = F_{1y}(t)/\eta_0,
\end{cases}
\tag{34}
$$

where $F_{1x}(t) \equiv (n\Phi_0/c)j_{1y}(t)$ and $F_{1y}(t) \equiv -(n\Phi_0/c)j_{1x}(t)$, where $j_{1x}(t) = j_1(t)\sin\alpha$, $j_{1y}(t) = j_1(t)\cos\alpha$, and $j_1(t) = j_1 \exp i\omega t$. We use the fact that $\tilde{F}_{0x} + F_{px} \equiv \tilde{F}_{px} = -d\tilde{U}_p(x)/dx$, where $\tilde{U}_p(x) \equiv U_p(x) - x\tilde{F}_{0x}$ is the effective PP, taking into account the driving force component along the x axis (see Fig. 4). In this case, the rest coordinate for a vortex is given by Eq. (30). The effective PP $\tilde{U}_p(x)$ can be expanded in series in the small difference $(x - x_0)$ like it was done in Sec. 3.2, and using Eq. (30) one gets

$$
\tilde{F}_{px}/\eta_0 = -\tilde{\omega}_p(x - x_0), \quad \text{where} \quad \tilde{\omega}_p = \omega_p\sqrt{1 - (\tilde{j}_{0y}/j_c)^2}, \quad \text{and} \quad \omega_p \equiv k_p/\eta_0.
\tag{35}
$$

Now it is possible to rewrite Eq. (34) as

$$
\begin{cases}
(\gamma + \tilde{\omega}_p/i\omega)v_{1x}(t) + \delta v_{1y}(t) = F_{1x}(t)/\eta_0, \\
-\delta v_{1x}(t) + (1/\gamma)v_{1y}(t) - \delta v_{1x}(t) = F_{1y}(t)/\eta_0.
\end{cases}
\tag{36}
$$

The solution of this system of equation is

$$
v_{1x}(t) = (F_{1x}/\gamma - \delta F_{1y})/(\eta_0 \Delta_\gamma), \qquad v_{1y}(t) = [\gamma F_{1y}(1 + \tilde{\omega}_{p\gamma}/i\omega) + \delta F_{1x}]/(\eta_0 \Delta_\gamma),
\tag{37}
$$

where $\tilde{\omega}_{p\gamma} \equiv \tilde{\omega}_p/\gamma$ and $\Delta_\gamma \equiv \Delta + \tilde{\omega}_{p\gamma}/i\omega$. From the relation $\mathbf{E}_1(t) = \hat{Z}\mathbf{j}_1(t)$, where the components of the ac impedance tensor $\hat{Z}$ are measured in the xy coordinate system (see Fig. 2), Eqs. (4), and (37), the longitudinal and transverse (with respect to the direction of $\mathbf{j}_1$) impedances $Z_\parallel$ and $Z_\perp$ are determined as

$$
\begin{cases}
Z_\parallel = (\rho_f/\gamma\Delta)[\gamma^2(\Delta - \delta^2 Z_1)\sin^2\alpha + Z_1\cos^2\alpha], \\
Z_\perp = (\rho_f/\gamma\Delta)\{\delta\gamma Z_1 + [Z_1 - \gamma^2(\Delta - \delta^2 Z_1)]\sin\alpha\cos\alpha]\},
\end{cases}
\tag{38}
$$

where $Z_1 \equiv \Delta/\Delta_\gamma = 1/(1 - i\omega_q/\omega)$, and

$$\omega_q = \tilde{\omega}_{p\gamma}/\Delta = [\omega_p/\gamma\Delta]\sqrt{1 - (\tilde{j}_{0y}/j_c)^2} = [\omega_p/\gamma\Delta]\sqrt{1 - (j_{0y}/j_c)^2(\cos\alpha + \delta\gamma\sin\alpha)^2} \quad (39)$$

The quantity ω_q in Eq. (39) a generalization to the case $\gamma \neq 1$ and $\epsilon \neq 0$, and is physically analogous to the depinning frequency ω_p introduced in Seq. 3.1 and dependent on the subcritical transport current. However, it should be emphasized that, unlike the depinning frequency ω_p which is independent on the **B**-inversion, the value of the ω_q changes with the replacement $n \to -n$, i.e., $\delta \to -\delta$ because the Hall effect is present.

Having separated even and odd parts in the impedance components, finally the experimentally deducible, field orientation-independent quantities are

$$\begin{cases} Z_\parallel^+ = (\rho_f/\gamma\Delta)[\gamma^2(\Delta - \delta^2 Z_1^+)\sin^2\alpha + Z_1^+\cos^2\alpha], \\[2mm] Z_\parallel^- = (\rho_f/\gamma\Delta)Z_1^-(\cos^2\alpha - \delta^2\gamma^2\sin^2\alpha), \\[2mm] Z_\perp^+ = (\rho_f/\gamma\Delta)\{Z_1^+ - \gamma^2(\Delta - \delta^2 Z_1^+)]\sin\alpha\cos\alpha\}, \\[2mm] Z_\perp^- = (\rho_f/\gamma\Delta)\{\delta\gamma Z_1^+ + Z_1^-(1 + \delta^2\gamma)\sin\alpha\cos\alpha\}, \end{cases} \quad (40)$$

4.3.3. Determination of the Hall constant from microwave measurements

Let us assume that $\gamma = 1$ and the Hall constant is arbitrary but satisfies the condition $\tilde{F}_{0x} < F_c$. Then $\check{Z}_1 \equiv Z_1(\gamma = 1) = 1/(1 - i\tilde{\omega}_q/\omega)$, where $\tilde{\omega}_q \equiv (\omega_p/\Delta)\sqrt{1 - (j_0/j_c)^2(\cos\alpha + \delta\sin\alpha)^2}$. In this case, the expressions for the real part of the longitudinal and transverse impedances $\mathrm{Re}\check{Z}_{\parallel,\perp}$ are determined by

$$\begin{cases} \mathrm{Re}\check{Z}_\parallel = (\rho_f/\Delta)[\mathrm{Re}\check{Z}_1 + \Delta(1 - \mathrm{Re}\check{Z}_1)\sin^2\alpha], \\[2mm] \mathrm{Re}\check{Z}_\perp = (\rho_f/\Delta)[\delta\mathrm{Re}\check{Z}_1 - \Delta(1 - \mathrm{Re}\check{Z}_1)\sin\alpha\cos\alpha], \end{cases} \quad (41)$$

where $\mathrm{Re}\check{Z}_1 = 1/[1 + (\tilde{\omega}_q/\omega)^2]$.

The culmination of this subsection is an analysis of the dependence $\mathrm{Re}\check{Z}_{\parallel,\perp}^\pm$ as a function of ω at large or small frequencies. If $\omega \to \infty$, one has $(\tilde{\omega}_q/\omega) \to 0$, i.e., $\mathrm{Re}\check{Z}_1 = 1$. Then in the main approximation with respect to $1/\omega$ one has $\mathrm{Re}\check{Z}_\parallel = \rho_f/\Delta$, (i.e., $\mathrm{Re}\check{Z}_\parallel^+ = \mathrm{Re}\check{Z}_\parallel$ and $\mathrm{Re}\check{Z}_\parallel^- = 0$), and $\mathrm{Re}\check{Z}_\perp = \rho_f\delta/\Delta$, i.e., $\mathrm{Re}\check{Z}_\perp^+ = 0$ and $\mathrm{Re}\check{Z}_\perp^- = \mathrm{Re}\check{Z}_\perp$. Thus, as the Hall constant ϵ increases (i.e., δ increases), the absorbed power P decreases as $\omega \to \infty$ ($P = P_0/\Delta$). Moreover, as $\omega \to \infty$, for any α, there is a relationship of the form $\delta = \mathrm{Re}\check{Z}_\perp^-/\mathrm{Re}\check{Z}_\parallel^-$. Now let $\omega \to 0$ (i.e., $\mathrm{Re}\check{Z}_1 = 1$). Then, in the main approximation with respect to ω, the Hall effect is unmeasurable, since $\mathrm{Re}\check{Z}_\parallel = \rho_f\sin^2\alpha$ and $P = P_0\sin^2\alpha$, while $\mathrm{Re}\check{Z}_\perp = -\rho_f\sin\alpha\cos\alpha$; i.e., δ has been canceled out of the results. It follows from this that $\rho_f = \mathrm{Re}\check{Z}_\parallel/\sin^2\alpha$; i.e., $\eta_0 = B\Phi_0\sin^2\alpha/\mathrm{Re}\check{Z}_\parallel c^2$.

Thus, the high-frequency limit $\omega \gg \bar{\omega}_q$ ($\omega \to \infty$) is needed to determine $\epsilon = \alpha_H / \eta_0$, whereas the low-frequency limit $\omega \ll \bar{\omega}_q$ ($\omega \to 0$) is sufficient to determine η_0. Because of the dependence $\bar{\omega}_q(j)$, appropriate measurements can be performed *even at one fixed frequency* $\omega \lesssim \omega_p$. Thus, for any ϵ and $\alpha = \pi/2$, the Hall constant in a periodic PP from microwave absorption by vortices is determined by

$$\alpha_H = (B\Phi_0/\bar{\rho}(0)c^2 \sqrt{\bar{\rho}(0)/\bar{\rho}(\infty) - 1}, \tag{42}$$

where $\bar{\rho}(0) = \rho_f \sin^2 \alpha$ for $\omega \to 0$ and $\bar{\rho}(\infty) = \rho_f/\Delta$ for $\omega \to \infty$.

4.3.4. Microwave absorption by vortices in a washboard pinning potential

By analogy with Sec. 4.2.4, the absorbed power is $P = (j_1^2/2)\mathrm{Re}Z_\| \equiv (j_1^2/2)\bar{\rho}$, where now

$$\bar{\rho} = (\rho_f/\gamma\Delta)\{\Delta\gamma^2 \sin^2 \alpha + [1 - (1 + \delta^2\gamma^2) \sin^2 \alpha]\mathrm{Re}Z_1\} \tag{43}$$

If $\delta = 0$ and $\gamma = 1$, Eq. (43) reduces to $\bar{\rho} = \rho_f(\sin^2 \alpha + \mathrm{Re}Z_1 \cos^2 \alpha)$ which has been dealt with previously [50]. Let $Z_1 \equiv 1 - iG_1$, where $G_1 = -(\omega_q/\omega)/(1 - i\omega_q/\omega)$, and consequently $\mathrm{Re}Z_1 = 1 - \mathrm{Re}(iG_1) = 1 + \mathrm{Im}G_1$. Then from Eq. (43) one has that

$$\bar{\rho} = (\rho_f/\gamma\Delta)\{\gamma^2 \sin^2 \alpha + \cos^2 \alpha + [1 - (1 + \delta^2\gamma^2) \sin^2 \alpha]\mathrm{Im}G_1\} \tag{44}$$

where $\mathrm{Im}G_1 = -1/[1 + (\omega/\omega_q)^2]$. When $\gamma = 1$, Eq. (44) reduces to Eq. (85) in our previous work [25]. Finally,

$$P = P_0\{1 + (\gamma^2 - 1) \sin \alpha + [1 - (1 + \delta^2\gamma^2) \sin^2 \alpha]\mathrm{Im}G_1\}/\gamma\Delta, \tag{45}$$

where $P_0 = \rho_f(j_1^2/2)/$ Unlike the case in which is no pinning (Sec. 4.2.4), the absorbed power P in this case not only depends on angle α, anisotropy parameter γ, and Hall constant ϵ, but also depends on frequency ω and current density j_0.

It is essential to point out, that in that case under consideration, the absorbed power contains both even and odd parts with regards to the change $\mathbf{B} \to -\mathbf{B}$; this is because of the dependence of $\mathrm{Im}G_1$ through ω_q on n [see Eq. (39)]. Thus, the experimentally observed $P(B)$ changes under the reversal of the direction of $\mathbf{B}$. Therefore, it is convenient to represent the absorbed power as $P(B) = P^+(B) + P^-(B)$, where $P^\pm(B) \equiv [P(B) \pm P(-B)]/2$ are moduli that do not change their quantities under inversion of $\mathbf{B}$.

5. Conclusion

5.1. Summary

The microwave absorbtion by vortices in a superconductor with a periodic (washboard-type) pinning potential in the presence of the Hall effect and viscosity anisotropy has been studied theoretically. Two groups of results have been discussed. First, it has been shown how the Gittleman-Rosenblum model can be generalized for the case when a subcritical dc current is superimposed on a weak ac current. It has been elucidated how the coordinate dependence of a pinning potential can be reconstructed from the microwave absorption data measured at a set of subcritical dc currents. Second, the dependences of the longitudinal and transverse dc resistivity and ac impedance tensors, as well as of the absorbed power on the subcritical constant current density j_0, the ac frequency ω, the dimensionless Hall parameter δ, the anisotropy coefficient γ, and the angle α between the direction of the collinear currents j_0 and $j_1(t)$ with respect to the channels of the WPP have been derived for the general case. The physics of the vortex motion in a pinning potential subjected to superimposed dc an ac drives has been elucidated. In particular, it has been shown that the results are most substantially affected not by the value of γ but by the value of the Hall parameter ϵ. At a constant subcritical current $j_0 < j_c$, it turns out that ϵ does not appear in the resistive responses, whereas, at a small ac current, two new effects result from the presence of ϵ, namely (i) a falloff of the absorbtion as ϵ increases at any subcritical current $j_0 < j_c$, and (ii) the appearance of an odd-in-field component $P^-(\omega)$ when $\alpha \neq 0; 90$, and this increases with increasing ϵ.

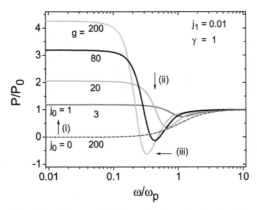

Figure 8. Dependence of the absorbed power P/P_0 on the dimensionless frequency ω/ω_p calculated on the basis of a stochastic model [25]. In the presence of a dc current $j_0 = 1$ and a series of dimensionless inverse temperatures $g = U_p/2T$, as indicated, $P(\omega)$ demonstrates (i) an enhanced power absorption by vortices at low frequencies, (ii) a pronounced temperature-dependent minimum at intermediate frequencies and (iii) a sign change at certain conditions [27]. Experimentally, $U_0 \simeq 1000 \div 5000$ K typically [10], and ω_p usually ensues in the microwave range [16, 30, 42]. In the limit of zero dc current $j_0 = 0$, the curves coincide with the well-known results of Coffey and Clem (dashed line) [49].

Our discussion has been limited by the consideration of $j_0 < j_c$ and $j_1 \to 0$ mainly for two reasons. First, at a steady-state dc current $j_0 > j_c$ the dissipated power $P_0 \sim \rho_f j_0^2$ may be significant and because of this the *overheating* of the sample is unavoidable and as a result the heat release in the film should be properly analyzed. We note, however, this difficulty can be overcome provided high-speed current sweeps [59] or short-pulse measurements [60] are employed. The second reason is that at $j_0 > j_c$ a *running* mode in the vortex dynamics appears [25], when the vortex moves in a tilted WPP with instantaneous velocity oscillating with frequency ω_i. Due to the presence of the two frequencies, i.e., *intrinsic* ω_i and *external* ω, the problem of their synchronization arises. Though an analytical treatment of this issue could be presented, to work out a clear physical picture for this problem would be more complicate.

5.2. Extension of the theory for non-zero temperature

Finally, let us compare the results presented in the chapter with the analogous but more general results obtained by the authors [25] on the basis of a stochastic model for arbitrary temperature T and densities j_0 and j_1. In that work, the Langevin equation (1), supplemented with a thermofluctuation term, has been *exactly* solved for $\gamma = 1$ in terms of a matrix continued fraction [52] and, depending on the WPP's tilt caused by the dc current, two substantially different modes in the vortex motion have been predicted. In more detail, at low temperatures and relatively high frequencies in a *nontilted* pinning potential each pinned vortex is *confined* to its pinning potential well during the *ac* period. In the case of superimposed strong ac and dc driving currents a *running* state of the vortex may appear when it can visit several (or many) potential wells during the ac period. As a result, two branches of new findings have been elucidated [25, 27]. First, the influence of an ac current on the usual $E_0(j_0)$ and ratchet $E_0(j_1)$ CVCs has been analyzed. Second, the influence of a dc current on the ac nonlinear impedance response and nonlinear power absorption has been investigated. In particular, the appearance of Shapiro-like steps in the usual CVC and the appearance of phase-locking regions in the ratchet CVC has been predicted. At the same time, it has been shown that an anomalous power absorption in the ac response is expected at close-to critical currents $j_0 \simeq j_c$ and relatively low frequencies $\omega \lesssim \omega_p$. Figure 8 shows the main predictions of these works. Namely, predicted are (i) an enhanced power absorption at low frequencies, (ii) a temperature- and current-dependent minimum at intermediate frequencies. (iii) At substantially low temperatures, the absorption can acquire negative values which physically corresponds to the generation by vortices. However, a more general and formally precise solution of the problem in terms of a matrix-continued fraction does not allow the main physical results of the problem to be investigated in the form of explicit analytical functions of the main physical quantities (j_0, j_1, ω, α, T, ϵ, and γ) which, we believe, has helped us greatly to elucidate the physics in the problem under consideration.

Acknowledgements

The authors are very grateful to Michael Huth for useful comments and critical reading. O.V.D. gratefully acknowledges financial support by the Deutsche Forschungsgemeinschaft (DFG) through Grant No. DO 1511/2-1.

Author details

Valerij A. Shklovskij
Institute of Theoretical Physics, NSC-KIPT, 61108 Kharkiv
Physical Department, Kharkiv National University, 61077 Kharkiv, Ukraine

Oleksandr V. Dobrovolskiy
Physikalisches Institut, Goethe-University, 60438 Frankfurt am Main, Germany

6. References

[1] L. V. Shubnikov, V. I. Khotkevich, Yu. D. Shepelev, and Yu. N. Ryabinin. Magnetic properties of superconductors and alloys. *Zh. Eksper. Teor. Fiz.*, 7:221–237, 1937.

[2] L. V. Shubnikov, V. I. Khotkevich, Yu. D. Shepelev, and Yu. N. Ryabinin. Magnetic properties of superconductors and alloys. *Ukr. J. Phys.*, 53:42–52, 2008.

[3] A. A. Abrikosov. On the magnetic properties of second kind superconductors. *Sov. Phys. JETP.*, 5:1174–1182, 1957.

[4] Ernst Helmut Brandt. The flux-line lattice in superconductors. *Rep. Progr. Phys.*, 58:1465–1594, 1995.

[5] A. V. Silhanek, J. Van de Vondel, and V. V. Moshchalkov. *Guided Vortex Motion and Vortex Ratchets in Nanostructured Superconductors*, chapter 1, pages 1–24. Springer-Verlag, Berlin Heidelberg, 2010.

[6] G. Blatter, M. V. Feigel'man, V. B. Geshkenbein, A. I. Larkin, and V. M. Vinokur. Vortices in high-temperature superconductors. *Rev. Mod. Phys.*, 66:1125–1388, Oct 1994.

[7] P. W. Anderson. Theory of flux creep in hard superconductors. *Phys. Rev. Lett.*, 9:309–311, Oct 1962.

[8] Ivar Giaever. Magnetic coupling between two adjacent type-ii superconductors. *Phys. Rev. Lett.*, 15:825–827, Nov 1965.

[9] A.T. Fiory, A. F. Hebard, and S. Somekh. Critical currents associated with the interaction of commensurate fluxline sublattices in a perforated al film. *Appl. Phys. Lett.*, 32:73–75, Jan 1978.

[10] Oleksiy K. Soroka, Valerij A. Shklovskij, and Michael Huth. Guiding of vortices under competing isotropic and anisotropic pinning conditions: Theory and experiment. *Phys. Rev. B*, 76:014504, Jul 2007.

[11] O. V. Dobrovolskiy, M. Huth, and V. A. Shklovskij. Anisotropic magnetoresistive response in thin nb films decorated by an array of co stripes. *Supercond. Sci. Technol.*, 23(12):125014, 2010.

[12] Valerij A. Shklovskij and Oleksandr V. Dobrovolskiy. Influence of pointlike disorder on the guiding of vortices and the hall effect in a washboard planar pinning potential. *Phys. Rev. B*, 74:104511, Sep 2006.

[13] O. K. Soroka. *Vortex dynamics in superconductors in the presence of anisotropic pinning*. PhD thesis, Johannes Gutenberg University, 2005.

[14] A. K. Niessen and C. H. Weijsenfeld. Anisotropic pinning and guided motion of vortices in type-ii superconductors. *J. Appl. Phys.*, 40:384–394, 1969.

[15] R Wördenweber, J. S. K. Sankarraj, P. Dymashevski, and E. Holmann. Anomalous hall effect studied via guided vortex motion. *Physica C*, 434:1010–104, 2006.

[16] E. Silva, N. Pompeo, S. Sarti, and C. Amabile. *Vortex State Microwave Response in Superconducting Cuprates*, chapter 1, pages 201–243. Nova Science, Hauppauge, NY, 2006.

[17] C. S. Lee, B. Janko, I. Derenyi, and A. L. Barabasi. Reducing vortex density in superconductors using the 'ratchet effect'. *Nature*, 400:337–340, May 1999.

[18] B. L. T. Plourde. Nanostructured superconductors with asymmetric pinning potentials: Vortex ratchets. *IEEE Trans. Appl. Supercond.*, 19:3698 – 3714, Oct 2009.

[19] Peter Hänggi and Fabio Marchesoni. Artificial brownian motors: Controlling transport on the nanoscale. *Rev. Mod. Phys.*, 81:387–442, Mar 2009.

[20] R. Wördenweber, P. Dymashevski, and V. R. Misko. Guidance of vortices and the vortex ratchet effect in high-T_c superconducting thin films obtained by arrangement of antidots. *Phys. Rev. B*, 69:184504, May 2004.

[21] A. Crisan, A. Pross, D. Cole, S. J. Bending, R. Wördenweber, P. Lahl, and E. H. Brandt. Anisotropic vortex channeling in YBaCuO thin films with ordered antidot arrays. *Phys. Rev. B*, 71:144504, Apr 2005.

[22] J. E. Villegas, K. D. Smith, Lei Huang, Yimei Zhu, R. Morales, and Ivan K. Schuller. Switchable collective pinning of flux quanta using magnetic vortex arrays: Experiments on square arrays of co dots on thin superconducting films. *Phys. Rev. B*, 77:134510, Apr 2008.

[23] J. I. Martin, Y. Jaccard, A. Hoffmann, J. Nogues, J. M. George, J. L. Vicent, and I. K. Schuller. Fabrication of submicrometric magnetic structures by electron-beam lithography. *J. Appl. Phys.*, 434:411–416, 1998.

[24] V. A. Shklovskij, A. K. Soroka, and A. A. Soroka. Nonlinear dynamics of vortices pinned to unidirectional twins. *J. Exp. Theor. Phys.*, 89:1138–1153, 1999.

[25] Valerij A. Shklovskij and Oleksandr V. Dobrovolskiy. ac-driven vortices and the hall effect in a superconductor with a tilted washboard pinning potential. *Phys. Rev. B*, 78:104526, Sep 2008.

[26] Valerij A. Shklovskij and Vladimir V. Sosedkin. Guiding of vortices and ratchet effect in superconducting films with asymmetric pinning potential. *Phys. Rev. B*, 80:214526, Dec 2009.

[27] Valerij A. Shklovskij and Oleksandr V. Dobrovolskiy. Frequency-dependent ratchet effect in superconducting films with a tilted washboard pinning potential. *Phys. Rev. B*, 84:054515, Aug 2011.

[28] Valerij A. Shklovskij. Guiding of vortices and the hall conductivity scaling in a bianisotropic planar pinning potential. *Phys. Rev. B*, 65:092508, Feb 2002.

[29] M. Huth, K.A. Ritley, J. Oster, H. Dosch, and H. Adrian. Highly ordered Fe and Nb stripe arrays on facetted $\alpha - al_2o_3$ $(10\bar{1}0)$. *Adv. Func. Mat.*, 12(5):333–338, 2002.

[30] Jonathan I. Gittleman and Bruce Rosenblum. Radio-frequency resistance in the mixed state for subcritical currents. *Phys. Rev. Lett.*, 16:734–736, Apr 1966.

[31] D. D. Morrison and R. M. Rose. Controlled pinning in superconducting foils by surface microgrooves. *Phys. Rev. Lett.*, 25:356–359, Aug 1970.

[32] D. Jaque, E. M. Gonzalez, J. I. Martin, J. V. Anguita, and J. L. Vicent. Anisotropic pinning enhancement in Nb films with arrays of submicrometric Ni lines. *Appl. Phys. Lett.*, 81:2851 – 2854, Oct 2002.

[33] Y Yuzhelevski and G. Jung. Artificial reversible and programmable magnetic pinning for high-T_c superconducting thin films. *Physica C: Superconductivity*, 314(314):163 – 171, 1999.

[34] O. V. Dobrovolskiy, E. Begun, M. Huth, and V. A. Shklovskij. private communication. 2012.

[35] T. A. Friedmann, M. W. Rabin, J. Giapintzakis, J. P. Rice, and D. M. Ginsberg. Direct measurement of the anisotropy of the resistivity in the ab plane of twin-free, single-crystal, superconducting $YBa_2Cu_3O_{7-\delta}$. *Phys. Rev. B*, 42:6217–6221, Oct 1990.

[36] Y. Matsuda, N. P. Ong, Y. F. Yan, J. M. Harris, and J. B. Peterson. Vortex viscosity in $YBa_2Cu_3O_{7-\delta}$ at low temperatures. *Phys. Rev. B*, 49:4380–4383, Feb 1994.

[37] P. Berghuis, E. Di Bartolomeo, G. A. Wagner, and J. E. Evetts. Intrinsic channeling of vortices along the *ab* plane in vicinal $YBa_2Cu_3O_{7-\delta}$ films. *Phys. Rev. Lett.*, 79:2332–2335, Sep 1997.

[38] H. Pastoriza, S. Candia, and G. Nieva. Role of twin boundaries on the vortex dynamics in YBa$_2$Cu$_3$O$_{7-\delta}$. *Phys. Rev. Lett.*, 83:1026–1029, Aug 1999.

[39] G. D'Anna, V. Berseth, L. Forró, A. Erb, and E. Walker. Scaling of the hall resistivity in the solid and liquid vortex phases in twinned single-crystal YBa$_2$Cu$_3$O$_{7-\delta}$. *Phys. Rev. B*, 61:4215–4221, Feb 2000.

[40] V. A. Shklovskij. Hot electrons in metals at low temperatures. *Journal of Low Temperature Physics*, 41:375–396, 1980. 10.1007/BF00117947.

[41] A. I. Bezuglyj and V. A. Shklovskij. Effect of self-heating on flux flow instability in a superconductor near T$_c$. *Physica C: Superconductivity*, 202:234 – 242, 1992.

[42] B. B. Jin, B. Y. Zhu, R. Wördenweber, C. C. de Souza Silva, P. H. Wu, and V. V. Moshchalkov. High-frequency vortex ratchet effect in a superconducting film with a nanoengineered array of asymmetric pinning sites. *Phys. Rev. B*, 81:174505, May 2010.

[43] C. Song, M. P. DeFeo, and B. L. T. Yu, K. Plourde. Reducing microwave loss in superconducting resonators due to trapped vortices. *Appl. Phys. Lett.*, 95:232501, 2009.

[44] N. S. Lin, T. W. Heitmann, K. Yu, B. L. T. Plourde, and V. R. Misko. Rectification of vortex motion in a circular ratchet channel. *Phys. Rev. B*, 84:144511, Oct 2011.

[45] N. Pompeo, E. Silva, S. Sarti, C. Attanasio, and C. Cirillo. New aspects of microwave properties of nb in the mixed state. *Physica C: Superconductivity*, 470(19):901 – 903, 2010.

[46] D. Janjušević, M. S. Grbić, M. Požek, A. Dulčić, D. Paar, B. Nebendahl, and T. Wagner. Microwave response of thin niobium films under perpendicular static magnetic fields. *Phys. Rev. B*, 74:104501, Sep 2006.

[47] V. A. Shklovskij and O. V. Dobrovolskiy. private communication. 2012.

[48] Jonathan I. Gittleman and Bruce Rosenblum. The pinning potential and high-frequency studies of type-ii superconductors. *J. Appl. Phys.*, 39(6):2617–2621, 1968.

[49] Mark W. Coffey and John R. Clem. Unified theory of effects of vortex pinning and flux creep upon the rf surface impedance of type-ii superconductors. *Phys. Rev. Lett.*, 67:386–389, Jul 1991.

[50] V. A. Shklovskij and Dang Thi Bich Hop. Effect of the transport current on microwave absorption by vortices in type-ii superconductors. *Low Temp. Phys.*, 35(5):365 – 369, 2009.

[51] V. A. Shklovskij and Dang Thi Bich Hop. The hall effect and microwave absorption by vortices in an anisotropic superconductor with a periodic pinning potential. *Low Temp. Phys.*, 36(71):71 – 80, 2010.

[52] W. T. Coffey, J. L. Déjardin, and Yu. P. Kalmykov. Nonlinear impedance of a microwave-driven josephson junction with noise. *Phys. Rev. B*, 62:3480–3487, Aug 2000.

[53] Yasunori Mawatari. Dynamics of vortices in planar pinning centers and anisotropic conductivity in type-ii superconductors. *Phys. Rev. B*, 56:3433–3437, Aug 1997.

[54] Yasunori Mawatari. Anisotropic current-voltage characteristics in type-ii superconductors with planar pinning centers. *Phys. Rev. B*, 59:12033–12038, May 1999.

[55] G. Grüner. The dynamics of charge-density waves. *Rev. Mod. Phys.*, 60:1129–1181, Oct 1988.

[56] A. A. Golubov, M. Yu. Kupriyanov, and E. Il'ichev. The current-phase relation in josephson junctions. *Rev. Mod. Phys.*, 76:411–469, Apr 2004.

[57] A. M. Campbell and J. E. Evetts. Flux vortices and transport currents in type ii superconductors. *Adv. Phys.*, 21(90):199–428, 1972.

[58] J. M. Harris, Y. F. Yan, O. K. C. Tsui, Y. Matsuda, and N. P. Ong. Hall angle evidence for the superclean regime in 60 k $YBa_2Cu_3O_{6+y}$. *Phys. Rev. Lett.*, 73:1711–1714, Sep 1994.

[59] A. Leo, G. Grimaldi, R. Citro, A. Nigro, S. Pace, and R. P. Huebener. Quasiparticle scattering time in niobium superconducting films. *Phys. Rev. B*, 84:014536, Jul 2011.

[60] Manlai Liang and Milind N. Kunchur. Vortex instability in molybdenum-germanium superconducting films. *Phys. Rev. B*, 82:144517, Oct 2010.

Chapter 3

Effective Models of Superconducting Quantum Interference Devices

R. De Luca

Additional information is available at the end of the chapter

1. Introduction

It is well known that the electrodynamic properties of SQUIDs (Superconducting Quantum Interference Devices) are obtained by means of the dynamics of the Josephson junctions in these superconducting system (Barone & Paternò, 1982; Likharev, 1986; Clarke & Braginsky, 2004). Due to the intrinsic macroscopic coherence of superconductors, r. f. SQUIDs have been proposed as basic units (qubits) in quantum computing (Bocko et al., 1997). In the realm of quantum computing non-dissipative quantum systems with small (or null) inductance parameter and finite capacitance of the Josephson junctions (JJs) are usually considered (Crankshaw & Orlando, 2001). The mesoscopic non-simply connected classical devices, on the other hand, are generally operated and studied in the overdamped limit with negligible capacitance of the JJs and small (or null) values of the inductance parameter. Nonetheless, r. f. SQUIDs find application in a large variety of fields, from biomedicine to aircraft maintenance (Clarke & Braginsky, 2004), justifying actual scientific interest in them.

As for d. c. SQUIDs, these systems can be analytically described by means of a single junction model (Romeo & De Luca, 2004). The elementary version of the single-junction model for a d. c. SQUID takes the inductance L of a single branch of the device to be negligible, so that $\beta = LI_J/\Phi_0 \approx 0$, where Φ_0 is the elementary flux quantum and I_J is the average value of the maximum Josephson currents of the junctions. In this way, the Josephson junction dynamics is described by means of a nonlinear first-order ordinary differential equation (ODE) written in terms of the phase variable ϕ, which represents the average of the two gauge-invariant superconducting phase differences, ϕ_1 and ϕ_2, across the junctions in the d. c. SQUID. By considering a device with equal Josephsons junction in each of the two symmetric branches, the dynamical equation of the variable ϕ can be written as follows (Barone & Paternò, 1982):

$$\frac{d\phi}{d\tau} + (-1)^n \cos \pi \psi_{ex} \sin \phi = \frac{i_B}{2} \tag{1}$$

where n is an integer denoting the number of fluxons initially trapped in the superconducting interference loop, $\tau = 2\pi R I_j t / \Phi_0 = t / \tau_\phi$, R being the intrinsic resistive junction parameter, $\psi_{ex} = \Phi_{ex} / \Phi_0$ is the externally applied flux normalized to Φ_0 and $i_B = I_B / I_j$, is the bias current normalized to I_j. In what follows we shall consider zero-field cooling conditions, thus taking $n = 0$. Eq. (1) is similar to the non-linear first-order ODE describing the dynamics of the gauge-invariant superconducting phase difference across a single overdamped JJ with field-modulated maximum current I_{JF} ($I_{JF} = |\cos \pi \psi_{ex}|$) in which a normalized bias current $i_B/2$ flows. This strict equivalence comes from the hypothesis that the total normalized flux $\psi = \Phi / \Phi_0$ linked to the interferometer loop can be taken to be equal to ψ_{ex}. However, being

$$\psi = \psi_{ex} + \beta(i_1 - i_2), \tag{2}$$

we may say that the above hypothesis may be stated merely by means of the following identity: $\beta = 0$. Therefore, for finite values of the parameter β, Eq. (1) is not anymore valid and the device behaves as if the equivalent Josephson junction possessed a non-conventional current-phase relation (CPR). In fact, for small finite values of β, one can see that the following model may be adopted (Romeo & De Luca, 2004):

$$\frac{d\phi}{d\tau} + X_{ex} \sin \phi + \pi \beta Y_{ex}^2 \sin 2\phi = \frac{i_B}{2} \tag{3}$$

where $X_{ex} = \cos \pi \psi_{ex}$ and $Y_{ex} = \sin \pi \psi_{ex}$. A second-order harmonic in ϕ thus appears in addition to the usual $\sin \phi$ term. The $\sin 2\phi$ addendum, however, arises solely from electromagnetic coupling between the externally applied flux and the system, as described by Eq. (2), when $\beta \neq 0$. Therefore, the non-conventional CPR of the equivalent JJ in the SQUID model cannot be considered as a strict consequence of an intrinsic non-conventional CPR of the single JJs. The Josephson junctions in the device, in fact, could behave in the most classical way, obeying strictly to the Josephson current-phase relation; the interferometer, however, would still show the additional $\sin 2\phi$ term for finite values of β. In order to understand how the reduction in the dimensional order of the dynamical equations is possible, it is noted that the quantities $\tau_\phi = \Phi_0 / 2\pi R I_j$ and $\tau_\psi = L/R$, denoting the characteristic time scales of the variables ϕ and of the number of fluxons ψ in the superconducting SQUID loop, respectively, are intimately linked to the parameter β, since $\tau_\psi / \tau_\phi = 2\pi \beta$. In this way, for constant applied magnetic fields, the flux dynamics for small values of β can be considered very fast with respect to the equivalent junction dynamics given in Eq. (3). As a consequence, the superconducting phase ϕ can be assumed to be adiabatic and the equation of motion for ψ in terms of the quasi-static variable ϕ can be solved by perturbation analysis. When the information for ψ is substituted back into the effective dynamical equation for ϕ, Eq. (3) is finally obtained.

The single-junction model can be adopted also when dealing with more complex systems, as one-dimensional Josephson junction arrays. In fact, by assuming small inductance values in the N current loops of a one-dimensional array containing $N+1$ identical overdamped Josephson junctions, the dynamical equations for the gauge-invariant superconducting phase differences can be reduced to a single non linear differential equation (Romeo & De Luca, 2005). The resulting time-evolution equation is seen to be similar to the single-junction dynamical equation with an appropriately defined current-phase relation. As specified before, the critical current, the I-V characteristics and the flux-voltage curves of the array can be determined analytically by means of the effective model. Furthermore, a one dimensional array of N cells of 0- and π-junctions in parallel can be considered (De Luca, 2011). In this case, by assuming that junctions parameters and effective loop areas alternate as one moves along the longitudinal direction of the array, going from 0- to π-junctions, an effective single junction model for the system can be derived. It can be shown that, by this model, interference patterns of the critical current as a function of the applied magnetic flux can be analytically found and compared with existing experiments (Scharinger et al., 2010).

Finally, a single-junction model for a d.c. SQUID is derived when we consider the effect of rapidly varying applied fields whose frequency ω is comparable with τ_ψ^{-1}. By letting the applied magnetic flux have, in addition to a constant term A, an a. c. component, we can find, by similar reasoning as in the case of a constant applied magnetic flux, the effective reduced single-junction model for the system. In particular, for $\beta = 0$, the critical current of the device is seen to depend on A, and on the frequency ω and the amplitude B of the a. c. component of the applied magnetic flux in a closed analytic form. From the analysis of the voltage vs. applied flux curves it can be argued that the quantities ω and B can play the role of additional control parameters in the device.

The work will thus be organized as follows. In Section 2 the derivation of an effective single-junction model for a symmetric d. c. SQUID containing two equal junctions will be briefly reviewed. In Section 3 the extension to this model to Josephson junction arrays with equal junctions in all branches will be considered. In Section 4 the case of the alternate presence of 0- and π-junctions in the array is considered, the system being similar to multifacets Josepshon junctions. In Section 5 the effective single-junction model for a d. c. SQUID in the presence of rapidly varying field is derived. Finally, in Section 6 conclusions are drawn and further investigations are suggested.

2. Two-junction quantum interference devices

Let us consider a symmetric two junction interferometer with equal junctions of negligible capacitance, as shown in fig. 1. The dynamical equations for the variables ϕ and ψ characterizing this system, can be written in the following form (Romeo & De Luca, 2004):

$$\frac{d\phi}{d\tau} + (-1)^n \cos(\pi\psi)\sin\phi = \frac{i_B}{2}; \qquad \text{(a)}$$

$$\pi\frac{d\psi}{d\tau} + (-1)^n \sin(\pi\psi)\cos\phi + \frac{\psi}{2\beta} = \frac{\psi_{ex}}{2\beta}. \quad \text{(b)}$$

(4)

where n is the number of initially trapped fluxons in the superconducting loop. Let us consider a new time variable $\theta = \dfrac{\tau}{2\pi\beta} = \dfrac{R}{L}t$ and write the solutions for ϕ a and ψ in the following form: $\phi(\beta,\tau) \approx \phi_{0,\beta}(\tau) + \beta\phi_{1,\beta}(\tau)$; $\psi(\beta,\tau) \approx \psi_0(\theta) + \beta\psi_1(\theta)$. For simplicity, set $n=0$.

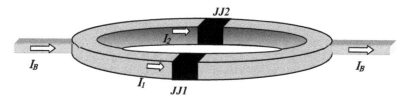

Figure 1. Schematic representation of a two-junction quantum interferometer

This approach allows us not only to account for the regular part of the solution, as seen in (Grønbech-Jensen et al., 2003) and in (Romeo & De Luca, 2004), but also to consider its singular part. Moreover, as we shall see, the role of the two time variables will become evident in what follows, since one time scale is defined for Eq. (4a) and one for Eq. (4b). Consider then $\phi(\beta,\tau)$ and $\psi(\beta,\tau)$ to be bounded, differentiable functions, and expand the sine and cosine functions appearing in Eqs. (4a-b) to first order in β. By then collecting all coefficients of identical power of β, we can obtain a system of equations for the functions $\phi_{k,\beta}(\tau)$ and $\psi_k(\theta)$, with $k = 0, 1$, describing the k-th order solutions for ϕ and ψ, respectively. These approximate solutions are determined according to the following sequential scheme. As a first step, we use Eq. (4b) to determine $\psi_0(\theta)$. We adopt the solution found and substitute it in Eq. (4a) to determine $\phi_{0,\beta}(\tau)$. The latter solution, on its turn, is substituted in Eq. (4b) to find $\psi_1(\theta)$ and, finally, this solution is used in Eq. (4a) to find $\phi_{1,\beta}(\tau)$. Note, therefore, that for defining first order solutions, knowledge of zero-th order solutions is required. Furthermore, we assume that the initial conditions are the following:

$$\phi(\beta,0) = \phi_{0,\beta}(0) + \beta\phi_{1,\beta}(0); \quad \psi(\beta,0) = \psi_0(0) + \beta\psi_1(0) . \tag{5}$$

As for initial conditions, from Eq. (4b) we may notice that $\psi_0(\tau) = \psi_{ex}$ for $\beta = 0$, in which case we cannot even define the time variable θ. This condition, however, is inherited by the function $\psi_0(\theta)$, since the following equalities are satisfied:

$$\psi_0\left(\frac{\tau}{\beta}\right)_{\beta=0} = \psi_0(\theta)_{\theta\to\infty} = \psi_{ex}. \tag{6}$$

Furthermore, we may also notice that $\psi_k(\theta)_{\theta=0} = \psi_k\left(\dfrac{\tau}{\beta}\right)_{\tau=0} = \psi_k(0)$, for $k = 0, 1$.

By the general procedure described above we get the following differential equations for the superconducting phase variables:

$$\frac{d\phi_{0,\beta}}{d\tau} + \cos\left(\pi\psi_0(\theta)\right)\sin\phi_{0,\beta}(\tau) = \frac{i_B}{2},$$

$$\frac{d\phi_{1,\beta}}{d\tau} + \phi_{1,\beta}(\tau)\cos\left(\pi\psi_0(\theta)\right)\cos\phi_{0,\beta}(\tau) = \pi\psi_1(\theta)\sin\left(\pi\psi_0(\theta)\right)\sin\phi_{0,\beta}(\tau);$$ (b)

(a)

(7)

and the following for the flux number variables:

$$\frac{d\psi_0}{d\theta} + \psi_0(\theta) = \psi_{ex},$$ (a)

$$\frac{d\psi_1}{d\theta} + \psi_1(\theta) = -2\sin\left(\pi\psi_0(\theta)\right)\cos\phi_{0,\beta}(2\pi\beta\theta).$$ (b)

(8)

In Eqs. (7a-b) and (8a-b) we may notice the appearance of two different time scales the first, $\tau_\psi = \frac{L}{R}$, linked to flux motion in and out the superconducting ring, the second, $\tau_\phi = \frac{\Phi_0}{2\pi RI_J}$, pertaining to the dynamics of the superconducting phase difference value ϕ. We have already noticed that $\frac{\tau_\psi}{\tau_\phi} = 2\pi\beta$, so that, for negligible values of this ratio, the system behaves effectively as if the adiabatic time evolution of the superconducting phase difference variable ϕ could be studied by taking asymptotic solutions of ψ. In this case, therefore, we may first let the flux variable evolve, so that a stationary magnetic state is reached, and then solve for the superconducting phase difference time evolution of the system. This is exactly what is done, under the assumptions of negligible value of the ratio $r = \frac{\tau_\psi}{\tau_\phi}$, in (Grønbech-Jensen et al., 2003) and in (Romeo & De Luca, 2004). However, if one were to acquire the regular solution for the system dynamics, even when considering the approximate solution for the variables ϕ and ψ, one would follow the more general perturbation analysis described above, where the ratio r might not a priori be considered as negligible. Furthermore, considering that this ratio is proportional to the perturbation parameter β, one might wish to generalize the procedure described above to higher order in β to acquire a wider range of validity of the analysis.

Despite the fact that the more general approach allows extension to higher order approximations of the perturbation solutions, we wish to limit our analysis to the study of the electrodynamic properties of a two-junction or a multi-junction quantum interferometer with very small parameter β. Therefore, while in the present section we shall only be concerned with a single time scale, namely τ_ϕ, by assuming that the transient of the flux variable rapidly vanishes ($\frac{\tau_\psi}{\tau_\phi} \ll 1$), in Section 5 we shall consider both evolution time, by introducing a rapidly varying external magnetic field.

By considering, for the time being, only the time scale τ_ϕ, the dynamical equations for the flux variable (Eqs. (8a-b)) give the following steady-state solution for ψ_0 and ψ_1:

$$\psi_0 = \psi_{ex}, \qquad \text{(a)}$$

$$\psi_1 = -2\sin(\pi\psi_{ex})\cos\phi_0 - 2\pi\frac{d\psi_{ex}}{d\tau}, \quad \text{(b)}$$

(9)

where the term $2\pi\dfrac{d\psi_{ex}}{d\tau}$ has been inserted in Eq. (9b), in order to correctly take account of first order contributions in β, and the subscript β in $\phi_{0,\beta}(\tau)$ has been elided, as it will be done for $\phi_{1,\beta}(\tau)$ from this point on, since these functions will not depend on β in this limit.

In order to obtain some preliminary results, we start by considering ψ_{ex} as constant. By the general procedure schematized above we get the following differential equations for the superconducting phase variables:

$$\frac{d\phi_0}{d\tau} + \cos(\pi\psi_{ex})\sin\phi_0(\tau) = \frac{i_B}{2}, \qquad \text{(a)}$$

$$\frac{d\phi_1}{d\tau} + \phi_1(\tau)\cos(\pi\psi_{ex})\cos\phi_0(\tau) = \pi\psi_1(\tau)\sin(\pi\psi_{ex})\sin\phi_0(\tau), \quad \text{(b)}$$

(10)

and the following for the flux number variables:

$$\psi_0 = \psi_{ex}, \qquad \text{(a)}$$

$$\psi_1(\tau) = -2\sin(\pi\psi_{ex})\cos\phi_0(\tau). \text{ (b)}$$

(11)

According to the scheme described above, by having already set $\psi_0 = \psi_{ex}$, we may now solve for $\phi_0(\tau)$ in Eq. (10a). Let us therefore briefly discuss how to obtain this solution. In the case $\left|\dfrac{i_B}{2\cos(\pi\psi_{ex})}\right| = \dfrac{1}{|a|} > 1$, which characterizes the running state of the junctions, we have

$$\phi_0(\tau) = 2\tan^{-1}\left[a + \sqrt{1-a^2}\,\tan\left(\gamma\tau + \tan^{-1}\xi_0\right)\right]. \ + 2k\pi \qquad (12)$$

where k is an integer, $\gamma = \dfrac{1}{2}\sqrt{\dfrac{i_B^2}{4} - \cos^2(\pi\psi_{ex})}$ and $\xi_0 = \dfrac{\tan\left(\dfrac{\phi_0(0)}{2}\right) - a}{\sqrt{1-a^2}}$.

On the other hand, in the case $\left|\dfrac{i_B}{2\cos(\pi\psi_{ex})}\right| = \dfrac{1}{|a|} < 1$, which characterizes the superconducting state of the junctions, we have

$$\phi_0(\tau) = 2\tan^{-1}\left[a + \sqrt{a^2-1}\,\tanh\left(\tilde{\gamma}\tau + \tanh^{-1}(\chi_0)\right)\right], \qquad (13)$$

where $\tilde{\gamma} = \dfrac{1}{2}\sqrt{\cos^2(\pi\psi_{ex}) - \dfrac{i_B^2}{4}}$ and $\chi_0 = \dfrac{\tan\left(\dfrac{\phi_0(0)}{2}\right) - a}{\sqrt{a^2 - 1}}$.

Finally, in the case $|a| = 1$, we have

$$\phi_0(\tau) = \operatorname{sgn}(a)\left[2\tan^{-1}\left(\frac{i_B\tau}{2} + \omega_0^{(\pm)}\right) - \frac{\pi}{2}\right], \tag{14}$$

where $\omega_0^{(\pm)} = \tan\left(\operatorname{sgn}(a)\dfrac{\phi_0(0)}{2} + \dfrac{\pi}{4}\right)$. Having found the time dependence of the variable ϕ_0,

ψ_1 can be found by Eq. (11b) by substitution. Finally, by knowledge of ψ_0, ϕ_0 and ψ_1, the function ϕ_1 can be found by Eq. (10b), which is a standard first order linear differential equation. Solutions for ϕ_0 are shown for $\cos(\pi\psi_{ex}) = 0.3$ in figs. 2a, 2b, and 2c for $i_B = 1.6$, $i_B = 0.4$, and $i_B = 0.6$, respectively, along with the solution obtained by numerically integrating Eqs. (4a-b). In figs. 2b-c the first order approximation of the solution is shown as a dotted line.

Solutions for ϕ_1 are shown in figs. 3a-c for $\cos(\pi\psi_{ex}) = 0.3$ and $i_B = 1.6$, $i_B = 0.4$ and $i_B = 0.6$, respectively. The above analysis thus leads to a solution in a closed form, to first order in the parameter β. Notice that in the case of time-dependent bias currents one should adopt a more general procedure.

As a simple application, let us calculate, to first order in the parameter β, the circulating current i_S in the circuit, normalized to I_J, given by (Barone & Paternò, 1982):

$$i_S = \frac{\psi - \psi_{ex}}{\beta}. \tag{15}$$

For an arbitrary value n, which represents the number of fluxons initially trapped in the superconducting ring, we have

$$i_S = \psi_1(\tau) = -2\sin(\pi\psi_{ex})\cos\phi_0(\tau). \tag{16}$$

Graphs of circulating currents are shown in figs. 4a, 4b, 4c for n even, $i_B = 2.2$ and for $\psi_{ex} = 0.1$ and 0.3, $\psi_{ex} = 0.5$, $\psi_{ex} = 0.7$ and 0.9, respectively. The period T of these curves is equal to the pseudo-period of $\phi_{0,\beta}$ which is given by the following expression in terms of ψ_{ex} and i_B:

$$T = \frac{2\pi}{\sqrt{\left(\dfrac{i_B}{2}\right)^2 - \cos^2(\pi\psi_{ex})}}. \tag{17}$$

Notice that the lowest value of the period is obtained for $\psi_{ex} = 0.5$ and that the curves for $\psi_{ex} = 0.1$ and 0.9 and for $\psi_{ex} = 0.3$ and 0.7, although having the same period, as it can be argued from Eq. (17), are not equal.

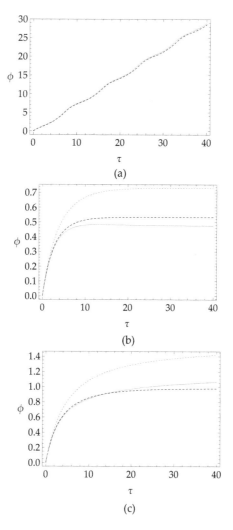

(a)

(b)

(c)

Figure 2. Average phase difference for $\cos(\pi\psi_{ex}) = 0.3$ and: a) $i_B = 1.6$; b) $i_B = 0.4$; c) $i_B = 0.6$. Dotted blue lines represent $\phi_0(\tau)$ for $\beta = 0.02$, full red lines represent $\phi(\tau)$ as calculated to first order for $\beta = 0.02$, and the dashed black lines represents the numerical solution of the complete system. In a) the first order approximation of the solution is not shown, for clarity reasons.

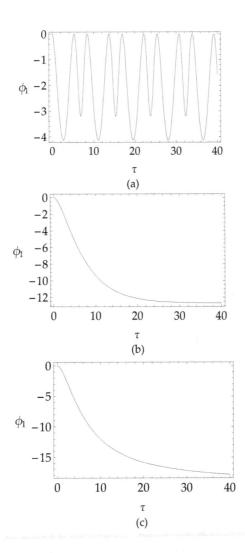

Figure 3. First order correction to $\phi_0(\tau)$ as calculated by the procedure described in the text by setting β=0.02, $\cos(\pi\psi_{ex})=0.3$ and: a) $i_B=1.6$; b) $i_B=0.4$; c) $i_B=0.6$.

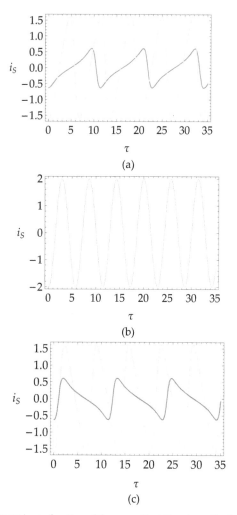

Figure 4. Circulating current i_S as a function of the normalized time for null values of the initially trapped flux and for $i_B = 2.2$ and: a) $\psi_{ex} = 0.1$ (orange), $\psi_{ex} = 0.3$ (cyan) ; b) $\psi_{ex} = 0.5$; c) $\psi_{ex} = 0.7$ (cyan), $\psi_{ex} = 0.9$ (orange).

The above results have been obtained for the magnetic response of the system in the presence of a constant applied flux. In the following we shall analyze the electrodynamic response of the two junction quantum interferometer in the presence of a time-dependent external flux. For this purpose, we shall take a sinusoidal forcing term, in such a way that $\psi_{ex}(t) = A + B\cos\omega t$, where A is the normalized d. c. component of the applied flux and B is the normalized amplitude of the a. c. signal.

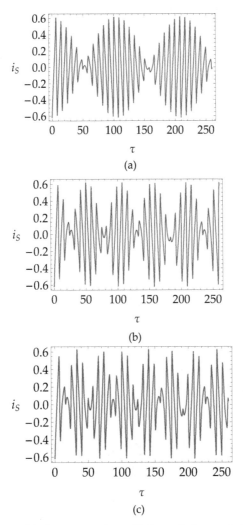

Figure 5. Circulating current i_S in the presence of an oscillating applied magnetic field shown as a function of the normalized time for $A=0$, $B=0.1$, $i_B = 2.5$, $\beta=0.01$ and for normalized frequencies equal to: a) $\tilde{\omega} = 0.03$; b) $\tilde{\omega} = 0.06$; c) $\tilde{\omega} = 0.09$. No initially trapped flux in the superconducting loop is present.

Now, since $t = \dfrac{\Phi_0}{2\pi RI_J}\tau$, we can write

$$\psi_{ex}(\tau) = A + B\cos\tilde{\omega}\tau, \tag{18}$$

where $\tilde{\omega} = \dfrac{\Phi_0}{2\pi RI_J}\omega$. By considering normalized frequencies of the order of τ_ϕ^{-1}, we

may still use the analysis described above. In Section 5 we shall relax the latter hypothesis. As also specified above, a different approach, which will be developed in Section 5, needs to be used when very rapidly varying applied fields are applied to the system. We shall assume that the normalized amplitude B of the oscillating signal is much less than one ($B<<1$). The perturbation analysis is then carried out in a way at all similar as done above.

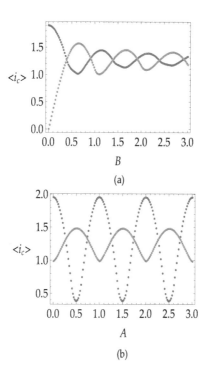

(a)

(b)

Figure 6. (a) Time average value $\langle i_c \rangle$ of the critical current i_c as a function of the amplitude B of the oscillating magnetic flux for A=0.1 (blue) and A=0.5 (orange). (b) Time average value $\langle i_c \rangle$ of the critical current i_c as a function of A for B=0.1 (blue) and B=0.5 (orange). Both figures are obtained by taking null values of the initially trapped flux.

We start by setting, by Eq. (9a) and (9b), $\psi_0(\tau) = \psi_{ex}(\tau) = A + B\cos(\tilde{\omega}\tau)$ and

$\psi_1 = -2\sin(\pi\psi_{ex}(\tau))\cos\phi_0 - 2\pi\dfrac{d\psi_{ex}}{d\tau}$ and solve the equations for the phase differences

$$\frac{d\phi_0}{d\tau} + \cos\left(\pi\psi_{ex}(\tau)\right)\sin\phi_0(\tau) = \frac{i_B}{2}, \tag{a}$$

$$\frac{d\phi_1}{d\tau} + \phi_1(\tau)\cos\left(\pi\psi_{ex}(\tau)\right)\cos\phi_0(\tau) = -\pi\sin^2\left(\pi\psi_{ex}(\tau)\right)\sin\left(2\phi_0(\tau)\right) \tag{b}$$

(19)

By noticing, however, that $\cos\left(\pi\psi_{ex}(\tau)\right) = \cos\left(\pi A + \pi B\cos\tilde{\omega}\tau\right) \approx \cos(\pi A) - \pi B\sin$, $(\pi A)\cos\tilde{\omega}\tau$ Eq. (19a) can be written in the following form:

$$\frac{d\phi_0}{d\tau} + \left(a - b\cos\tilde{\omega}\tau\right)\sin\phi_0(\tau) = \frac{i_B}{2}, \tag{20}$$

where $a = \cos(\pi A)$ and $b = \pi B\sin(\pi A)$. In Eq. (20) we find a perturbed solution in terms of the parameter b, so that, by setting $\phi_0(\tau) = \eta_0(\tau) + b\eta_1(\tau)$, we can write:

$$\frac{d\eta_0}{d\tau} + a\sin\eta_0(\tau) = \frac{i_B}{2}, \tag{a}$$

$$\frac{d\eta_1}{d\tau} + a\cos\eta_0(\tau)\eta_1(\tau) = \sin\eta_0(\tau)\cos(\tilde{\omega}\tau) \tag{b}$$

(21)

Notice then that the solutions to the above equations can be found by exactly the same procedure described for the case of constant applied fields. Once the solution for $\phi_0(\tau)$ is found, by substituting in Eq. (19b), the solution for $\phi_1(\tau)$ can be determined by solving a first order linear differential equation with time-dependent coefficients. Assuming thus $\phi_0(\tau) = \eta_0(\tau) + b\eta_1(\tau)$ to be a known expression, we can then write:

$$i_S = -2\sin\left(\pi\psi_{ex}(\tau)\right)\cos\left(\eta_0(\tau) + b\eta_1(\tau)\right) + 2\pi\tilde{\omega}B\sin\tilde{\omega}\tau \tag{22}$$

As before, the above expression is equal to $\psi_1(\tau)$ and represents the circulating current i_S in the circuit. In figs. 5a, 5b, and 5c the time dependence of the current i_S for normalized frequency values $\tilde{\omega} = 0.03, 0.06, 0.09$, respectively, and for $A=0$, $\beta=0.01$ and $i_B = 2.5$ is represented. In these graphs we notice that the oscillating patterns, which we have already detected in the constant applied field case, are modulated by the externally applied oscillating signal.

Another important quantity to be measured in these systems is the critical current i_c, defined as the maximum value of the current bias i_B which can be injected in the two junction interferometer without giving rise to dissipation. By considering the stationary case of Eq. (21a) we write:

$$i_B = 2\cos\left(\pi\psi_{ex}(\tau)\right)\sin\phi_0(\tau). \tag{23}$$

Therefore, we have

$$i_c = 2\left|\cos\left(\pi A + \pi B\cos(\tilde{\omega}\tau)\right)\right|. \tag{24}$$

Noticing that the time-averaged value <i_c> of the critical current does not depend on the normalized frequency, it can be calculated in terms of solely A and B, the results being shown in fig. 6a and fig. 6b for null values of the initially trapped flux. In particular, in fig. 6a <i_c> is shown as a function of the applied magnetic field amplitude B, for $A=0.1$ and $A=0.4$, while in fig. 6b, <i_c> vs. A curves are shown for $B=0.1$ and $B=0.2$. In the curves in fig. 6a we notice Fraunhofer-like oscillations, while ordinary cosinusoidal oscillations are present in fig. 6b.

For what seen above, the electrodynamic properties of a symmetric quantum interferometer containing two identical junctions with negligible capacitance can be studied by means of a perturbation approach in the parameter β, whose value gives the strength of the electromagnetic coupling between the two junction in the system. The analysis is rather similar to what done in other works in the literature (Grønbech-Jensen et al., 2003; Romeo & De Luca, 2004). However, in the present section we have presented a rather general procedure to obtain the solution to the problem to first order in the parameter β. Considering at first transient solutions, we have noticed that the function $\psi(\beta,\theta)$ governs fluxon dynamics, where θ is the ordinary time t, normalized to the characteristic circuital time constant $\tau_\psi = \dfrac{L}{R}$. By this more general approach it becomes evident that the characteristic time constant $\tau_\phi = \dfrac{\Phi_0}{2\pi R I_J}$ of the dynamics of the average superconducting phase difference ϕ is different from the fluxon dynamics characteristic time τ_ψ, so that the asymptotic solution for the system, proposed in the analyses carried out in (Grønbech-Jensen et al., 2003) and in (Romeo & De Luca, 2004), acquires a more precise meaning in this context. Indeed, when the parameter β is sufficiently small to allow, for finite values of τ, an asymptotic evaluation of $\psi_0\left(\dfrac{\tau}{2\pi\beta}\right)$ and $\psi_1\left(\dfrac{\tau}{2\pi\beta}\right)$, the general solution given in the present work coincides with the asymptotic perturbed solution proposed in (Grønbech-Jensen et al., 2003) and in (Romeo & De Luca, 2004) in the limit of negligible junction capacitance.

The perturbation analysis has been first carried out for a constant applied magnetic flux. Successively, since it could be experimentally possible to force the system with a time-dependent magnetic field, it is noted that the perturbed solution for the flux number ψ, obtained for a sinusoidal magnetic flux, needs careful evaluation. In order to exhibit experimentally detectable quantities, the circulating current i_S is evaluated as a function of time, for different values of the frequency of the forcing field, whose a. c. component is assumed to be small. Finally, the time average <i_c> of the critical current of the device has been studied both as a function of the d. c. component A and of the small amplitude B of the oscillating part of the applied flux. In these curves two characteristic behaviors have been detected: A Fraunhofer-like pattern in <i_c> vs. B curves; independence of <i_c> from $\tilde{\omega}$. We shall see in Section 5 of the present Chapter that dependence from $\tilde{\omega}$ appears in a more general case, i. e., when $\tilde{\omega} \approx 1$ and the coefficient B is not assumed to be small.

3. Multi-junction quantum interference devices

In this section we shall consider the one-dimensional Josephson junction array (1D-JJA) represented in fig. 7, consisting of $N+1$ identical overdamped junctions connected in parallel. In this figure we notice that the bias current I_B is evenly applied to the two external branches of the array. By assuming perfectly identical overdamped Josephson junctions with resistive parameter R and maximum Josephson current I_J, we take the inductance L of the horizontal upper branches to be such that $\beta = \dfrac{LI_J}{\Phi_0} \ll 1$, where Φ_0 is the elementary flux quantum. This parameter has been defined as each single loop could be compared to a two-junction quantum interference device. By fluxoid quantization, the normalized magnetic flux $\Psi_k = \dfrac{\Phi_k}{\Phi_0}$ linked to the k-th cell of the array is related to the superconducting phase differences ϕ_{k-1} and ϕ_k across the two junctions in the cell, so that:

$$\phi_{k-1} - \phi_k + 2\pi\Psi_k = 2\pi n_k, \tag{25}$$

where n_k is an integer and $k = 1, 2, ..., N$.

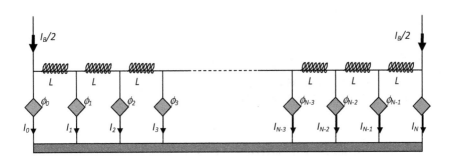

Figure 7. Schematic representation of a one-dimensional array of equal Josephson junctions.

When an external magnetic field $\mathbf{H}$, orthogonal to the plane of the array, is applied to the system so that the normalized geometric applied flux through each cell is $\Psi_{ex} = \dfrac{\Phi_{ex}}{\Phi_0} = \dfrac{\mu_0 H S_0}{\Phi_0}$, where S_0 is the cell area, we may set

$$\Psi_k = \beta \sum_{m=1}^{k} i_m - \beta \frac{i_B}{2} + \Psi_{ex}, \tag{26}$$

for $k = 1, 2, ..., N$, where $i_k = \dfrac{I_k}{I_J}$ is the normalized current flowing in the k-th branch

and $i_B = \dfrac{I_B}{I_J}$. The dynamical equation for each Josephson junction in the array is written

by means of the resistively shunted junction (RSJ) model (Barone & Paternò, 1982) as follows:

$$\frac{d\phi_k}{d\tau} + \sin \phi_k = i_k, \tag{27}$$

where $k = 0, 1, 2, ..., N$ and $\tau = \dfrac{2\pi R I_J}{\Phi_0} t$. Eqs. (25-27) can be used to define the dynamics of the

gauge-invariant superconducting phase difference ϕ_k in terms of the forcing parameters

Ψ_{ex} and $i_B = \dfrac{I_B}{I_J} = \sum_{k=0}^{N} i_k$. In addition, the instantaneous voltage $v(\tau)$ of the system can be

obtained by setting:

$$v(\tau) = \frac{1}{N+1} \sum_{k=0}^{N} \frac{d\phi_k}{d\tau}. \tag{28}$$

Define now the partial sum s_n ($1 \le n \le N$) of the normalized fluxes as follows:

$$s_n = \sum_{k=1}^{n} \Psi_k. \tag{29}$$

By fluxoid quantization (Eq. (25)), setting all n_k's to zero (under the hypothesis of zero initially trapped flux in the array), we can write:

$$\phi_k = \phi_0 + 2\pi s_k, \tag{30}$$

for $k = 1, 2, ..., N$, so that the dynamical equations (Eq. (27)) can be rewritten as follows:

$$\frac{d\phi_0}{d\tau} + \sin \phi_0 = i_0, \qquad \text{(a)}$$

$$2\pi \frac{ds_n}{d\tau} + \sin(\phi_0 + 2\pi s_n) = i_n - i_0 + \sin \phi_0 \quad \text{(b)} \tag{31}$$

where $n = 1, 2, ..., N$ in the second equation. Expressing now, by means of Eq. (26), the currents i_0 and i_n in terms of the forcing terms Ψ_{ex} and i_B and of the partial sums of the flux variables s_n, we may finally rewrite Eqs. (31a-b) as follows:

$$\frac{d\phi_0}{d\tau} + \sin\phi_0 = \frac{i_B}{2} + \frac{s_1 - \Psi_{ex}}{\beta}, \tag{a}$$

$$2\pi\frac{ds_n}{d\tau} + \sin\left(\phi_0 + 2\pi s_n\right) - \sin\phi_0 - \frac{s_{n+1} - 2s_n + s_{n-1}}{\beta} = \frac{\Psi_{ex} - s_1}{\beta} - \frac{i_B}{2}, \tag{b}$$

$$2\pi\frac{ds_N}{d\tau} + \sin\left(\phi_0 + 2\pi s_N\right) - \sin\phi_0 + \frac{s_N - s_{N-1}}{\beta} = \frac{\Psi_{ex}}{\beta} - \frac{s_1 - \Psi_{ex}}{\beta} \tag{c}$$

(32)

where $n = 1, 2, ..., N - 1$ in Eq. (32b).

The above analysis has been carried out essentially to write the dynamical equations in terms of the effective superconducting phase difference ϕ_0 and of the N partial sums s_n.

We shall now develop a reduction of these variables to one, by assuming small values of the parameter β. Therefore, start by considering the dynamical equations of the system as written in Eqs. (32a-b). For small values of the parameter β we can set:

$$s_n \approx s_n^{(0)} + \beta s_n^{(1)}. \tag{33}$$

By substituting the above expression in Eq. (32b) and, after having multiplied both members by β, by setting the coefficients of β^m ($m = 0, 1$) equal to zero, we obtain, for $n = 1, 2, ..., N$:

$$s_n^{(0)} = n\Psi_{ex}. \tag{34}$$

For the first order corrections, on the other hand, we need to solve the following set of equations:

$$\begin{cases} y_1 - y_0 = s_2^{(1)} - 3s_1^{(1)} - \dfrac{i_B}{2} \\[2mm] y_2 - y_0 = s_3^{(1)} - 2s_2^{(1)} - \dfrac{i_B}{2} \\[2mm] y_3 - y_0 = s_4^{(1)} - 2s_3^{(1)} + s_2^{(1)} - s_1^{(1)} - \dfrac{i_B}{2} \\[2mm] .... \\[2mm] y_{N-1} - y_0 = s_N^{(1)} - 2s_{N-1}^{(1)} + s_{N-2}^{(1)} - s_1^{(1)} - \dfrac{i_B}{2} \\[2mm] y_N - y_0 = s_{N-1}^{(1)} - s_N^{(1)} - s_1^{(1)} \end{cases} \tag{35}$$

where $y_n = \sin\left(\phi_0 + 2\pi n\Psi_{ex}\right)$, $n = 0, 1, 2, ..., N$. By solving for $s_1^{(1)}$, which is the quantity required in Eq. (32a), we have:

$$s_1^{(1)} = -\frac{N-1}{N+1}\frac{i_B}{2} - \frac{1}{N+1}\sum_{k=1}^{N}\left(y_k - y_0\right), \tag{36}$$

Substitution of the above results into Eq. (32a) gives:

$$\frac{d\phi_0}{d\tau} + \frac{1}{(N+1)} \sum_{n=0}^{N} y_n = \frac{i_B}{(N+1)}, \tag{37}$$

The above differential equation represents the effective model for the 1D-JJA represented in Fig. 7, containing $N+1$ identical over-damped junctions connected in parallel. We notice that the reduced model in Eq. (37), even being of first order in β, does not explicitly contain a β^1 term, given that Eq. (32a) contains β^{-1} terms.

We can now explicitly perform the sum in Eq. (37), so that we write:

$$\frac{d\phi_0}{d\tau} + \frac{B_{N+1}}{(N+1)} \sin\left(\phi_0 + N\pi\Psi_{ex}\right) = \frac{i_B}{(N+1)}, \tag{38}$$

where the absolute value of $B_{N+1} = \dfrac{\sin\left((N+1)\pi\Psi_{ex}\right)}{\sin\left(\pi\Psi_{ex}\right)}$ is the normalized critical current of the device in this approximation, as already known from the literature (Likharev, 1986). We start by finding the voltage , as defined in Eq. (28), under the assumption that Eqs. (32b-c) can be replaced by the effective model given by first-order perturbation analysis above. Therefore, we may set:

$$v(\tau) = \frac{d\phi_0}{d\tau} = \frac{i_B}{(N+1)} - \frac{B_{N+1}}{(N+1)} \sin\left(\phi_0 + N\pi\Psi_{ex}\right). \tag{39}$$

In this way, we can find the I-V characteristics by simply integrating Eq. (39), recalling the well-known procedure for a single overdamped junction. Indeed, noticing that

$$\langle v \rangle = \frac{1}{T} \int_0^T \frac{d\phi_0}{d\tau} d\tau = \frac{2\pi}{T}, \tag{40}$$

where T is the period for the instantaneous voltage curve and a pseudo-period for the superconducting phase difference ϕ_0 (i.e., $\phi_0(\tau+T) = \phi_0(\tau) + 2\pi$), we first find the expression for ϕ_0 from Eq. (39), for $i_B > |B_{N+1}|$ and k integer, so that

$$\phi_0(\tau) + N\pi\Psi_{ex} = 2\tan^{-1}\left[\frac{B_{N+1}}{i_B} - \frac{\sqrt{i_B^2 - B_{N+1}^2}}{i_B}\tan\left(\frac{1}{2(N+1)}\sqrt{i_B^2 - B_{N+1}^2}\,\tau\right)\right] + 2k\pi . \tag{41}$$

The pseudo-period T of the above solution can be found by inspection, so that:

$$T = \frac{2\pi(N+1)}{\sqrt{i_B^2 - B_{N+1}^2}}. \tag{42}$$

Therefore, by Eqs. (40) and (42), the I-V characteristics are given by the following expression:

$$i_B = \sqrt{(N+1)^2 \langle v \rangle^2 + B_{N+1}^2}, \tag{43}$$

where only the positive branch has been chosen.

In Figs. 8a-b B_{N+1} vs. Ψ_{ex} curves are shown. In particular, in Fig. 8a the number of JJ's in the system $(N+1)$ is taken to be equal to 10, while in Fig. 8b it is set equal to 15. We notice that, when the normalized flux approaches the first zero in the B_{N+1} vs. Ψ_{ex} curve for $(N+1)=10$, namely $\Psi_{ex}=1/10$, the system goes toward a purely resistive behaviour. In Fig. 8b, the same resistive behaviour is expected for $\Psi_{ex}=1/15$, given that the B_{N+1} vs. Ψ_{ex} curve for $(N+1)=15$ reaches its first zero exactly at $\Psi_{ex}=1/15$.

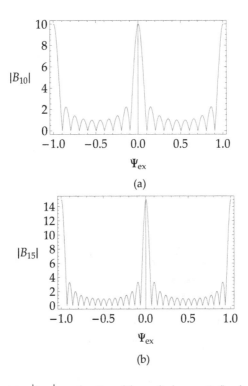

Figure 8. Critical current $i_C=|B_{N+1}|$ as a function of the applied magnetic flux for an array of 10 (a) and 15 (b) Josephson junctions.

In Figs. 9a-b *I-V* characteristics for different externally applied flux values are shown. In Fig. 9a the number of JJ's is taken to be equal to 10, while in Fig. 9b it is set equal to 15. Starting from Fig. 9a, we notice that, as the normalized flux approaches the first zero in the B_{N+1} vs. Ψ_{ex} curve for $(N+1)=10$, namely $\Psi_{ex}=1/10$, the behaviour of the system indeed becomes purely resistive. In Fig. 8b, the same resistive behaviour is obtained at $\Psi_{ex}=1/15$, as predicted.

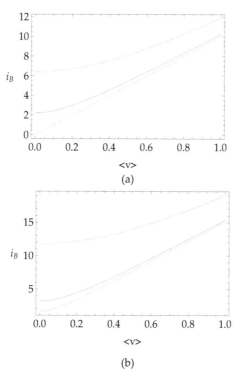

Figure 9. (a) *I-V* characteristics of an array of 10 Josephson junctions for the following values of the applied magnetic flux Ψ_{ex} : 0.05 (blue); 0.075 (cyan); 0.10 (orange); 0.15 (red). (b) *I-V* characteristics of an array of 15 Josephson junctions for the following values of the applied magnetic flux Ψ_{ex} : 0.025 (blue); 0.05 (cyan); 0.075 (orange); 0.10 (red).

For fixed bias current values, the voltage versus flux curves can be obtained by Eq. (43) and is given by the following:

$$\langle v \rangle = \frac{\sqrt{i_B^{\,2} - \left[B_{N+1} \left(\Psi_{ex} \right) \right]^2}}{\left(N+1 \right)}.$$

(44)

The above expression is similar to the homologous d. c. SQUID $\langle v \rangle$ vs. Ψ_{ex} curves.

In Figs. 10a and 10b we report the $\langle v \rangle$ vs. Ψ_{ex} curve for $(N+1) = 10$ and for $(N+1) = 15$, respectively. The normalized bias current values are $i_B = 4.0$, 10.0, 16.0 (from bottom to top) for Fig. 10a and $i_B = 4.0$, 10.0, 15.0 (from bottom to top) for Fig. 10b. Notice that the zero-voltage state regions on the Ψ_{ex} -axis disappear at values of the bias current greater or equal to the critical current of the system. In Fig. 10a, indeed, we see that disappearance of zero-voltage states occurs, for increasing normalized flux values, exactly at $i_B = 10.0$. On the other hand, in Fig. 10b, the same behaviour is detected at $i_B = 15.0$. Notice also that, for relatively high bias currents, the Ψ_{ex} intervals in which finite-voltage states are present tend to be flatter than in the case of a low number of Josephson junctions. This feature is more evident in Fig. 10b, where the region of zero-voltage states is narrower than in Fig. 10a. Therefore, for high enough values of the the number of Josephson junction in the array and of the bias current, away from integer values of the normalized applied flux, the interval in which finite-voltage states spread can be approximated by a horizontal segment.

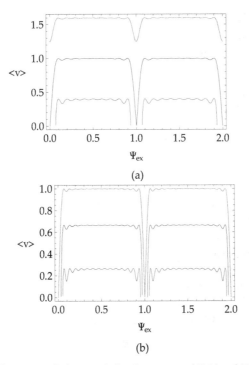

Figure 10. Average voltage vs. applied magnetic flux for an array of 10 (a) and 15 (b) Josephson junctions for three different values of the normalized bias current i_B. In (a) $i_B = 4.0$, 10.0, 16.0 (from bottom to top). In (b) $i_B = 4.0$, 10.0, 15.0 (from bottom to top).

In conclusion, by considering the dynamical equations of one-dimensional arrays containing $N+1$ identical overdamped Josephson junctions, the system of $N+1$ nonlinear first-order ordinary differential equation equations can be broken into two coupled subsystems, one consisting of only one equation for the superconducting phase of one junction in the array (arbitrarily chosen to be the first), the second describing the time evolution of N opportunely defined normalized flux variables.

When a solution of the latter N equations is found, by means of a perturbative approach to first order in the parameter β, the dynamical properties of the system are described by a single time-evolution equation. In this way, we may affirm that, for small values of β, the system may be described by an equivalent single-junction model, where the maximum Josephson current is appropriately defined in terms of the normalized applied flux Ψ_{ex}. When we compare our present analysis to equivalent studies carried out for a two-junction interferometer, we realize that the degree of approximation of the present model in β is one order less than the first-order perturbation analysis carried out for the simplest two-junction system. This is a consequence of the approach followed in the present work, where we had to appropriately define partial sums of flux variables in order to separate the dynamical equations into two subsystems. When we refer to the SQUID case, then, we might state that the present analysis corresponds exactly to the $\beta = 0$ limit. Further work is therefore needed to carry out more detailed information on the system behaviour for finite values of β.

Even though part of the present analysis reproduces known results, as, for example, the expression for the maximum Josephson current, it still represents a simple way of approaching the problem of the electrodynamic response of one-dimensional arrays of overdamped Josephson junctions by an equivalent single-junction model. In fact, by starting with the simple representation of the dynamics of the system given in Eq. (38), all the results obtained for a single Josephson junction can be reproduced for an array of $N+1$ equal overdamped Josephson junctions. In addition, in case the solutions of the normalized flux variable would be extended to second order in the parameter β, following the same analysis as in the present section, effects due to finite β values in the electrodynamic properties of the system would be detected. Finally, considering that the present analysis has been carried out in the absence of flux fluctuations, its extension to noise effects can be obtained by means of already known results obtained for a single overdamped Josephson junction (Ambegaokar & Halperin, 1969; Bishop & Trullinger, 1978). In this case, however, care must be taken in considering the stochastic terms on all branches of the array.

4. Parallel connections of $N \times$ (0-π) overdamped Josephson junctions

In the previous section we have considered an array of $N+1$ overdamped 0-junctions, without considering the possibility of inserting π-junctions in the system. We briefly recall that π-junctions (Bulaevskii et al., 1977; Geshkenbein et al., 1987; Baselman et al., 1999; Ryazanov et al., 2001 M. Weides et al., 2006), when compared to 0-junctions, possess an intrinsic phase difference exactly equal to π. By inserting a 0-junction and a π-junction in the same superconducting loop, π-SQUIDs may be realized. These non-conventional SQUIDs can be fabricated either by exploiting the symmetry properties of d-wave superconductors (Chesca, 1999; Schultz et al., 2000) or by utilizing both s-wave and d-wave superconductors (Wollman et al., 1993; Smilde et al., 2004). A π-SQUID can thus be viewed as an elementary cell of a $N\times$(0-π) one-dimensional array of overdamped Josephson junctions shown in fig. 11.

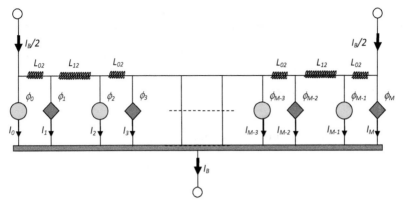

Figure 11. Schematic representation of a Josephson junction array with alternate parameters. In the branches of the parallel connections overdamped 0-junctions (green circles) and π-junctions (red diamonds) are alternately present.

Therefore, π-SQUIDs can be viewed as the building block of discretized models of multifacets Josephson junctions (MJJs) (Scharinger et al., 2010), in which the critical current density alternates between two opposite values along the junction length. However, even though some characteristic features of MJJs can be qualitatively reproduced by N×(0-π) one-dimensional arrays, one should bear in mind that the latter are, in general, less complex systems than MJJs.

For conventional arrays of overdamped Josephson junctions we have already shown that, for small enough values of the characteristic parameter β a series solution for the magnetic flux variable can be found by perturbation analysis. In this way, the multi-junction interferometer model reduces to a single non-linear ordinary differential equation. The same perturbation approach will be proposed again in the present section to derive the equivalent single-junction model of N×(0-π) one-dimensional arrays of overdamped Josephson junctions. Therefore, we start by considering the model system represented in fig. 11, consisting of identical overdamped junctions connected in parallel. In this system one half of the bias current I_B is evenly applied to the two external branches of the array. This condition is obtained, for example, by injecting the current by means of a superconducting bar of width w equal to the length of the array. In order to have well focused bias, the penetration length of the superconducting bar would be much smaller than its width w.

4.1. The homogeneous case

Consider, as a first approach to the problem, the loop areas S_k, the junctions resistive parameters R_k and maximum Josephson currents I_{Jk}, and the inductances L_k in the horizontal upper branches to be all equal. In this way one may write $S_k = S_0$, $R_k = R_0$, $I_{Jk} = I_{J0}$, $L_k = L_0$, for all allowed values of k. Define, as in the previous section in this case,

$\beta = \dfrac{L_0 I_{J0}}{\Phi_0}$. Consider, finally, as also represented in fig. 11, that 0-junctions (green circles) occupy even positions ($k = 0, 2, 4, .., M - 1$) and π-junctions (diamonds) occupy odd positions ($k = 1, 3, .., M$). The reader will notice very many similarities with the analysis in the previous section. However, due to the higher degree of complexity of the problem, we need to proceed step by step.

By fluxoid quantization, the normalized magnetic flux $\Psi_k = \dfrac{\Phi_k}{\Phi_0}$ linked to the k-th cell of the array is related to the superconducting phase differences ϕ_{k-1} and ϕ_k across the two junctions in the cell as follows:

$$\phi_{k-1} - \phi_k + 2\pi\Psi_k = 2\pi\left[n_k + \frac{(-1)^k - 1}{4} \right],\tag{45}$$

where n_k is an integer and $k = 1, 2, ..., M$ (notice that the number of loops is one less the number of junctions). If an external magnetic field $\mathbf{H}$, orthogonal to the plane of the array, is applied to the system, the normalized geometric applied flux through each cell now is $\Psi_{ex} = \dfrac{\Phi_{ex}}{\Phi_0} = \dfrac{\mu_0 H S_0}{\Phi_0}$. By considering Φ_k as the sum of the applied and induced flux, we may write:

$$\Psi_k = \beta\left(\sum_{m=0}^{k-1} i_m - \frac{i_B}{2} \right) + \Psi_{ex},\tag{46}$$

for $k = 1, 2, ..., M$, where $i_k = \dfrac{I_k}{I_{J0}}$ is the normalized current flowing in the k-th branch and $i_B = \dfrac{I_B}{I_{J0}}$. The dynamical equation for each Josephson junction in the array can be written by means of the RSJ model (Barone & Paternò, 1982), so that:

$$\frac{d\phi_k}{d\tau} + \sin\phi_k = i_k,\tag{47}$$

where $k = 0, 1, 2, ..., M$ and $\tau = \dfrac{2\pi R_0 I_{J0}}{\Phi_0} t$. Eqs. (45-47) can be used to define the dynamics of the gauge-invariant superconducting phase difference ϕ_k in terms of the forcing parameters Ψ_{ex} and $i_B = \dfrac{I_B}{I_J} = \sum_{k=0}^{M} i_k$. Define now the partial sum s_n ($1 \le n \le M$) of the normalized fluxes as follows:

$$s_n = \sum_{k=1}^{n} \Psi_k. \tag{48}$$

By fluxoid quantization, setting all n_k's to zero in Eq. (45) under the hypothesis of zero initially trapped flux in the array, we can write:

$$\phi_k = \phi_0 + 2\pi s_k + \pi \left[\frac{1-(-1)^k}{2} \right], \tag{49}$$

for $k = 1, 2, ..., M$, so that the dynamical equations can be rewritten as follows:

$$\frac{d\phi_0}{d\tau} + \sin\phi_0 = i_0, \tag{a}$$

$$2\pi \frac{ds_{2n}}{d\tau} + \sin(\phi_0 + 2\pi s_{2n}) = i_{2n} - i_0 + \sin\phi_0, \; (k = 2n) \tag{b} \tag{50}$$

$$2\pi \frac{ds_{2n-1}}{d\tau} - \sin(\phi_0 + 2\pi s_{2n-1}) = i_{2n-1} - i_0 + \sin\phi_0, \; (k = 2n-1) \tag{c}$$

where $n = 1, 2, ..., \frac{M-1}{2}$ in (50b), and $n = 1, 2, ..., \frac{M+1}{2}$ in (50c). Expressing now, by means of Eq. (46), the currents i_0, i_{2n}, and i_{2n-1} in terms of the forcing terms Ψ_{ex} and i_B and of the partial sums of the flux variables s_k defined in Eq. (48), we may finally rewrite Eqs. (50a-c) as follows:

$$\frac{d\phi_0}{d\tau} + \sin\phi_0 = \frac{i_B}{2} + \frac{s_1 - \Psi_{ex}}{\beta}, \tag{a}$$

$$2\pi \frac{ds_{2n}}{d\tau} + \sin(\phi_0 + 2\pi s_n) - \sin\phi_0 - \frac{s_{2n+1} - 2s_{2n} + s_{2n-1}}{\beta} = \frac{\Psi_{ex} - s_1}{\beta} - \frac{i_B}{2}, \tag{b}$$

$$2\pi \frac{ds_{2n-1}}{d\tau} - \sin(\phi_0 + 2\pi s_{2n-1}) - \sin\phi_0 - \frac{s_{2n} - 2s_{2n-1} + s_{2n-2}}{\beta} = \frac{\Psi_{ex} - s_1}{\beta} - \frac{i_B}{2}, \tag{c} \tag{51}$$

$$2\pi \frac{ds_M}{d\tau} - \sin(\phi_0 + 2\pi s_M) - \sin\phi_0 + \frac{s_M - s_{M-1}}{\beta} = \frac{\Psi_{ex}}{\beta} - \frac{s_1 - \Psi_{ex}}{\beta}, \tag{d}$$

where now $n = 1, 2, ..., \frac{M-1}{2}$ in Eqs. (51b) and (51c) and where it has been set $s_0 = 0$.

Considering the dynamical equations of the system as written in Eqs. (51a-d), for small values of the parameter β we may assume that the solution, to first order in this perturbation parameter, can be written as follows:

$$s_n \approx s_n^{(0)} + \beta s_n^{(1)}. \tag{52}$$

By substituting the above expression in Eqs. (51b-d) and, after having multiplied both members by β, by setting to zero the coefficients of β^m ($m = 0, 1$), we obtain:

$$s_n^{(0)} = n\Psi_{ex}. \tag{53}$$

For the first order corrections, on the other hand, we need to solve the following set of equations:

$$\begin{cases} -y_1 - y_0 = s_2^{(1)} - 3s_1^{(1)} - \dfrac{i_B}{2} \\[2mm] y_2 - y_0 = s_3^{(1)} - 2s_2^{(1)} - \dfrac{i_B}{2} \\[2mm] -y_3 - y_0 = s_4^{(1)} - 2s_3^{(1)} + s_2^{(1)} - s_1^{(1)} - \dfrac{i_B}{2} \\[2mm] \cdots \\[2mm] -y_{M-2} - y_0 = s_{M-1}^{(1)} - 2s_{M-2}^{(1)} + s_{M-1}^{(1)} - s_1^{(1)} - \dfrac{i_B}{2} \\[2mm] y_{M-1} - y_0 = s_M^{(1)} - 2s_{M-1}^{(1)} + s_{M-2}^{(1)} - s_1^{(1)} - \dfrac{i_B}{2} \\[2mm] -y_M - y_0 = s_{M-1}^{(1)} - s_M^{(1)} - s_1^{(1)} \end{cases} \tag{54}$$

where $y_n = \sin(\phi_0 + 2\pi n\Psi_{ex})$, $n = 0, 1, 2, ..., M$. By solving for $s_1^{(1)}$, which is the required quantity in Eq. (51a), we have:

$$s_1^{(1)} = -\frac{M-1}{M+1}\frac{i_B}{2} - \frac{1}{M+1}\sum_{k=1}^{M}\left[(-1)^k y_k - y_0\right], \tag{55}$$

Substitution of the above result into Eq. (51a) gives:

$$\frac{d\phi_0}{d\tau} + \frac{1}{(M+1)}\sum_{k=0}^{M}(-1)^k \sin(\phi_0 + 2\pi k\Psi_{ex}) = \frac{i_B}{(M+1)}. \tag{56}$$

It is now possible to explicitly calculate the finite sum in Eq. (56) to get, in terms of the number $N = \dfrac{M+1}{2}$ of the individual (0-π) cells:

$$\frac{d\phi_0}{d\tau} - \frac{A_{2N}}{2N}\cos\left[\phi_0 + (2N-1)\pi\Psi_{ex}\right] = \frac{i_B}{2N}, \tag{57}$$

where $A_{2N} = \dfrac{\sin(2N\pi\Psi_{ex})}{\cos(\pi\Psi_{ex})}$. The above equation represents the equivalent single-junction model for the homogeneous array consisting of N cells, each one containing one 0-junction and one π-junction.

4.2 The non-homogeneous case

Consider, next, a non-homogeneous array with alternating 0-π Josephson junctions. We shall take the parameters of all 0-junctions equal. The parameters of π-junctions, even being equal among them, are assumed to be different from those of the 0-junctions. In this case we can omit some of the calculations, having already treated the problem in detail in the previous subsection.

Considering again the 1D-JJA represented in fig. 11, we now assume that the loop areas S_k, the junctions resistive parameters R_k and maximum Josephson currents I_{Jk}, and the inductances L_k of the horizontal upper branches alternate in their values, as we go along the array. In this way we write

$$
\begin{aligned}
S_k &= \begin{cases} S_0 & k \text{ even} \\ S_1 = \sigma S_0 & k \text{ odd} \end{cases} \quad \text{(a)} \\[4pt]
R_k &= \begin{cases} R_0 & k \text{ even} \\ R_1 = \alpha S_0 & k \text{ odd} \end{cases} \quad \text{(b)} \\[4pt]
I_{Jk} &= \begin{cases} I_{J0} & k \text{ even} \\ I_{J1} = \varepsilon I_{J0} & k \text{ odd} \end{cases} \quad \text{(c)} \\[4pt]
L_k &= \begin{cases} L_0 & k \text{ even} \\ L_1 = \sigma S_0 & k \text{ odd} \end{cases} \quad \text{(d)}
\end{aligned}
\tag{58}
$$

for all allowed values of k. In this way, the additional parameters α, ε, and σ are implicitly defined. As before, we set $\tau = \dfrac{2\pi R_0 I_{J0}}{\Phi_0} t$ and $\beta = \dfrac{L_0 I_{J0}}{\Phi_0}$; therefore, we may define $\beta_1 = \dfrac{L_1 I_{J1}}{\Phi_0} = \varepsilon \alpha \beta$.

Fluxoid quantization give the same relation as in Eq. (45) between the superconducting phases and the normalized magnetic flux Ψ_k linked to the k-th cell of the array. However, Eq. (46) is modified as follows

$$
\Psi_k = \begin{cases} \beta \left(\displaystyle\sum_{m=0}^{k-1} i_m - \dfrac{i_B}{2} \right) + \Psi_{ex} & k \text{ even} \\[14pt] \beta \sigma \varepsilon \left(\displaystyle\sum_{m=0}^{k-1} i_m - \dfrac{i_B}{2} \right) + \sigma \Psi_{ex} & k \text{ odd} \end{cases}
\tag{59}
$$

for $k = 1, 2, ..., M$. The equations of the motion for the superconducting phases can now be written as follows:

$$
\begin{aligned}
\frac{d\phi_{2n}}{d\tau} + \sin\phi_{2n} &= i_{2n}, \quad &\text{(a)} \\[6pt]
\frac{d\phi_{2n-1}}{d\tau} + \varepsilon\alpha \sin\phi_{2n-1} &= \alpha i_{2n-1}, \quad &\text{(b)}
\end{aligned}
\tag{60}
$$

where n runs over all allowed k-values also in all following equations.

By adopting a first-order perturbation analysis in the parameter β, we set:

$$s_{2n} \approx s_{2n}^{(0)} + \beta s_{2n}^{(1)} = n(\sigma+1)\Psi_{ex} + \beta s_{2n}^{(1)}, \qquad \text{(a)}$$

$$s_{2n-1} \approx s_{2n-1}^{(0)} + \beta s_{2n-1}^{(1)} = \left[n(\sigma+1)-1\right]\Psi_{ex} + \beta s_{2n-1}^{(1)}, \qquad \text{(b)} \tag{61}$$

By substituting the above expression in Eqs. (49) to obtain the superconducting phases ϕ_k we then write the differential equations (60a-b) in terms of the sole phase variable ϕ_0 and of the $2N$ normalized flux variables s_k. By following the same steps as in the previous section, we obtain the following equivalent single-junction model for the superconducting array:

$$\frac{d\tilde{\phi}_0}{d\tau} + \frac{\alpha \tilde{A}_N}{(\alpha+1)N}\left[\sin\tilde{\phi}_0 - \varepsilon\sin\left(\tilde{\phi}_0 + 2\pi\sigma\Psi_{ex}\right)\right] = \frac{\alpha i_B}{(\alpha+1)N}, \tag{62}$$

Where here $\tilde{A}_N = \dfrac{\sin\left(\pi N(\sigma+1)\Psi_{ex}\right)}{\sin\left(\pi(\sigma+1)\Psi_{ex}\right)}$ and $\tilde{\phi}_0 = \phi_0 + \pi(N-1)(\sigma+1)\Psi_{ex}$. Naturally, for α, ε,

and σ all equal to one, the ordinary differential equation (62) reduces to Eq. (57).

4.3. Critical current

In order to find the critical current of arrays with alternating 0-π Josephson junctions, we proceed as follows. First, consider the homogeneous array described in Section 4.1. We look for the maximum value of the bias current i_B which can be injected in the system at zero voltage ($\frac{d\phi_0}{d\tau} = 0$ in Eq. (57)). Therefore, by maximizing with respect to ϕ_0 the following expression

$$i_B = -A_{2N}\cos\left[\phi_0 + (2N-1)\pi\Psi_{ex}\right], \tag{63}$$

we can express the critical current of the homogeneous device as follows:

$$i_c = \left|A_{2N}\right| = \left|\frac{\sin\left(2N\pi\Psi_{ex}\right)}{\cos\left(\pi\Psi_{ex}\right)}\right|. \tag{64}$$

In order to understand the origin of the patterns we are going to show for the non homogeneous case, let us consider the result in Eq. (64) as the product of the envelop function

$$E\left(\Psi_{ex}\right) = \sqrt{1+\varepsilon^2 - 2\varepsilon\cos\left(2\pi\sigma\Psi_{ex}\right)}, \tag{65}$$

($\varepsilon = 1$, $\sigma = 1$ in this case) and of the rapidly oscillating function

$$F\left(\Psi_{ex}\right) = \left|\frac{\sin\left(N\pi(\sigma+1)\Psi_{ex}\right)}{\sin\left(\pi(\sigma+1)\Psi_{ex}\right)}\right|, \tag{66}$$

with $\sigma = 1$, giving rise to secondary peaks in the overall curves. The functions $N \cdot E(\Psi_{ex})$ and $2F(\Psi_{ex})$ are separately shown in fig. 12a for $\varepsilon = 1, \sigma = 1$ and $N=10$. The factors in front of these functions are chosen in such a way that they can attain the same maximum value of the product $E(\Psi_{ex}) \cdot F(\Psi_{ex})$.

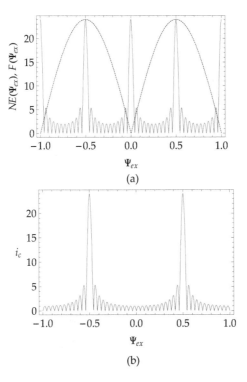

(a)

(b)

Figure 12. (a) Envelope function $NE(\Psi_{ex})$ as a function of $2F(\Psi_{ex})$. (b) Critical current as a function of the normalized applied flux for a homogeneous parallel array with $N=12$, $\varepsilon=1$, $\sigma=1$.

One notices that regular primary peaks of $F(\Psi_{ex})$ appear at integer and half integer values of Ψ_{ex}. By multiplying the functions $E(\Psi_{ex})$ and $F(\Psi_{ex})$ as reported in fig. 12b, we notice that primary peaks positioned at half-integer Ψ_{ex} values are left unchanged, while those appearing at integer Ψ_{ex} values are depressed to zero.

By proceeding in the same way for the non-homogeneous case, starting from Eq. (62) we find that the critical current of the non homogeneous array described in Section 3 can be written as follows:

$$i_c = E(\Psi_{ex})F(\Psi_{ex}).$$ (67)

Notice that the resulting pattern does not depend on α, which can be absorbed, in the effective dynamical equation, by a rescaling of the normalized time τ. Notice also that the periodicity of the envelop function $E\left(\Psi_{ex}\right)$ may now not be commensurable with that of the rapidly oscillating term $\left|\tilde{A}_N\right|$. Consider, in fact, the case reported in figs. 13a and 13b, where the pattern is shown, as a full line, for a non homogeneous array with $N=12$, $\varepsilon=1.3$ and $\sigma=1.5$.

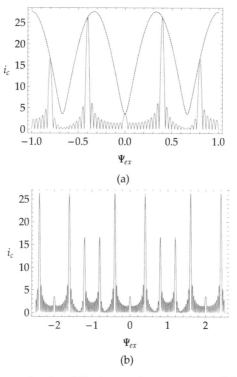

(a)

(b)

Figure 13. Critical current as a function of Ψ_{ex} for a non-homogeneous parallel array with $N=12$, $\varepsilon=1.3$, $\sigma=1.5$ in the following ranges: (a) [-1.25, 1.25]; (b) [-2.5, 2.5]. In (a) the envelope function $NE\left(\Psi_{ex}\right)$ is shown as a dotted line.

In fig. 13a the envelop curve $N\cdot E\left(\Psi_{ex}\right)$ is shown as a dotted line. In the interference pattern in fig. 13a we may notice the finiteness of the small peak at $\Psi_{ex}=0$ due to the non-vanishing value of $E(0)=\left|\varepsilon-1\right|$. Also, notice the reduction in height of the primary peaks close to half-integer values as compared to those in fig. 12b. Finally, notice the appearance of two extra peaks of equal height in between two successive primary peaks. In fig. 13b, where the range of Ψ_{ex} is increased, we notice that this feature repeats over a period $\Delta\Psi_{ex}=2$, as it can be directly confirmed by inspection of Eqs. (65-67). Differently from fig. 13a and 13b, when we consider an array with $N=12$, $\varepsilon = 1.3$ and $\sigma = \sqrt{2}$, as in figs. 14a-c, the fundamental periods of the functions $E(\Psi_{ex})$ and $F(\Psi_{ex})$ are incommensurable, so that the overall pattern is not periodic. This feature can be detected by gradually increasing the range of the pattern, as it is done in fig. 14a and fig. 14b. In this way, when looking for the conditions giving periodicity, one realizes that the ratio $\dfrac{\sigma+1}{\sigma}$ needs to be a rational number. Therefore, when the parameter σ is a rational number, one has periodicity in the i_c vs. Ψ_{ex} curves; otherwise, irregular patterns, like those shown in figs. 14a-b, are found.

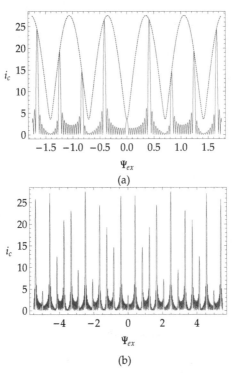

(a)

(b)

Figure 14. Critical current as a function of Ψ_{ex} for a non-homogeneous parallel array with $N=12$, $\varepsilon=1.3$, $\sigma = \sqrt{2}$ in the following ranges: (a) [-1.75, 1.75]; (b) [-5.5, 5.5]. In (a) the envelope function $NE(\Psi_{ex})$ is shown as a dotted line.

We have seen that a reduced single-junction model can be adopted to describe the overall dynamics of a one-dimensional arrays with alternating parameters of $N\times(0\text{-}\pi)$ Josephson junctions. This effective model is very useful, since it allows to obtain the critical current vs. normalized magnetic flux curves in closed analytic form. The interference patterns are seen to be qualitatively similar to recently obtained experimental results on multifacets Josephson junctions (Scharinger et al., 2010) in which the critical current density alternates many times between two opposite values along the junction length. Discrete Josephson junction arrays, even presenting some analogies with the latter devices, are much too simple systems to describe the complete behaviour of MJJs. As a matter of fact, shielding current effects is not taken into account by analysis carried out in this section in the lowest order approximation in β. Nevertheless, the analytic results obtained for the interference patterns shed some light on the causes of the presence or absence of periodicity and on the nature of primary and secondary peaks in the i_c vs. Ψ_{ex} curves. We finally notice that the analysis relies on the choice of an even number of Josephson junction in the array. Different patterns are expected for an odd number of alternating junctions in the array, depending also on which type (0 or π) of junctions is predominant. Further work is therefore necessary to address this problem in its fullest extent.

5. Quantum interferometers in the presence of rapidly varying fields

In Section 2 we have analyzed the effective model for a two-junction quantum interferometer in the presence of an oscillating magnetic flux under the hypothesis that the frequency of oscillation is comparable with the inverse of the characteristic time evolution τ_ϕ of the superconducting time variable ϕ. In this way, the quasi-static approach described in Section 2 has been proven to be applicable. In the present section, on the other hand, we shall consider rapidly varying externally applied fluxes, whose frequency ω is comparable with τ_ψ^{-1}, so that $\omega \gg \tau_\phi^{-1}$.

Let us therefore consider an externally applied flux having d. c. component A and a. c. amplitude B, so that

$$\psi_{ex}(t) = A + B \sin \omega t \tag{68}$$

where $\omega \approx \tau_\psi^{-1} \gg \tau_\phi^{-1}$ is the frequency of the sinusoidal term. Let us again consider Eqs. (4a-b) rewritten, for $n=0$, as follows:

$$\frac{d\phi}{d\tau} + \cos \pi\psi \sin \phi = \frac{i_B}{2}, \qquad \text{(a)}$$

$$\pi \frac{d\psi}{d\tau} + \sin \pi\psi \cos \phi = \frac{\psi_{ex} - \psi}{2\beta} \quad \text{(b)} \tag{69}$$

where the normalization $\tau = \dfrac{2\pi RI_J}{\Phi_0} t = \dfrac{t}{\tau_\phi}$ prescribes a τ-dependence of the externally

applied flux as written in Eq. (18) with a normalized frequency $\tilde{\omega} = \dfrac{\Phi_0}{2\pi RI_J} \omega = \omega\tau_\phi$. We

therefore need to consider again all steps in Section 2, having care to integrate opportunely the right hand side term of (69b), in order to obtain a solution for Ψ in terms of the superconducting phase by perturbation analysis on β for arbitrary values of A and B. In this way, the effective dynamics for ϕ can be found when the solution for Ψ is substituted in the cosine term of (69a).

Start by considering the cosine term in (69b) as a quasi-static quantity (i. e., it does not vary appreciably over an interval of time of the order of τ_ψ). This hypothesis is confirmed by what already stated in the previous section; i. e., while the variables ϕ varies on a characteristic time interval $\Delta\tau_\phi$ the variable Ψ varies within a time interval $\Delta\tau_\psi = 2\pi\beta\Delta\tau_\phi \ll \Delta\tau_\phi$. Within the former time interval $\Delta\tau_\phi$ it is then possible to choose a subinterval, of the order of $\Delta\tau_\psi$ in which the variable ϕ does not vary appreciably. We can thus solve (69b), by perturbation analysis, by first setting $\tau = 2\pi\beta\theta$ and by rewriting it as follows:

$$\frac{d\psi}{d\theta} + 2\beta\sin\pi\psi\cos\phi + \psi(\theta) = \psi_{ex}(\theta). \tag{70}$$

By now setting

$$\psi(\theta) = \psi_0(\theta) + \beta\psi_1(\theta) \tag{71}$$

we again find the ODEs for ψ_0 and ψ_1 in (8a) and (8b). Recall that an ODE of the type

$$\frac{df}{d\theta} + f(\theta) = g(\theta), \tag{72}$$

has solution $f(\theta) = e^{-\theta}\int^\theta g(x)e^x dx$. By now considering (68), by taking the non-decaying solutions of the system of ordinary differential equations (8a-b) and by considering the non-vanishing solution of (8a) for ψ_0 at large values of θ, we have

$$\psi_0 = A + \frac{B}{1+\tilde{\omega}^2}(\cos\tilde{\omega}\theta + \tilde{\omega}\sin\tilde{\omega}\theta). \tag{73}$$

Having found the solution to (8a), we can find the solution to (8b) by the same type of reasoning. After some rather long calculations one finds

$$\psi_1 = -2h(\theta)\cos\phi, \tag{74}$$

where

$$h(\theta) = \sum_{n=-\infty}^{+\infty}\sum_{k=-\infty}^{+\infty} g_{n,k}\left[\sin\alpha_{n,k}(\theta) - \tilde{\omega}\cos\alpha_{n,k}(\theta)\right], \tag{75}$$

with $g_{n,k} = \dfrac{J_n(\gamma)J_n(\gamma\tilde{\omega})}{1+(n+k)^2\tilde{\omega}^2}$, $J_n(x)$ being the Bessel function of order n, and

$\alpha_{n,k}(\theta) = \pi A + (n+k)\tilde{\omega}\theta + n\dfrac{\pi}{2}$. Therefore, the SQUID dynamics can be described, to first order in β, by the following ODE:

$$\frac{d\phi}{d\tau} + X_0(\theta)\sin\phi + \pi\beta h(\theta)Y_0(\theta)\sin 2\phi = i_B/2m, \tag{76}$$

where $X_0(\theta) = \cos\pi\psi_0$ and $Y_0(\theta) = \sin\pi\psi_0$.

Equation (76) thus represents the differential equation describing the dynamics of the superconducting phase difference ϕ in a d. c. SQUID in the presence of a time-varying externally applied flux, whose frequency ω is considered to be comparable with τ_ψ^{-1}, in such a way that $\tilde{\omega} = \omega\tau_\psi \approx 1$. For slowly varying fields ($\tilde{\omega} \ll 1$) one can readily verify from (8a) and (8b) that $\psi_0(\theta) \to \psi_{ex}(\theta)$ and $\psi_1 \to -2\sin\psi_{ex}\cos\phi$, respectively. In this way, the dynamics described by the quasi-static d. c. SQUID in (3) holds.

For $\tilde{\omega} = \omega\tau_\psi \approx 1$, time evolution of the two variables, ϕ and ψ, still occurs with two completely different time scales. In fact, as already stated, one has $\tau_\psi = 2\pi\beta\tau_\phi \ll \tau_\phi$. Therefore, flux motion is very fast with respect to the dynamics of the phase variable ϕ. The only difference, here, is that the externally applied flux $\psi_{ex}(\theta)$ is able to follow this fast dynamics. Having carefully solved (8a) and (8b), and having found the effective single-junction dynamical equation for a d. c. SQUID, we can determine the effective time-averaged equation, by taking the time average over the fast variable ψ. Therefore, we may write

$$\frac{d\phi}{d\tau} + \langle X_0(\theta)\rangle\sin\phi + \pi\beta\langle h(\theta)Y_0(\theta)\rangle\sin 2\phi = i_B/2, \tag{77}$$

where the symbol <x> stands for the time average of the variable x. Equation (77) can thus be considered an effective single-junction model for a d. c. SQUID in the presence of a rapidly varying magnetic field ($\tilde{\omega} = \omega t_L \approx 1$). The average values $\langle X_0(\theta)\rangle$ and $\langle h(\theta)Y_0(\theta)\rangle$ can be calculated as follows. First of all, set

$$X_0(\theta) = \mathrm{Re}\{\exp(i\pi\psi_0)\}. \tag{78}$$

Let us next express the exponential of a cosine and a sine terms in $\exp(i\pi\psi_0)$ through the following Bessel function identities

$$e^{ia\cos x} = \sum_{n=-\infty}^{+\infty} i^n J_n(a)e^{inx}, \quad \text{(a)}$$

$$e^{ia\sin x} = \sum_{n=-\infty}^{+\infty} J_n(a)e^{inx}. \quad \text{(b)} \tag{79}$$

In this way, (78) becomes

$$\langle X_0(\theta)\rangle = \mathrm{Re}\left\{\sum_{n,k} J_n(\gamma)J_k(\gamma\tilde{\omega})\langle e^{i\alpha_{n,k}(\theta)}\rangle\right\}. \tag{80}$$

It is now easy to show that

$$\langle e^{i\alpha_{n,k}(\theta)}\rangle = e^{i\pi A}i^n\delta_{n,-k}, \tag{81}$$

where the symbol $\delta_{n,m}$ is the Kronecker delta. By inserting (81) in (80), we have

$$\langle X_0(\theta)\rangle = \cos\pi A\rho(\tilde{\omega},B), \tag{82}$$

with

$$\rho(\tilde{\omega},B) = J_0(\gamma)J_0(\gamma\tilde{\omega}) + 2\sum_{n=1}^{+\infty}(-1)^n J_{2n}(\gamma)J_{2n}(\gamma\tilde{\omega}). \tag{83}$$

Proceeding in a similar way in finding the effective coefficient of the $\sin 2\phi$ term, we find:

$$\langle h(\theta)Y_0(\theta)\rangle = \frac{1}{2}\sum_{n,m,l=-\infty}^{+\infty}\xi_{n,m,l}g_{n,m}J_l(\gamma)J_{n+m-l}(\gamma\tilde{\omega}). \tag{84}$$

where

$$\xi_{n,m,l} = \begin{cases} (-1)^{n-l} & \text{for } (n-l) \text{ even} \\ -(n+k)(-1)^{\frac{n-l-1}{2}} & \text{for } (n-l) \text{ odd} \end{cases}. \tag{85}$$

Having expressed the effective time-averaged terms in (77) in a closed analytic form, we can understand the effect of a high-frequency field on the electrodynamic behaviour a d. c. SQUID with extremely small value of the parameter β, for instance. The critical current i_c of the device in this case ($\beta = 0$) can be expressed as follows:

$$i_C(\tilde{\omega},A,B) = 2|\cos\pi A||\rho(\tilde{\omega},B)|, \tag{86}$$

where the quantity $|\rho(\tilde{\omega},B)|$ is the extra-factor modifying the usual form of a d. c. SQUID, expressed, in terms of a constant applied flux A, as $2|\cos\pi A|$. The above expression can be derived by inspection from Eq. (77), by setting $\beta = 0$ and getting the maximum stationary value for i_B with respect to ϕ.

In fig. 15a-b we thus show the critical current i_c in terms of the a. c. component B of the applied flux. In particular, in fig. 15a, we report the i_c vs. B curves for various values of the d. c. component A of the applied flux. normalized frequency $\tilde{\omega}$. In fig. 15b, on the other hand, the i_c vs. B curves are shown for various values of the normalized frequency $\tilde{\omega}$. From figs. 15a-b we notice Fraunhofer-like patterns of the critical current as shown as a function of the a. c. amplitude B.

The particular shape of these patterns in figs. 15a-b depends on the value of the normalized frequency $\tilde{\omega}$. When the d. c. component A and the time-varying portion of the magnetic flux attain fixed values, and we let the normalized frequency vary with continuity, the $\tilde{\omega}$ critical current of the device is the same as that of the quantity $|\rho(\tilde{\omega},B)|$. The i_c vs. $\tilde{\omega}$ curves are represented in fig. 16 for A=0.1 and for various values of B. In this respect, we notice that, while for small fixed B values the quantity $|\rho(\tilde{\omega},B)|$ is always increasing for increasing values of $\tilde{\omega}$ (see orange line in fig. 16) for values of B approaching 1.0, this character is lost (see brown and cyan lines in fig. 16).

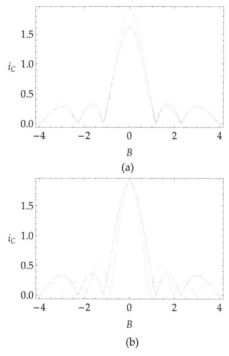

Figure 15. Critical current i_c as a function of the oscillating amplitude B. (a) The normalized frequency is fixed at $\tilde{\omega} = 1.0$ and various values of the d. c. component A are represented: A=0.1 (orange); A=0.2 (brown); A=0.3 (cyan). (b) The d. c. component A is fixed at A=0.1 and the normalized frequency attains the following values: $\tilde{\omega} = 0.5$ (orange); $\tilde{\omega} = 1.0$ (brown); $\tilde{\omega} = 1.5$ (cyan).

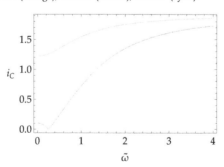

Figure 16. Critical current i_c as a function of the normalized frequency $\tilde{\omega}$ for A=0.1 and for various values of the oscillating amplitude B: B=0.4 (orange); B=0.8 (brown); B=1.2 (cyan).

We can now calculate, for $\beta=0$, the flux-voltage curves (v vs. A) by the following well known expression (Barone & Paternò, 1982):

$$v = \left\langle \frac{d\phi}{d\tau} \right\rangle = \sqrt{\frac{i_B^{\,2}}{4} - \cos^2(\pi A)\rho^2(\tilde{\omega}, B)}, \qquad (87)$$

for $i_B > 2\left|\cos(\pi A)\rho(\tilde{\omega}, B)\right|$. The important feature in this expression is that these curves depend both on B and $\tilde{\omega}$ though the extra-factor $\left|\rho(\tilde{\omega}, B)\right|$. Because of this dependence, the amplitude of the v vs. A curves can be varied and B and $\tilde{\omega}$ can be viewed as control parameters. In figs. 17a-b we thus report the v vs. A curves for $i_B = 2.5$, for $\tilde{\omega} = 0$ (a) and $\tilde{\omega} = 1.0$ (b), and for various values of B. In particular, in fig. 17a we notice that the amplitude of the v vs. A curves obtained at $B=0$ and $\tilde{\omega} = 0$ (orange line) decreases as we let B increase to 0.15 first (brown line) and to 0.30 next (cyan line). The same decreasing behavior is detected in fig. 17b for $\tilde{\omega} = 0.1$ when B increases from 0.0 (orange line) to 0.15 (brown line) and to 0.30 (cyan line).

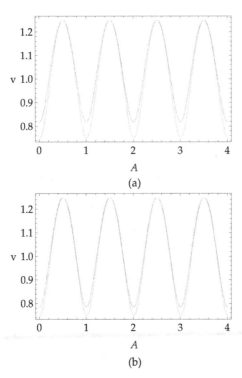

Figure 17. Voltage v versus the d. c. component A of the applied flux for $i_B = 2.5$ and $B=0.0$ (orange); $B=0.15$ (brown); $B=0.30$ (cyan). In (a) the normalized frequency values is $\tilde{\omega} = 0$, in (b) $\tilde{\omega} = 1.0$.

In fig. 18a-b, finally, by fixing the value of B first to 0.5 (a) and then to 1.0 (b), we notice that the amplitude of the v vs. A curves increases for increasing values of $\tilde{\omega}$, as shown for $\tilde{\omega} = 0.5$ (orange line), $\tilde{\omega} = 1.0$ (brown line), and $\tilde{\omega} = 1.5$ (cyan line).

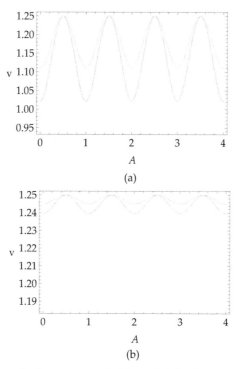

(a)

(b)

Figure 18. Voltage v versus the d. c. component A of the applied flux for $i_B = 2.5$ and $\tilde{\omega} = 0.5$ (orange); $\tilde{\omega} = 1.0$ (brown); $\tilde{\omega} = 1.5$ (cyan). In (a) the a. c. component of the applied magnetic flux is B=0.50, in (b) B=1.0.

6. Conclusion

We have studied the dynamical properties of quantum interferometers consisting of single or multiple superconducting loops, each containing two Josephson junctions.

A symmetric quantum interferometer containing two identical junctions with negligible capacitance has been considered first. The analysis of the system has been carried out by means of a perturbation approach in the parameter β, whose value gives the strength of the electromagnetic coupling between the two junction in the system. We have noticed that the flux-number function $\psi(\beta, \theta)$ governs fluxon dynamics, where θ is the laboratory time t normalized to the characteristic circuital time constant $\tau_\psi = \dfrac{L}{R}$. By this general approach it becomes evident that the characteristic time constant $\tau_\phi = \dfrac{\Phi_0}{2\pi RI_J}$ of the dynamics of the average superconducting phase difference ϕ is different from the characteristic time τ_ψ. The perturbation analysis has been carried out for both a constant and a time-dependent applied magnetic flux. The circulating current i_S and the time average of the critical current $<i_c>$ as a function of the d. c. and a. c. components of the applied flux are evaluated in the adiabatic limit, assuming that the oscillation frequency ω of the applied flux is much less then τ_ψ^{-1}. In this limit the Fraunhofer-like pattern in $<i_c>$ vs. B curves are shown to be independent from the normalized frequancy $\tilde\omega = \omega\tau_\psi$.

Next, the dynamical equations of one-dimensional arrays containing $N+1$ identical overdamped Josephson junctions are considered. It has been noticed that the system of $N+1$ nonlinear first-order ordinary differential equations can be broken into two coupled subsystems, one consisting of only one equation for the superconducting phase of one junction in the array (arbitrarily chosen to be ϕ_0), the second describing the time evolution of N opportunely defined normalized flux variables. When a solution of the latter N equations is found, by means of a perturbative approach to first order in the parameter β, the dynamical properties of the system are described by a single time-evolution equation for ϕ_0. In this way, we may affirm that, for small values of β, the system may be described by an equivalent single-junction model, where the maximum Josephson current is appropriately defined. The analysis represents a simple way of approaching the problem of the electrodynamic response of one-dimensional arrays of overdamped Josephson junctions by an equivalent single-junction model.

The same approach is followed for one-dimensional arrays with alternating parameters of $N\times(0\text{-}\pi)$ Josephson junction. Even in this case an effective single-junction model can be adopted to describe the overall dynamics of the system. This model is very useful, since it allows us to obtain the critical current vs. normalized magnetic flux curves in closed analytic form. The interference patterns are seen to be qualitatively similar to recently obtained experimental results on multifacets Josephson junctions (Scharinger et al., 2010) in which the critical current density alternates many times between two opposite values along the junction length. The analytic results for the interference patterns clarify the presence or absence of periodicity and the nature of primary and secondary peaks in these curves. Further investigation on the dependence of i_c on different distributions of the JJs in the array can be of interest.

Finally, by allowing the magnetic flux, applied to a two-junction superconducting quantum interference device, to have an a. c. component in addition to a constant term A, we derive the effective reduced single-junction model describing the dynamics of the average superconducting phase difference ϕ of the two junctions in the device. The difference between this case and the one previously treated is that the alternating flux now varies with a frequency ω of the same order of magnitude of τ_ψ^{-1}, so that the adiabatic approach does not apply. The single-junction model for the system is again obtained by perturbation analysis to first order in the parameter β. Averaging of the rapidly varying quantities in the differential equation for ϕ gives the effective dynamics of the two junctions in the system. In particular, for $\beta = 0$, the critical current of the device is seen to depend on A, on the frequency $\tilde{\omega}$ and the amplitude B of the a. c. component of the applied magnetic flux in a closed analytic form. From the analysis of the voltage vs. applied flux curves it can be argued that the quantities $\tilde{\omega}$ and B can play the role of additional control parameters in the device. Further work in extending the present analysis to finite values of β is necessary. Experimental work confirming the predictions of the present analysis needs to be performed. As far as non-normalized quantities are concerned, for direct experimental confirmation of the present results, we finally notice that the junction dynamics evolves with characteristic frequencies the order of 1 THz. Therefore one needs to run the experiment with very rapidly oscillating signals (10 THz or more) in such a way that normalized frequencies of $\tilde{\omega} = 1.0$ can be achieved.

Author details

R. De Luca

Dipartimento di Fisica "E. R. Caianiello", Università degli Studi di Salerno, Italy

7. References

Ambegaokar, V. & Halperin, B. I. (1969). Voltage Due to Thermal Noise in the dc Josephson Effect. *Phys. Rev. Lett.*, Vol.22, No.25, (June 1969), pp. 1364-1366, ISSN 0031-9007

Barone, A. & Paternò, G. (1982). *Physics and Applications of the Josephson Effect*, John Wiley & Sons, ISBN 0471014699, New York, USA

Baselmans, J. J.; Morpurgo, A. F.; Van Wees, B. J. & Klapwijk, M. (1999). Vortices with Half Magnetic Flux Quanta in "Heavy Fermion" Superconductors . *Nature*, Vol.397, No.6714, (January 1999), pp. 43-45, ISSN 0028-0836

Bishop, A. R. & Trullinger, S. E. (1978). Josephson Junction Threshold Viewed as a Critical Point. *Phys. Rev. B*, Vol.17, No.5, (March 1978), pp. 2175-2182, ISSN 1098-0121

Bocko, M. F.; Herr, A. M. & Feldman, M. J. (1997). Prospect for Quantum Coherence Computation Using Superconducting Electronics. *IEEE Transactions on Applied Superconductivity*, Vol.7, No.2, (June 1997), pp. 3638-3641, ISSN 1051-8223

Bulaevkii, L. N.; Kuzii, V. V. & Sobyanin, A. A. (1977). Superconducting System with Weak Coupling to the Current in the Ground State. *JETP Lett.*, Vol.25, No.7, (April 1997), pp. 290-294, ISSN 0021-3640

Chesca, B. (1999). Magnetic field dependencies of the critical current and of the resonant modes of dc SQUIDs fabricated from superconductors with s+id$_{x^2-y^2}$ order parameter symmetries. *Ann. Phys.(Leipzig)*, Vol.8, No.6, (September 1999), pp. 511-522, ISSN 1521-3889

Clarke, J. & Braginsky, A. I. (2004). *The SQUID Handbook*, Vol. I, Wiley-VCH, ISBN 3-527-40229-2, Weinheim, Germany

Crankshaw, D. S. & Orlando, T. P. (2001). Inductance Effect in the Persistent Current Qubit. *IEEE Transactions on Applied Superconductivity*, Vol.11, No.1, (March 2001), pp. 1006-1009, ISSN 1051-8223

De Luca, R. (2011). Quantum interference in parallel connections of N×(0–π) overdamped Josephson junctions. *Supercon. Sci. Technol.*, Vol.24, No.6, (June 2011), pp. 065026 1-5, ISSN 0953-2048

Geshkenbein, V. B.; Larkin, A. I. & Barone, A. (1987). Vortices with Half Magnetic Flux Quanta in "Heavy Fermion" Superconductors . *Phys. Rev. B*, Vol.36, No.1, (July 1987), pp. 235-238, ISSN 1098-0121

Grønbech-Jensen, N.; Thompson, D. B.; Cirillo, M. & Cosmelli, C. (2003). Thermal escape from zero-voltage states in hysteretic superconducting interferometers. *Phys. Rev. B*, Vol.67, No.22, (June 2003), pp. 224505 1-6, ISSN 1098-0121

Likharev, K. K. (1986). *Dynamics of Josephson Junctions and Circuits*, Gordon and Breach, ISBN 2881240429, Amsterdam, The Netherlands

Romeo, F. & De Luca, R. (2004). Effective Non-Sinusoidal Current-Phase Relations in Conventional d. c. SQUIDs. *Physics Letters A*, Vol.328, No.4-5, (August 2004), pp. 330-334, ISSN 0375-9601

Romeo, F. & De Luca, R. (2005). A reduced model for one-dimensional arrays of overdamped Josephson junctions. *Physica C*, Vol.432, No.3-4, (November 2005), pp. 159-166, ISSN 0921-4534

Ryazanov, V. V.; Oboznov, V. A.; Rusanov, A. Yu.; Veretennikov, A. V.; Golubov, A. A. & Aarts, J. (2001). Coupling of Two Superconductors Through a Ferromagnet: Evidence for a π Junction. *Phys. Rev. Lett.*, Vol.86, No.11, (March 2001), pp. 2427-2430, ISSN 0031-9007

Scharinger, S.; Gürlich, C.; Mints, R. G.; Weides, M.; Kohlstedt, H.; Goldobin, E.; Koelle, D. & Kleiner, R. (2010). Interference patterns of multifacet 20x(0-π) Josephson junctions with ferromagnetic barrier. *Phys. Rev. B*, Vol.81, No.17, (May 2010), pp. 174535 1-4, ISSN 1098-0121

Schultz, R. R.; Chesca, B.; Goetz, B.; Schneider, C. W.; Schmehl, A.; Bielefeldt, H.; Hilgenkamp, H.; Mannhart, J. & Tsuei, C. C. (2000). Design and Realization of an all d-wave dc π Superconducting Quantum Interference Device. *Appl. Phys. Lett.*, Vol.76, No.7, (February 2000), pp. 912-914, ISSN 0003-6951

Smilde, H. J. H.; Ariando; Blank, D. H. A.; Hilgenkamp, H. & Rogalla, H. (2004). π SQUIDs based on Josephson Contacts Between High-Tc and Low-Tc Superconductors . *Phys. Rev. B*, Vol.70, No.2, (July 2004), pp. 024519 1-12, ISSN 1098-0121

Weides, M.; Kemmler, M.; Goldobin, E.; Koelle, D., Kleiner, R.; Kohlstedt, H. & Buzdin, A. (2006). High quality ferromagnetic 0 and π Josephson tunnel junctions, *Appl. Phys. Lett.*, Vol.89, No.12, (September 2006), pp. 122511 1-3, ISSN 0003-6951

Wollman, D. A.; Van Harlingen, D. J.; Lee, W. C.; Ginsberg, D. M. & Leggett, A. J. (1993). Experimental Determination of the Superconducting Pairing State in YBCO from the phase coherence of YBCO-Pb dc SQUIDs. *Phys. Rev. Lett.*, Vol.71, No.13, (September 1993), pp. 2134-2137, ISSN 0031-9007

dc Josephson Current Between an Isotropic and a d-Wave or Extended s-Wave Partially Gapped Charge Density Wave Superconductor

Alexander M. Gabovich, Suan Li Mai,
Henryk Szymczak and Alexander I. Voitenko

Additional information is available at the end of the chapter

1. Introduction

The discovery and further development of superconductivity is extremely interesting because of its pragmatic (practical) and purely academic reasons. At the same time, the superconductivity science is very remarkable as an important object for the study in the framework of the history and methodology of science, since all the details are well documented and well-known to the community because of numerous interviews by participants including main heroes of the research and the fierce race for higher critical temperatures of the superconducting transition, T_c. Moreover, the whole science has well-documented dates, starting from the epoch-making discovery of the superconducting transition by Heike Kamerlingh-Onnes in 1911 [1–7], although minor details of this and, unfortunately, certain subsequent discoveries in the field were obscured [8–11]. As an illustrative example of a senseless dispute on the priority, one can mention the controversy between the recognition of Bardeen-Cooper-Schrieffer (BCS) [12] and Bogoliubov [13] theories.

If one looks beyond superconductivity, it is easy to find quite a number of controversies in different fields of science [14, 15]. Recent attempts [16–18] to contest and discredit the Nobel Committee decision on the discovery of graphene by Andre Geim and Kostya Novoselov [19, 20] are very typical. The reasons of a widespread disagreement concerning various scientific discoveries consist in a continuity of scientific research process and a tense competition between different groups, as happened at liquefying helium and other cryogenic gases [9, 21–24] and was reproduced in the course of studying graphite films [25, 26]. At the same time, the authors and the dates of major discoveries and predictions in the science of superconductivity are indisputable, fortunately to historians and teachers.

Macroscopic manifestations of the superconducting state and diverse properties of the plethora of superconductors are consequences of main fundamental features: (i) zero resistivity found already by Kamerlingh-Onnes (sometimes the existence of persistent currents discovered by him in 1914 is considered more prominent and mysterious [27]), (ii) expulsion of a weak magnetic field (the Meissner effect [28]), and (iii) the Josephson effects [29–37], i.e. the possibility of dc or ac super-currents in circuits, containing thin insulating or normal-metal interlayers between macroscopic superconducting segments. Of course, the indicated properties are interrelated. For instance, a macroscopic superconducting loop with three Josephson junctions can exhibit a superposition of two states with persistent currents of equal magnitudes and opposite polarity [38].

We note that those findings, reflecting a cooperative behavior of conducting electrons (later interpreted in terms of a quantum-mechanical wave function [12, 39–43]), had to be augmented by the observed isotope dependence of T_c [44, 45] in order that the first successful semi-microscopic (it is so, because the declared electron-phonon interaction was, in essence, reduced to the phenomenological four-fermion contact one) BCS theory of superconductivity [12] would come into being. Sometimes various ingenious versions of the BCS theory, explicitly taking into account the momentum and energy dependences of interaction matrix elements, as well as the renormalization of relevant normal-state properties by the superconducting reconstruction of the electron spectrum [46–50], are called "the BCS theory". Nevertheless, such extensions of the initial concept, explicitly related to Ref. [12] and results obtained therein, are inappropriate. This circumstance testifies that one should be extremely accurate with scientific terms, since otherwise it may lead to reprehensible misunderstandings [51].

Whatever be a theory referred to as "the BCS one" or as "the theory of superconductivity" [52], we still lack a true consistent microscopic picture scenario (scenarios?) of superconducting pairing in different various classes of superconductors. As a consequence, all existing superconducting criteria [53–72] are empirical rather than microscopic, although based on various relatively well-developed theoretical considerations. Hence, materials scientists must rely on their intuition to find new promising superconductors [73–78], although bearing also in mind a deep qualitative theoretical reasoning [43, 79–83].

It is no wonder that unusual transport properties of superconductors together with their magnetic-field sensibility led to a number of practically important applications. Namely, features (i) and (ii) indicated above made it possible to manufacture large-scale power cables, fly-wheel energy storage devices, bearings, high field magnets, fault current limiters, superconductor-based transformers, levitated trains, motors and power generators [84–93]. At the same time, the Josephson (weak-coupling) feature (iii) became the basis of small-scale superconducting electronics [88, 94–98], which also uses the emergence of half-integer magnetic flux quantization in circuits with superconducting currents [99, 100]. Smartly designed SQUID devices with several Josephson junctions and a quantized flux serve as sensible detectors of magnetic field and electromagnetic waves, which, in their turn, are utilized in industry, research, and medicine [95–98, 101]. Recently oscillatory effects inherent to superfluid ^{3}He [102–104] and ^{4}He [103–105], which are similar to the Josephson one, were used to construct superfluid helium quantum interference devices (SHeQUIDs) [106].

High-T_c oxide superconductors found in 1986 [107] and including large families of materials with $T_c \leq 138$ K [108–112] extended the application domain of superconductivity, because, first, liquid-nitrogen temperatures were achieved and, second, the predominant $d_{x^2-y^2}$- order parameter symmetry (at least in hole-doped oxides) made possible applications in electronics and quantum computation more diverse [37, 113–122].

While studying high-T_c cuprates, superconductivity was shown to compete with charge density waves (CDWs), so that the observed properties in the superconducting state must be modified by CDWs [123–128]. It should concern Josephson currents phenomenon too [129–134], although this topic has not been properly developed so far.

Of course, other superconducting materials found after the discovery of high-T_c oxide materials are also very remarkable, because of their non-trivial electron spectra, so that Josephson currents through junctions involving those materials should possess interesting features. We mean, in particular, MgB$_2$ with $T_c \leq 40$ K [135] and a multiple energy-gap structure [136, 137], as well as Fe-based pnictides and chalcogenides with $T_c \leq 56$ K and concomitant spin density waves (SDWs) suspected to have deep relations with superconductivity in those materials [78].

In this paper, we present our theoretical studies of dc Josephson currents between conventional superconductors and partially CDW-gapped materials with an emphasis on cuprates, although the gross features of the model can be applied to other CDW superconductors as well. The next Section 2 contains the justification of the approach and the formulation of the problem, whereas numerical results of calculations, as well as the detailed discussion, are presented in Section 3. Section 4 contains some general conclusions concerning dc Josephson currents across junctions involving partially gapped CDW superconductors.

A more involved case of Josephson junctions between two CDW superconductors with various symmetries of superconducting pairing will be treated elsewhere.

2. Theoretical approach

2.1. d-wave versus s-wave order parameter symmetry

Coherent properties of Fermi liquids in the paired state are revealed by measurements of dc or ac Josephson tunnel currents between two electrodes possessing such properties. The currents depend on the phase difference between superconducting order parameters of the electrodes involved [30, 31, 119]. Manifestations of the coherent pair tunneling are more complex for superconductors with anisotropic order parameters than for those with an isotropic energy gap. In particular, it is true for d-wave superconductors, where the order parameter changes its sign on the Fermi surface (FS) [119, 138–143]. As was indicated above, high-T_c oxides are usually considered as such materials, where the $d_{x^2-y^2}$ pairing is usually assumed at least as a dominating one [117, 144–152]. However, conventional s-wave contributions were also detected in electron tunneling experiments [153–160] and, probably, in nuclear magnetic resonance (NMR) and nuclear quadrupole resonance measurements [161]. Therefore, only a minority of researchers prefer to accept the isotropic s-wave (or extended s-wave) nature of superconductivity in cuprates [162–175]. Notwithstanding the existing fundamental controversies, the d-wave specificity of high-T_c oxide superconductivity has already been used in technical devices [95, 116, 118–120, 122].

2.2. Pseudogaps as a manifestation of non-superconducting gapping

In addition to the complex character of superconducting order parameter, cuprates reveal another intricacy of their electron spectrum. Namely, the pseudogap is observed both below and above T_c [176–180]. Here, various phenomena manifesting themselves in resistive, magnetic, optical, photoemission (ARPES), and tunnel (STM and break-junction) measurements are considered as a consequence of the "pseudogap"-induced depletion in the electron density of states, in analogy to what is observed in quasi-one-dimensional compounds above the mean-field phase-transition temperature [181, 182].

Notwithstanding large theoretical and experimental efforts, the pseudogap nature still remains unknown [126–128, 133, 178, 183–201]. Namely, some researchers associate them with precursor order parameter fluctuations, which might be either of a superconducting or some other competing (CDWs, SDWs, etc.) origin. Another viewpoint consists in relating pseudogaps to those competing orderings, but treating them, on the equal footing with superconductivity, as well-developed states that can be made allowance for in the mean field approximation, fluctuation effects being non-crucial. We believe that the available observations support the latter viewpoint (see, e.g., recent experimental evidences of CDW formation in various cuprates [202–205]). Moreover, although undoped cuprates are antiferromagnetic insulators [206], the CDW seems to be a more suitable candidate responsible for the pseudogap phenomena, which competes with Cooper pairing in doped high-T_c oxide samples [123–127], contrary to what is the most probable for iron-based pnictides and chalcogenides [78, 207]. Nevertheless, the type of order parameter competing with Cooper pairing in cuprates is not known with certainty. For instance, neutron diffraction studies of a number of various high-T_c oxides revealed a nonhomogeneous magnetic ordering (usually associated with SDWs) in the pseudogap state [208, 209].

2.3. Superconducting order parameter symmetry scenarios

Bearing in mind all the aforesaid, we present here the following scenarios of dc Josephson tunneling between a non-conventional partially gapped CDW superconductor and an ordinary s-wave one. The Fermi surface (FS) of the former is considered two-dimensional with a $d_{x^2-y^2}$-, d_{xy}- or extended s-wave (with a constant order parameter sign) four-lobe symmetry of superconducting order parameter and a CDW-related doping-dependent dielectric order parameter. The CDWs constitute a system with a four-fold symmetry emerging inside the superconducting lobes in their antinodal directions for cuprates (the $d_{x^2-y^2}$-geometry of the superconducting order parameter, see Figure 1) or in the nodal directions for another possible configuration allowed by symmetry (the d_{xy}-geometry of the superconducting order parameter). (Below, for the sake of brevity, when considering the extended s-wave geometries for the superconducting order parameter, we use the corresponding mnemonic notations $s^{ext}_{x^2-y^2}$ and s^{ext}_{xy}.) Thus, the CDW order parameter Σ competes with its superconducting counterpart Δ over the whole area of their coexistence, which gives rise to an interesting phenomena of temperature- (T-) reentrant Σ [126–128, 210, 211]. In this paper, the main objective of studies are the angular dependences, which might be observed in the framework of the adopted model. Of course, any admixture of Cooper pairing with a symmetry different from $d_{x^2-y^2}$-one [148, 154, 160, 212, 213] may alter the results. Moreover, the superconducting order parameter symmetry might be doping-dependent [214]. To obtain some insight into such more cumbersome situations, we treat here the pure isotropic s-wave case as well. Other possibilities for predominantly d-wave superconductivity coexisting with CDWs lie somewhere between those pure s- and d- extremes.

2.4. Formulation of the problem

The dc Josephson critical current through a tunnel junction between two superconductors, whatever their origin, is given by the general equation [30, 35]

$$I_c(T) = 4eT \sum_{\mathbf{pq}} \left| \widetilde{T}_{\mathbf{pq}} \right|^2 \sum_{\omega_n} \mathsf{F}^+_{\mathrm{HTSC}}(\mathbf{p};\omega_n) \mathsf{F}_{\mathrm{OS}}(\mathbf{q};-\omega_n), \tag{1}$$

Here, $\widetilde{T}_{\mathbf{pq}}$ are matrix elements of the tunnel Hamiltonian corresponding to various combinations of FS sections for superconductors taken on different sides of tunnel junction, $\mathbf{p}$ and $\mathbf{q}$ are the transferred momenta, $e > 0$ is the elementary electrical charge, $\mathsf{F}_{\mathrm{HTSC}}(\mathbf{p};\omega_n)$ and $\mathsf{F}_{\mathrm{OS}}(\mathbf{q};-\omega_n)$ are Gor'kov Green's functions for d-wave (CDW gapped!) and ordinary s-wave superconductor, respectively, and the internal summation is carried out over the discrete fermionic "frequencies" $\omega_n = (2n+1)\pi T$, $n = 0,\pm1,\pm2,\ldots$. The external summation should take into account both the anisotropy of electron spectrum $\xi(\mathbf{p})$ in a superconductor in the manner suggested long time ago for all kinds of anisotropic superconductors [215], the directionality of tunneling [216–220], and the concomitant dielectric (CDW) gapping of the nested FS sections [129].

Hereafter, we shall assume that the ordinary superconductor has the isotropic order parameter $\Delta^*(T)$. At the same time, the superconducting order parameter of the high-T_c CDW superconductor has the properly rotated (see Figure 1) pure d-wave form $\Delta(T)\cos[2(\theta-\gamma)]$, the angle θ being reckoned from the normal $\mathbf{n}$ to the junction plane and γ is a tilt angle between $\mathbf{n}$ and the bisectrix of the nearest positive lobe. Note that, for the s^{ext}-symmetry, the gap profile is the same as in the d-case, but the signs of all lobes are identical rather than alternating (for definiteness, let this sign be positive).

The dielectric order parameter $\Sigma(T)$ corresponds to the checkerboard system of mutually perpendicular CDWs (observed in various high-T_c oxides [221–223]). In the adopted model, it is nonzero inside four sectors, each of the width 2α, with their bisectrices rotated by the angle β with respect to the bisectrices of superconducting order parameter lobes [126–128, 210, 211]. Actually, we shall assume β to be either 0 or $\pi/4$. Since the nesting vectors are directed along the $\mathbf{k}_x$- and $\mathbf{k}_y$-axes in the momentum space [126, 224], the adopted choice corresponds to the choice between $d_{x^2-y^2}$- and d_{xy}-symmetry. Another possible, unidirectional CDW geometry is often observed in cuprates as well [225–227]. It can be treated in a similar way, but we shall not consider it in this work.

Note also that, in agreement with previous studies [216–220, 228], the tunnel matrix elements $\widetilde{T}_{\mathbf{pq}}$ in Eq. (1) should make allowance for the tunnel directionality (the angle-dependent probability of penetration through the barrier) [140, 229, 230]. We factorize the corresponding directionality coefficient $w(\theta)$. The weight factor $w(\theta)$ effectively disables the FS outside a certain given sector around $\mathbf{n}$, thus governing the magnitude and the sign of the Josephson

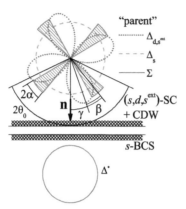

Figure 1. Geometry of the junction between a conventional s-wave superconductor (s-BCS) and a d-, s-extended (s^{ext}) or s-superconductor partially gapped by charge density waves (CDWs, induced by dielectric, i.e. electron-hole, pairing). The angle α denotes the half-width of each of four angular sectors at the Fermi surface, where the CDW gap appears. The gap profiles for the parent CDW insulator (Σ), s-(Δ_s), d- (Δ_d), and s-extended (Δ_{sext}) superconductors, and conventional superconductor (Δ^*) are shown. β is a misorientation angle between the nearest superconducting lobe and CDW-gapped sector, γ is a tilt angle of superconducting lobe with respect to the junction plane determined by the normal $\mathbf{n}$, θ_0 is a measure of tunneling directionality (see explanations in the text).

tunnel current. Specifically, we used the following model for $w(\theta)$:

$$w(\theta) = \exp\left[-\left(\frac{\tan\theta}{\tan\theta_0}\right)^2\right], \tag{2}$$

where θ_0 is an angle describing the effective width of the directionality sector. We emphasize that, for tunneling between two anisotropic superconductors, two different coefficients $w(\theta)$ associated with $\mathbf{p}$- and $\mathbf{q}$-distributions in the corresponding electrodes come into effect [216].

In accordance with the previous treatment of partially gapped s-wave CDW superconductors [123–125, 129, 130, 132, 231–234] and its generalization to their d-wave counterparts [126–128, 210, 211, 235] and in line with the basic theoretical framework for unconventional superconductors [236, 237], the anomalous Gor'kov Green's functions for high-T_c oxides are assumed to be different for angular sectors with coexisting CDWs and superconductivity (d sections of the FS) and the "purely superconducting" rest of the FS (nd sections)

$$F_{HTSC,nd}(\mathbf{p};\omega_n) = \frac{\Delta(T)\cos[2(\theta - \gamma)]}{\omega_n^2 + \Delta^2(T)\cos^2[2(\theta - \gamma)] + \xi_{nd}^2(\mathbf{p})}, \tag{3}$$

$$F_{HTSC,d}(\mathbf{p};\omega_n) = \frac{\Delta(T)\cos[2(\theta - \gamma)]}{\omega_n^2 + \Delta^2(T)\cos^2[2(\theta - \gamma)] + \Sigma^2(T) + \xi_d^2(\mathbf{p})}. \tag{4}$$

Here, we explicitly took into account a possible angle deviation γ of the Δ-lobe direction, which is governed by the crystal lattice geometry, from the normal $\mathbf{n}$ to the junction plane; the latter is created artificially and, generally speaking, can be not coinciding with a crystal facet. The concomitant rotation of the CDW sectors is made allowance for implicitly. The quasiparticle spectra $\xi_d(\mathbf{p})$ and $\xi_{nd}(\mathbf{p})$ correspond to "hot" and "cold" spots of the cuprate FS, respectively (see, e.g., Refs. [176, 238–240]).

Substituting Eqs. (2), (3), and (4) into Eq. (1) and carrying out standard transformations [30, 35], we obtain

$$I_c(T) = \frac{\Delta(0)\,\Delta^*(0)}{2eR_N} i_c(T), \tag{5}$$

$$i_c(T) = \frac{1}{2\pi} \int_{\theta_d} w(\theta) \cos\left[2\left(\theta - \gamma\right)\right] P\left[\Delta^*(T), \sqrt{\Sigma^2 + \Delta^2(T)\cos^2\left[2\left(\theta - \gamma\right)\right]}\right] d\theta$$
$$+ \frac{1}{2\pi} \int_{\theta_{nd}} w(\theta) \cos\left[2\left(\theta - \gamma\right)\right] P\left[\Delta^*(T), |\Delta(T)\cos 2\left(\theta - \gamma\right)|\right] d\theta. \tag{6}$$

Here, R_N is the normal-state resistance of the tunnel junction, determined by $\left|\widetilde{T}_{\mathbf{pq}}\right|^2$ without the factorized multiplier $w(\theta)$, the integration is carried out over the CDW-gapped and CDW-free FS sections (the FS-arcs θ_d and θ_{nd}, respectively, in the two-dimensional problem geometry), $\Delta^*(T)$ is the order parameter of the ordinary isotropic superconductor, whereas the function $P(\Delta_1, \Delta_2)$ is given by the expression [129, 215]

$$P(\Delta_1, \Delta_2) = \int_{\min\{\Delta_1,\Delta_2\}}^{\max\{\Delta_1,\Delta_2\}} \frac{dx\,\tanh\frac{x}{2T}}{\sqrt{\left(x^2 - \Delta_1^2\right)\left(\Delta_2^2 - x^2\right)}}. \tag{7}$$

Modified Eqs. (3)-(6) turn out valid for the calculation of dc Josephson current through a tunnel junction between an ordinary s-wave superconductor and a partially gapped CDW superconductor with an extended s-symmetry of superconducting order parameter [142, 241]. For this purpose, it is enough to substitute the cosine functions in Eqs. (3)-(6) by their absolute values.

At $w(\theta) \equiv 1$ (the absence of tunnel directionality), $\Sigma \equiv 0$ (the absence of CDW-gapping), and putting $\cos 2(\theta - \gamma) \equiv 1$ (actually, it is a substitution of an isotropic s-superconductor for the d-wave one), Eq. (6) expectedly reproduces the famous Ambegaokar–Baratoff result for tunneling between s-wave superconductors [30, 31, 35, 242].

Note that, in Eq. (6), the directionality is made allowance for only by introducing the angular function $w(\theta)$ reflecting the angle-dependent tunnel-barrier transparency. On the other hand, the tunneling process, in principle, should also take into account the factors $\left|\mathbf{v}_{g,nd} \cdot \mathbf{n}\right|$ and $\left|\mathbf{v}_{g,d} \cdot \mathbf{n}\right|$, responsible for extra directionality [140, 219, 230], where $\mathbf{v}_{g,nd} = \nabla\xi_{nd}$ and $\mathbf{v}_{g,d} = \nabla\xi_d$ are the quasiparticle group velocities for proper FS sections. Those factors can be considered as proportional to a number of electron attempts to penetrate the barrier [139]. They were introduced decades ago in the framework of general problem dealing with tunneling in heterostructures [243–245]. Nevertheless, we omitted here the group-velocity-dependent multiplier, since it requires that the FS shape should be specified, thus going beyond the applied semi-phenomenological scheme, as well as beyond similar semi-phenomenological approaches of other groups [138, 139, 141, 236, 246]. We shall take the additional directionality factor into account in subsequent publications, still being fully aware of the phenomenological nature of both $|\mathbf{v}_g \cdot \mathbf{n}|$ and $w\,(\theta)$ functions.

It is well known [143] that, in the absence of directionality, the Josephson tunneling between d- and s-wave superconductors is weighted-averaged over the FS, with the cosine multiplier in Eq. (6) playing the role of weight function. In this case, the Josephson current has to be strictly equal to zero. However, it was found experimentally that the dc Josephson current between $Bi_2Sr_2CaCu_2O_{8+\delta}$ and Pb [155], $Bi_2Sr_2CaCu_2O_{8+\delta}$ and Nb [247], $YBa_2Cu_3O_{7-\delta}$ and PbIn [248], $Y_{1-x}Pr_xBa_2Cu_3O_{7-\delta}$ and Pb [153] differ from zero. Hence, either a subdominant s-wave component of the superconducting order parameter does exist in cuprate materials, as was discussed above, or the introduction of directionality is inevitable to reconcile any theory dealing with tunneling of quasiparticles from (to) high-T_c oxides and the experiment.

We restrict ourselves mostly to the case $T = 0$, when formula (7) is reduced to elliptic functions [30, 249], although some calculations will be performed for $T \neq 0$ as well. The reason consists in the smallness of T_c for conventional s-wave superconductors (in our case, it is Nb, see below) as compared to T_c of anisotropic d-wave oxides. Hence, all effects concerning T-dependent interplay between Δ and Σ including possible reentrance of $\Sigma(T)$ [126–128, 210, 211, 235] become insignificant in the relevant T-range cut off by the s-wave-electrode order parameter. On the contrary, in the symmetrical case, when one studies tunneling between different high-T_c-oxide grains, T-dependences of the Josephson current are expected to be very interesting. This more involved situation will be investigated elsewhere.

3. Results and discussion

3.1. Total currents

In what follows, we shall consider in parallel the dc Josephson currents between a more or less conventional (weak-coupling BCS s-wave) Nb with a zero-T energy gap $\Delta^*(0) = 1.4\,meV$ and $T_c = 9.2\,K$ [247] and either a $d_{x^2-y^2}$- or a d_{xy}- superconductor ($\beta = 0$ and $\pi/4$, respectively). The latter is also possible from the symmetry viewpoint, but have not yet been found among existing classes of CDW superconductors.

The dependences of the dimensionless current $i_c(T = 0)$ on the tilt angle γ are shown in Figure 2(a) for $\alpha = 15°$ and various values of the parameter θ_0 describing the degree of directionality. Since $T = 0$, there is no need to solve the equation set for $\Sigma(T)$ and $\Delta(T)$ for partially CDW-gapped s-wave [233] or d-wave [128] superconductors self-consistently. Instead, for definiteness, we chose the experimental values $\Sigma(0) = 36.3\,meV$ and $\Delta(0) = 28.3\,meV$ appropriate to slightly overdoped $Bi_2Sr_2CaCu_2O_{8+\delta}$ samples [250] as input parameters. The half-width α of each of the four CDW sectors was *rather arbitrarily* chosen as 15°. In fact, it is heavily dependent on the doping extent and cannot be unambiguously extracted even from the most precise angle-resolved photoemission spectra (ARPES) [200, 251, 252]. Thus, hereafter we consider the parameter of dielectric FS gapping α as a *phenomenological* one on the same footing as the tunneling directionality parameter θ_0.

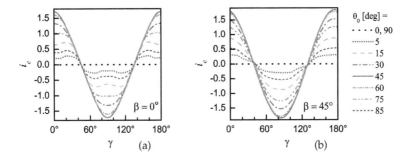

Figure 2. (a) Zero-temperature ($T = 0$) dependences on γ of the dimensionless dc Josephson current i_c for the tunnel junction between an s-wave superconductor and a CDW $d_{x^2-y^2}$-wave one ($\beta = 0°$) for various θ_0's. The specific gap values for electrodes correspond to the experimental data for Nb ($\Delta^*(T = 0) = 1.4$ meV) and $Bi_2Sr_2CaCu_2O_{8+\delta}$ ($\Sigma(T = 0) = 36.3$ meV and $\Delta(T = 0) = 28.3$ meV). The calculation parameter $\alpha = 15°$. See further explanations in the text. (b) The same as in panel (a), but for a CDW d_{xy}-superconductor ($\beta = 45°$).

It is evident that, if the sector θ_0 of effective tunneling equals zero, the Josephson current vanishes. It is also natural that, in the case of d-wave pairing and the absence of tunneling directionality ($\theta_0 = 90°$), the Josephson current disappears due to the exactly mutually compensating contributions from superconducting order parameter lobes with different signs [119, 138, 143]. Intermediate θ_0's correspond to non-zero Josephson tunnel current of either sign (conventional 0- and π-junctions [120, 122, 253]) except at the tilt angle $\gamma = 45°$, when $i_c = 0$. In this connection, one should recognize that the energy minimum *for non-conventional anisotropic superconductors* can occur, in principle, at any value of the order parameter phase [254]. As is seen from Figure 2(a), the existence of CDWs in cuprates ($\alpha \neq 0$, $\Sigma \neq 0$) influences the γ-dependences of i_c, which become non-monotonic for θ_0 close to α demonstrating a peculiar resonance between two junction characteristics. The effect appears owing to the actual $d_{x^2-y^2}$- pattern with the coinciding bisectrices of CDW sectors and superconducting lobes ($\beta = 0°$). This circumstance may ensure the finding of CDWs (pseudogaps) by a set of relatively simple transport measurements.

At the same time, for the hypothetical d_{xy} order parameter symmetry ($\beta = 45°$, Figure 2(b)), when hot spots lie in the nodal regions, the dependences $i_c(\gamma)$ become asymmetrical relative to $\gamma = 90°$ and remain monotonic as for CDW-free d-wave superconductors.

The role of superconducting-lobe and CDW (governed by the crystalline structure) orientation with respect to the junction plane (the angle γ) is most clearly seen for varying α, which is shown in Figure 3. The indicated above "resonance" between θ_0 and α is readily seen in Figure 3(a). One also sees that the Josephson current amplitude is expectedly reduced with the increasing α, since CDWs suppress superconductivity [123–127, 255]. For $\beta = 45°$ (Figure 3(b)), the curves $i_c(\gamma)$ are non-symmetrical, and their form is distorted by CDWs relative to the case of "pure" superconducting d-wave electrode.

The dependence of i_c on the CDW-sector width, i.e. the degree of dielectric FS gapping, is a rapidly dropping one, which is demonstrated in Figures 4(a) (for $\beta = 0°$, i.e. for $d_{x^2-y^2}$ or $s^{ext}_{x^2-y^2}$ symmetries) and 4(b) (for $\beta = 45°$, i.e. for d_{xy} or s^{ext}_{xy} symmetries) calculated for

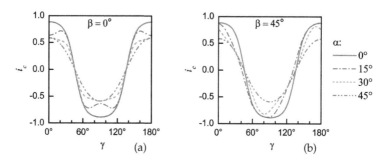

Figure 3. The same as in Figure 2, but for $\theta_0 = 15°$ and various α's.

$\gamma = 0°$ and $\theta_0 = 15°$. Indeed, for cuprates, where the directions of superconducting lobes and CDW sectors coincide, an extending CDW-induced gap reduces the electron density of states available to superconducting pairing until α becomes equal to θ_0 (see Figure 4(a)). A further increase of the pseudogapped FS arc has no influence on i_c, since it falls outside the effective tunneling sector. We note that the α-dependence of i_c for cuprates can be, in principle, non-linearly mapped onto the doping dependence of the pseudogap [200, 251, 252]. It is remarkable that, qualitatively, the results are the same for the extended s-symmetry (denoted as s^{ext}) of the superconducting order parameter and are very similar to those for the assumed s-wave order parameter (curves marked by s).

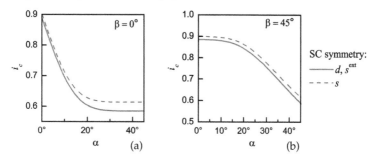

Figure 4. Dependences $i_c(\alpha)$ for $\gamma = 0°$ and $\theta_0 = 15°$ for d-, s-extended, and s-symmetries of superconducting order parameter.

At the same time, if the CDW sectors are rotated in the momentum space by 45° with respect to the superconducting lobes and/or the directional-tunneling θ_0-cone (see Figure 4(b)), the dependences $i_c(\alpha)$ are very weak at small α and become steep for $\alpha > \theta_0$. This result is true for the d_{xy}-, rotated extended s-, and isotropic s-symmetries of the superconducting order parameter coexisting with its dielectric counterpart.

One sees from Figure 4 that, for small $\theta_0 = 15°$, the d- and extended s-order parameters result in the same $i_c(\alpha)$. Of course, it is no longer true for larger θ_0, when contributions from different lobes into the total Josephson current start to compensate each other for d-wave superconductivity, whereas no compensation occurs for the extended s-wave scenario. To make sure that this assertion is valid, we calculated the dependences $i_c(\theta_0)$ for $\gamma = 0°$, $\alpha = 15°$, and $\beta = 0°$ and $45°$. The results are presented in Figure 5. Indeed, for $\theta_0 \geq 30°$, the curves corresponding to d-wave and extended s-wave superconductors come apart, as it has to be. Thus, Josephson currents between isotropic and CDW d-wave superconductors, similarly to the CDW-free case, are non-zero only because the tunneling is non-isotropic.

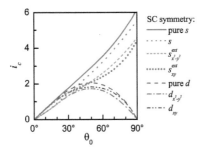

Figure 5. Dependences $i_c(\theta_0)$ for $\gamma = 0°$, $\beta = 0°$, and $\alpha_0 = 15°$ for various symmetries of superconducting order parameter.

It is instructive to compare the tilt-angle-γ dependences of the Josephson currents i_c for possible superconducting order parameter symmetries, which are considered, in particular, for cuprates. The results of calculations are displayed in Figure 7 for $\alpha = \theta_0 = 15°$. For an s-wave CDW-free superconductor, $i_c(\gamma) = $ const. The reference curve $i_c(\gamma)$ for a CDW-free $d_{x^2-y^2}$-wave superconductor (Figure 7(a)) is periodic and alternating. CDWs distort both curves. Namely, the CDW $d_{x^2-y^2}$-wave superconductor demonstrates a non-monotonic behavior of $i_c(\gamma)$, as was indicated above, whereas $i_c(\gamma)$ for the s-wave CDW superconductor becomes a periodic dependence of a constant sign. The curve $i_c(\gamma)$ for the extended s-wave CDW superconductor has a different form than in the s-wave case, although being qualitatively similar. The presented data demonstrate that CDWs can significantly alter angle dependences often considered as a smoking gun,when determining the actual order parameter symmetry for cuprates or other like materials.

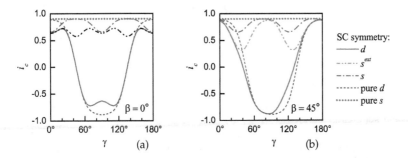

Figure 6. The same as in Figure 2, but for $\theta_0 = 15°$ and various symmetries of superconducting order parameter.

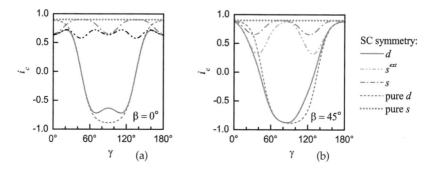

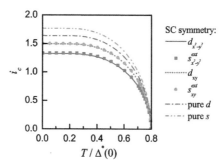

Figure 7. The same as in Figure 2, but for $\theta_0 = 15°$ and various symmetries of superconducting order parameter.

The results for $\beta = 45°$ (Figure 7(b)) differ quantitatively from their counterparts found for $\beta = 0°$, but qualitative conclusions remain the same.

As was indicated above, the temperature behavior of i_c between ordinary superconductors and cuprates is determined by the order parameter dependence $\Delta^*(T)$ for the material with much lower T_c, Nb in our case. This is demonstrated in Figure 8 for d-, extended s- and s-wave CDW high-T_c superconductors. One sees that all curves $i_c(T)$ are similar, differing only in magnitudes.

Figure 8. Dependences $i_c(T)$ for $\gamma = 0°$, $\theta_0 = 30°$, $\alpha_0 = 15°$ and various symmetries of superconducting order parameter.

3.2. Analysis of current components

In Figure 9, the dependences $i_c(\gamma)$ resolved into d and nd components are shown for CDW d-wave superconductors with $\beta = 0°$, $\alpha = 15°$, and various θ_0's. Note that the order parameter *amplitudes* at $T = 0$ are the same throughout the paper! It comes about that, for a narrow directionality cone θ_0, the contribution of the nested (d) FS sections has quite a different tilt (γ) angle behavior as compared to their nd counterparts. All that gives rise to a non-monotonic pattern seen, e.g., in Figure 2(a).

dc Josephson Current Between an Isotropic and a d-Wave or Extended s-Wave Partially Gapped
Charge Density Wave Superconductor

105

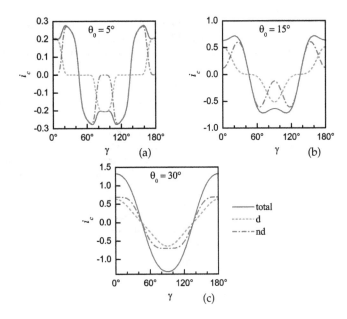

Figure 9. Dependences of i_c and its d and nd components on γ for $d_{x^2-y^2}$ order parameter symmetry, $\alpha_0 = 15°$, and various θ_0's (panel a to c).

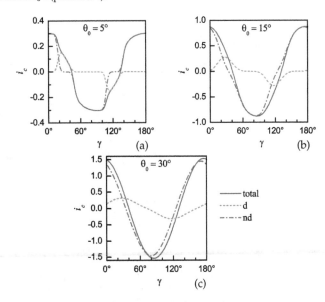

Figure 10. The same as in Figure 9, but for d_{xy} order parameter symmetry.

In Figure 10, the same dependences as in Figure 9 are shown, but for $\beta = 45°$. One sees that, whatever complex is the γ-angle behavior of d contribution to the overall tunnel currents between a d_{xy}-superconductor and Nb, the CDW influence is much weaker in governing the dependences $i_c(\gamma)$.

It is illustrative to carry out the same analysis in the scenario, when the high-T_c CDW superconductor is assumed to be an extended s-wave one, i.e. when the sign of superconducting order parameter is the same for all lobes. In the case $\beta = 0°$, the corresponding results can be seen in Figure 11, where the γ-dependences of d and nd components of i_c, as well as the total $i_c(\gamma)$ dependences, are depicted for the same parameter set as in Figure 9. We see that the d and nd contributions oscillate with the varying γ almost in antiphase, remaining, nevertheless, positive. For large $\theta_0 = 30°$ (Figure 11(c)), oscillations largely compensate each other making the curve $i_c(\gamma)$ almost flat, which mimics the behavior appropriate to CDW-free isotropic s-wave superconductors. However, we emphasize that this, at the first glance, dull result obtained for a relatively wide CDW sector is actually a consequence of a peculiar superposition involving the periodic dependences of d and nd components on γ with rapidly varying amplitudes.

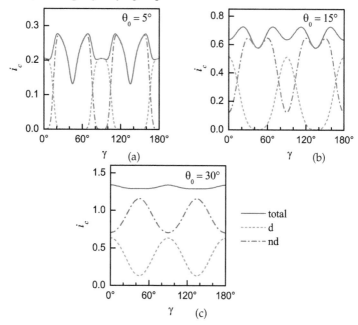

Figure 11. The same as in Figure 9, but for $s^{ext}_{x^2-y^2}$ order parameter symmetry.

The same plots as in Figure 11 were calculated for $\beta = 45°$ and depicted, in Figure 12. Here, the directionality angle θ_0 is the main factor determining the amplitude of i_c, the role of CDWs being much weaker than in the case $\beta = 0°$. It is natural, because now CDW-gapping is concentrated in the nodal regions.

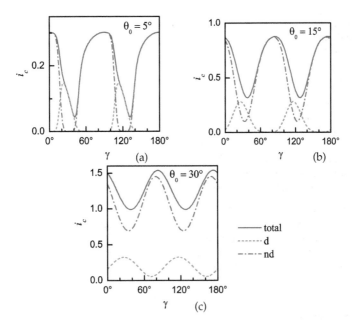

Figure 12. The same as in Figure 11, but for s_{xy}^{ext} order parameter symmetry.

4. Conclusions

The results obtained confirm that the dc Josephson current, probing coherent superconducting properties [30, 31, 33, 37, 119, 256–258], is always suppressed by the electron-hole CDW pairing, which, in agreement with the totality of experimental data, is assumed here to compete with its superconducting electron-electron (Cooper) counterpart [129, 130, 132, 259–262]. We emphasize that, as concerns the quasiparticle current, the results are more ambiguous. In particular, the states on the FS around the nodes of the d-wave superconducting order parameter are engaged into CDW gapping [126–128, 210, 211, 235, 263], so that the ARPES or tunnel spectroscopy feels the overall energy gaps being larger than their superconducting constituent.

Our examination demonstrates that the emerging CDWs should distort the dependences $i_c(\gamma)$, whatever is the symmetry of superconducting order parameter. It is easily seen that, for equal (or almost equal) θ_0 and α, CDWs make the $i_c(\gamma)$ curves non-monotonic and quantitatively different from their CDW-free counterparts. In particular, i_c values are conspicuously smaller for $\Sigma \neq 0$. The required resonance between θ_0 and α can be ensured by the proper doping, i.e. a series of samples and respective tunnel junctions should be prepared with attested tilt angles γ, and the Josephson current should be measured for them. Of course, such measurements could be very cumbersome, although they may turn out quite realistic to be performed.

At the same time, when an s-wave contribution to the actual order parameter in a cuprate sample is dominant up to the complete disappearance of the d-wave component, the $i_c(\gamma)$ dependences for junctions involving CDW superconductors are no longer constant as in the CDW-free case. This prediction can be verified for CDW superconductors with *a fortiori* s-wave order parameters (such materials are quite numerous [123–128]).

In this paper, our approach was purely theoretical. We did not discuss unavoidable experimental difficulties to face with in fabricating Josephson junctions necessary to check the results obtained here. We are fully aware that the emerging problems can be solved on the basis of already accumulated knowledge concerning the nature of grain boundaries in high-T_c oxides [37, 115–119, 122, 264–268]. Note that required junctions can be created at random in an uncontrollable fashion using the break-junction technique [250]. This method allows to comparatively easily detect CDW (pseudogap) influence on the tilt-angle dependences.

To summarize, measurements of the Josephson current between an ordinary superconductor and a d-wave or extended s-wave one (e.g., a high-T_c oxide) would be useful to detect a possible CDW influence on the electron spectrum of the latter. Similar studies of iron-based superconductors with doping-dependent spin density waves (SDWs) would also be of benefit (see, e.g., recent Reviews [78, 269–275]), since CDW and SDW superconductors have similar, although not identical, properties [123–125].

Acknowledgements

AIV is grateful to Kasa im. Józefa Mianowskiego, Fundacja na Rzecz Nauki Polskej, and Fundacja Zygmunta Zaleskiego for the financial support of his visits to Warsaw. The work was partially supported by the Project N 8 of the 2012-2014 Scientific Cooperation Agreement between Poland and Ukraine.

Author details

Gabovich Alexander M. and Voitenko Alexander I.
Institute of Physics, National Academy of Sciences of Ukraine, 46, Nauka Ave., Kyiv 03028, Ukraine

Mai Suan Li and Szymczak Henryk
Institute of Physics, Polish Academy of Sciences, 32/46, Al. Lotników, PL-02-668 Warsaw, Poland

5. References

[1] Kamerlingh-Onnes H (1911) Further experiments with liquid helium. C. On the change of electric resistance of pure metals at very low temperatures etc. IV. The resistance of pure mercury at helium temperatures. Communs Phys. Lab. Univ. Leiden. 120: 3–5.

[2] Editorial (2011) The super century. Nature Mater. 10: 253.

[3] Editorial (2011) A very cool birthday. Nature Phys. 7: 271.

[4] Cho A (2011) Superconductivity's smorgasbord of insights: A movable feast. Science 332: 190–192.

[5] Larbalestier D, Canfield PC (2011) Superconductivity at 100 – Where we've been and where we're going. Mater. Res. Soc. Bull. 36: 590–593.

[6] van der Marel D, Golden M (2011) Heike's heritage. Nature Phys. 7: 378–379.
[7] Grant PM (April 2011) Down the path of least resistance. Phys. World 24: 18–22.
[8] de Nobel J (September 1996) The discovery of superconductivity. Phys. Today 49: 40–42.
[9] van Delft D (2007) Freezing Physics. Heike Kamerlingh-Onnes and the Quest for Cold.
     Amsterdam: Koninklijke Nederlandse Akademie van Wetenschappen.
[10] van Delft D, Kes P (September 2010) The discovery of superconductivity. Phys. Today
     63: 38–43.
[11] van Delft D, Kes P (2011) The discovery of superconductivity. Europhys. News 42:
     21–25.
[12] Bardeen J, Cooper LN, Schrieffer JR (1957) Theory of superconductivity. Phys. Rev. 108:
     1175–1204.
[13] Bogoliubov NN (1958) On the new method in the theory of superconductivity. Zh. Éksp.
     Teor. Fiz. 34: 58–65.
[14] Burgin MS, Gabovich AM (March-April 1997) Why the discovery has not been made?
     Visn. Nats. Akad. Nauk Ukrainy: 55–60.
[15] Hudson RG (2001) Discoveries, when and by whom? Brit. J. Phil. Sci. 52: 75–93.
[16] de Heer WA (2011) Epitaxial graphene: A new electronic material for the 21st century.
     Mater. Res. Soc. Bull. 36: 632–639.
[17] de Heer WA, Berger C, Ruan M, Sprinkle M, Li X, Hu Y, Zhang B, Hankinson J,
     Conrad E (2011) Large area and structured epitaxial graphene produced by confinement
     controlled sublimation of silicon carbide. Proc. Nat. Acad. Sci. USA 108: 16900–16905.
[18] de Heer WA (2012) The invention of graphene electronics and the physics of epitaxial
     graphene on silicon carbide. Phys. Scripta T146: 014004.
[19] Geim AK (2011) Nobel Lecture: Graphene: Random walk to graphene. Rev. Mod. Phys.
     83: 851–862.
[20] Novoselov KS (2011) Nobel Lecture: Graphene: Materials in the Flatland. Rev. Mod.
     Phys. 83: 837–849.
[21] Soulen RJ Jr (March 1996) James Dewar, his flask and other achievements. Phys. Today
     49: 32–38.
[22] Reif-Acherman S (2004) Heike Kamerlingh-Onnes: Master of experimental technique
     and quantitative research. Phys. Perspect. 6: 197–223.
[23] van Delft D (March 2008) Little cup of helium, big science. Phys. Today 61: 36–42.
[24] Kubbinga H (2010) A tribute to Wróblewski and Olszewski. Europhys. News 41: 21–24.
[25] Dresselhaus MS (2012) Fifty years in studying carbon-based materials. Phys. Scripta
     T146: 014002.
[26] Geim AK (2012) Graphene prehistory. Phys. Scripta T146: 014003.
[27] Kitazawa K (2012) Superconductivity: 100th anniversary of its discovery and its future.
     Jpn. J. Appl. Phys. 51: 010001.
[28] Meissner W, Ochsenfeld R (1933) Ein neuer Effekt bei Eintritt der Supraleitfähigkeit.
     Naturwiss. 33: 787–788.
[29] Josephson BD (1962) Possible new effects in superconductive tunneling. Phys. Lett. 1:
     251–253.
[30] Kulik IO, Yanson IK (1972) Josephson Effect in Superconducting Tunnel Structures.
     Jerusalem: Israel Program for Scientific Translation.

[31] Waldram JR (1976) The Josephson effects in weakly coupled superconductors. Rep. Prog. Phys. 39: 751–821.

[32] Kulik IO, Omel'yanchuk AN (1978) Josephson effect in superconducting bridges: microscopic theory. Fiz. Nizk. Temp. 4: 296–311.

[33] Rogovin D, Scully M (1976) Superconductivity and macroscopic quantum phenomena. Phys. Rep. 25: 175–291.

[34] Likharev KK (1979) Superconducting weak links. Rev. Mod. Phys. 51: 101–159.

[35] Barone A, Paterno G (1982) The Physics and Applications of the Josephson Effect. New York: John Wiley and Sons.

[36] Likharev KK (1985) Introduction into Dynamics of Josephson Junctions. Moscow: Nauka. in Russian.

[37] Golubov AA, Kupriyanov M Yu, Il'ichev E (2004) The current-phase relation in Josephson junctions. Rev. Mod. Phys. 76: 411–469.

[38] van der Wal CH, ter Haar ACJ, Wilhelm FK, Schouten RN, Harmans CJPM, Orlando TP, Lloyd S, Mooij JE (2000) Quantum superposition of macroscopic persistent-current states. Science 290: 773–777.

[39] Rickayzen G (1971) Superconductivity and long-range order. Essays Phys. 3: 1–33.

[40] Eggington MA, Leggett AJ (1975) Is ODLRO a necessary condition for superfluidity? Collect. Phenom. 2: 81–87.

[41] Svidzinsky AV (1982) Spatially-Nonhomogeneous Problems of the Superconductivity Theory. Moscow: Nauka. in Russian.

[42] Pitaevskii L (2008) Phenomenology and microscopic theory: Theoretical foundations. In: Bennemann KH, Ketterson JB, editors. Superconductivity. Vol. 1: Conventional and Unconventional Superconductors. Berlin: Springer Verlag. pp. 27–71.

[43] Leggett AJ (2011) The ubiquity of superconductivity. Annu. Rev. Condens. Matter Phys. 2: 11–30.

[44] Maxwell E (1950) Isotope effect in the superconductivity of mercury. Phys. Rev. 78: 477.

[45] Reynolds CA, Serin B, Wright WH, Nesbitt LB (1950) Superconductivity of isotopes of mercury. Phys. Rev. 78: 487.

[46] Eliashberg GM (1960) Interaction of electrons with lattice vibrations in a superconductor. Zh. Éksp. Teor. Fiz. 38: 966–976.

[47] Eliashberg GM (1960) Temperature Green's functions of electrons in a superconductor. Zh. Éksp. Teor. Fiz. 39: 1437–1441.

[48] Kirzhnits DA, Maksimov EG, Khomskii DI (1973) The description of superconductivity in terms of dielectric response function. J. Low Temp. Phys. 10: 79–93.

[49] Carbotte JP (1990) Properties of boson-exchange superconductors. Rev. Mod. Phys. 62: 1027–1157.

[50] Carbotte JP, Marsiglio F (2003) Electron-phonon superconductivity. In: Bennemann KH, Ketterson JB, editors. The Physics of Superconductors. Vol. 1: Conventional and High-T_c superconductors. Berlin: Springer Verlag. pp. 233–345.

[51] Kuznetsov VI (1997) Concept and Its Strutures. Methodological Analysis. Kiev: Institute of Philosophy. in Russian.

[52] Anderson PW (1997) The Theory of Superconductivity in the High-T_c Cuprates. Princeton: Princeton Univeristy Press.

[53] Matthias BT (1955) Empirical relation between superconductivity and the number of valence electrons per atom. Phys. Rev. 97: 74–76.

[54] Kulik IO (1964) On the superconductivity criterion. Zh. Éksp. Teor. Fiz. 47: 2159–2167.

[55] Matthias BT (1973) Criteria for superconducting transition temperatures. Physica 69: 54–56.

[56] Stern H (1973) Trends in superconductivity in the periodic table. Phys. Rev. B 8: 5109–5121.

[57] Kulik IO (1974) Superconductivity and macroscopic stability criteria for electron-phonon systems. Zh. Éksp. Teor. Fiz. 66: 2224–2239.

[58] Stern H (1975) Superconductivity in transition metals and compounds. Phys. Rev. B 12: 951–960.

[59] Gualtieri DM (1975) The correlation between the superconducting critical temperature and the number of stable isotopes among superconducting elements. Solid State Commun. 16: 917–918.

[60] Kulik IO (1976) Spin fluctuations and superconductivity (properties of Pd–H solutions). Fiz. Nizk. Temp. 2: 486–499.

[61] Gabovich AM (1977) On the criterion of superconductivity of metals. Ukr. Fiz. Zh. 22: 2072–2074.

[62] Gabovich AM, Moiseev DP (1978) Superconductivity of metals with the allowance for ion core repulsion. Fiz. Nizk. Temp. 4: 1115–1124.

[63] Gabovich AM (1980) About superconductivity of polar semiconductors. Fiz. Tverd. Tela 22: 3231–3235.

[64] Gabovich AM, Moiseev DP (1981) Isotope effect in jellium and Brout models. Fiz. Tverd. Tela 23: 1511–1514.

[65] Chapnik IM (1984) Regularities in the occurrence of superconductivity. J. Phys. F 14: 1919–1921.

[66] Carbotte JP (1987) On criteria for superconductivity. Sci. Progr. 71: 327–350.

[67] Luo Q-G, Wang R-Y (1987) Electronegativity and superconductivity. J. Phys. Chem. Sol. 48: 425–430.

[68] Jou CJ, Washburn J (1991) Relationship between T_c and electronegativity differences in compound superconductors. Appl. Phys. A 53: 87–93.

[69] Hirsch JE (1997) Correlations between normal-state properties and superconductivity. Phys. Rev. B 55: 9007–9024.

[70] Buzea C, Robbie K (2005) Assembling the puzzle of superconducting elements: a review. Supercond. Sci. Technol. 18: R1–R8.

[71] Koroleva L, Khapaeva TM (2009) Superconductivity, antiferromagnetism and ferromagnetism in periodic table of D.I. Mendeleev. J. Phys.: Conf. Ser. 153: 012057.

[72] Taylor BJ, Maple MB (2009) Formula for the critical temperature of superconductors based on the electronic density of states and the effective mass. Phys. Rev. Lett. 102: 137003.

[73] Pickett WE (2008) The next breakthrough in phonon-mediated superconductivity. Physica C 468: 126–135.

[74] Blase X (2011) Superconductivity in doped clathrates, diamond and silicon. C. R. Physique 12: 584–590.

[75] Canfield PC (2011) Still alluring and hard to predict at 100. Nature Mater. 10: 259–261.

[76] Forgan T (April 2011) Resistance is futile. Phys. World 24: 33–38.

[77] Greene LH (April 2011) Taming serendipity. Phys. World 24: 41–43.

[78] Stewart GR (2011) Superconductivity in iron compounds. Rev. Mod. Phys. 83: 1589–1652.

[79] Geballe TH (2006) The never-ending search for high-temperature superconductivity. J. Supercond. 19: 261–276.

[80] Cohen ML (2006) Electron–phonon-induced superconductivity. J. Supercond. 19: 283–290.

[81] Pickett WE (2006) Design for a room-temperature superconductor. J. Supercond. 19: 291–297.

[82] Leggett AJ (2006) What DO we know about high T_c? Nature Phys. 2: 134–136.

[83] Maksimov EG, Dolgov OV (2007) A note on the possible mechanisms of high-temperature superconductivity. Usp. Fiz. Nauk 177: 983–988.

[84] Hassenzahl WV, Hazelton DW, Johnson BK, Komarek P, Noe M, Reis CT (2004) Electric power applications of superconductivity. Proc. IEEE 92: 1655–1674.

[85] Gourlay SA, Sabbi G, Kircher F, Martovetsky N, Ketchen D (2004) Superconducting magnets and their applications. Proc. IEEE 92: 1675–1687.

[86] Hull JR, Murakami M (2004) Applications of bulk high-temperature superconductors. Proc. IEEE 92: 1705–1718.

[87] Tsukamoto O (2004) Ways for power applications of high temperature superconductors to go into the real world. Supercond. Sci. Technol. 17: S185–S190.

[88] Malozemoff AP, Mannhart J, Scalapino D (April 2005) High-temperature cuprate superconductors get to work. Phys. Today 58: 41–47.

[89] Hassenzahl WV, Eckroad SEC, Grant PM, Gregory B, Nilsson S (2009) A high-power superconducting DC cable. IEEE Trans. Appl. Supercond. 19: 1756–1761.

[90] Shiohara Y, Fujiwara N, Hayashi H, Nagaya S, Izumi T, Yoshizumi M (2009) Japanese efforts on coated conductor processing and its power applications: New 5 year project for materials and power applications of coated conductors (M-PACC). Physica C 469: 863–867.

[91] Maguire JF, Yuan J (2009) Status of high temperature superconductor cable and fault current limiter projects at American Superconductor. Physica C 469: 874–880.

[92] Editorial (April 2011) Fantastic five. Phys. World 24: 23–25.

[93] Malozemoff AP (2011) Electric power grid application requirements for superconductors. Mater. Res. Soc. Bull. 36: 601–607.

[94] Mannhart J (1996) High-T_c transistors. Supercond. Sci. Technol. 9: 49–67.

[95] Koelle D, Kleiner R, Ludwig F, Dantsker E, Clarke J (1999) High-transition-temperature superconducting quantum interference devices. Rev. Mod. Phys. 71: 631–686.

[96] Winkler D (2003) Superconducting analogue electronics for research and industry. Supercond. Sci. Technol. 16: 1583–1590.

[97] Yang H-C, Chen J-C, Chen K-L, Wu C-H, Horng H-E, Yang SY (2008) High-T_c superconducting quantum interference devices: Status and perspectives. J. Appl. Phys. 104: 011101.

[98] Anders S, Blamire MG, Buchholz F-Im, Crete D-G, Cristiano R, Febvre P, Fritzsch L, Herr A, Il'ichev E, Kohlmann J, Kunert J, Meyer H-G, Niemeyer J, Ortlepp T, Rogalla H, Schurig T, Siegel M, Stolz R, Tarte E, ter Brake HJM, Toepfer H, Villegier J-C, Zagoskin

AM, Zorin AB (2010) European roadmap on superconductive electronics – status and perspectives. Physica C 470: 2079–2126.

[99] Deaver BS Jr, Fairbank WM (1961) Experimental evidence for quantized flux in superconducting cylinders. Phys. Rev. Lett. 7: 43–46.

[100] Doll R, Näbauer M (1961) Experimental proof of magnetic flux quantization in a superconducting ring. Phys. Rev. Lett. 7: 51–52.

[101] Foley CP, Hilgenkamp H (2009) Why NanoSQUIDs are important: an introduction to the focus issue. Supercond. Sci. Technol. 22: 064001.

[102] Borovik-Romanov AS, Bun'kov YuM, de Vaard A, Dmitriev VV, Makrotsieva V, Mukharskii YuM, Sergatskov DA (1988) Observation of a spin-current analog of the Josephson effect. Pis'ma Zh. Éksp. Teor. Fiz. 47: 400–403.

[103] Packard RE (1998) The role of the Josephson-Anderson equation in suprefluid helium. Rev. Mod. Phys. 70: 641–651.

[104] Varoquaux E (2001) Superfluid helium interferometry: an introduction. C. R. Acad. Sci. Paris 2: 531–544.

[105] Sukhatme K, Mukharsky Yu, Chui T, Pearson D (2001) Observation of the ideal Josephson effect in superfluid ^{4}He. Nature 411: 280–283.

[106] Sato Y, Packard RE (2012) Superfluid helium quantum interference devices: physics and applications. Rep. Prog. Phys. 75: 016401.

[107] Bednorz JG, Müller KA (1986) Possible high T_c superconductivity in the Ba-La-Cu-O system. Z. Phys. 64: 189–193.

[108] Hauck J, Mika K (1995) Classification of superconducting oxide structures. Supercond. Sci. Technol. 8: 374–381.

[109] Fisk Z, Sarrao JL (1997) The new generation high temperature superconductors. Annu. Rev. Mat. Sci. 27: 35–67.

[110] Hauck J, Mika K (1998) Structure families of superconducting oxides and intersitial alloys. Supercond. Sci. Technol. 11: 614–630.

[111] Cava RJ (2000) Oxide superconductors. J. Am. Ceram. Soc. 83: 5–28.

[112] Ott HR (2008) High-T_c superconductivity. In: Bennemann KH, Ketterson JB, editors. Superconductivity. Vol. 2: Novel Superconductors. Berlin: Springer Verlag. pp. 765–831.

[113] Mannhart J, Hilgenkamp H (1997) Wavefunction symmetry and its influence on superconducting devices. Supercond. Sci. Technol. 10: 880–883.

[114] Schulz RR, Chesca B, Goetz B, Schneider CW, Schmehl A, Bielefeldt H, Hilgenkamp H, Mannhart J, Tsuei CC (2000) Design and realization of an all d-wave dc π-superconducting quantum interference device. Appl. Phys. Lett. 76: 912–914.

[115] Schneider CW, Bielefeldt H, Goetz B, Hammerl G, Schmehl A, Schulz RR, Hilgenkamp H, Mannhart J (2001) Interfaces in high-T_c superconductors: fundamental insights and possible implications. Curr. Appl. Phys. 1: 349–353.

[116] Mannhart J, Chaudhari P (November 2001) High-T_c bicrystal grain boundaries. Phys. Today 54: 48–53.

[117] Hilgenkamp H, Mannhart J (2002) Grain boundaries in high-T_c superconductors. Rev. Mod. Phys. 74: 485–549.

[118] Tafuri F, Kirtley JR, Lombardi F, Medaglia PG, Orgiani P, Balestrino G (2004) Advances in high-T_c grain boundary junctions. Fiz. Nizk. Temp. 30: 785–794.

[119] Tafuri F, Kirtley JR (2005) Weak links in high critical temperature superconductors. Rep. Prog. Phys. 68: 2573–2663.

[120] Hilgenkamp H (2008) Pi-phase shift Josephson structures. Supercond. Sci. Technol. 21: 034011.

[121] Loder F, Kampf AP, Kopp T, Mannhart J (2009) Flux periodicities in loops of nodal superconductors. New J. Phys. 11: 075005.

[122] Kirtley JR (2010) Fundamental studies of superconductors using scanning magnetic imaging. Rep. Prog. Phys. 73: 126501.

[123] Gabovich AM, Voitenko AI (2000) Superconductors with charge- and spin-density waves: theory and experiment (Review). Fiz. Nizk. Temp. 26: 419–452.

[124] Gabovich AM, Voitenko AI, Annett JF, Ausloos M (2001) Charge- and spin-density-wave superconductors. Supercond. Sci. Technol. 14: R1–R27.

[125] Gabovich AM, Voitenko AI, Ausloos M (2002) Charge-density waves and spin-density waves in existing superconductors: competition between Cooper pairing and Peierls or excitonic instabilities. Phys. Rep. 367: 583–709.

[126] Gabovich AM, Voitenko AI, Ekino T, Li MS, Szymczak H, Pękała M (2010) Competition of superconductivity and charge density waves in cuprates: Recent evidence and interpretation. Adv. Condens. Matter Phys. 2010: 681070.

[127] Ekino T, Gabovich AM, Li MS, Pękała M, Szymczak H, Voitenko AI (2011) d-wave superconductivity and s-wave charge density waves: Coexistence between order parameters of different origin and symmetry. Symmetry 3: 699–749.

[128] Ekino T, Gabovich AM, Li MS, Pękała M, Szymczak H, Voitenko AI (2011) The phase diagram for coexisting d-wave superconductivity and charge-density waves: cuprates and beyond. J. Phys.: Condens. Matter 23: 385701.

[129] Gabovich AM, Moiseev DP, Shpigel AS, Voitenko AI (1990) Josephson tunneling critical current between superconductors with charge- or spin-density waves. Phys. Status Solidi B 161: 293–302.

[130] Gabovich AM (1992) Josephson and quasiparticle tunneling in superconductors with charge density waves. Fiz. Nizk. Temp. 18: 693–704.

[131] Gabovich AM, Voitenko AI (1997) Non-stationary Josephson effect for superconductors with charge-density waves: $NbSe_3$. Europhys. Lett. 38: 371–376.

[132] Gabovich AM, Voitenko AI (1997) Nonstationary Josephson effect for superconductors with charge-density waves. Phys. Rev. B 55: 1081–1099.

[133] Gabovich AM, Voitenko AI (1997) Josephson tunnelling involving superconductors with charge-density waves. J. Phys.: Condens. Matter 9: 3901–3920.

[134] Gabovich AM, Voitenko AI (2012) dc Josephson current for d-wave superconductors with charge density waves. Fiz. Nizk. Temp. 38: 414–422.

[135] Nagamatsu J, Nakagawa N, Muranaka T, Zenitani Y, Akimitsu J (2001) Superconductivity at 39 K in magnesium diboride. Nature 410: 63–64.

[136] Daghero D, Gonnelli RS (2010) Probing multiband superconductivity by point-contact spectroscopy. Supercond. Sci. Technol. 23: 043001.

[137] Putti M, Grasso G (2011) MgB_2, a two-gap superconductor for practical applications. Mater. Res. Soc. Bull. 36: 608–613.

[138] Tanaka Y (1994) Josephson effect between s wave and $d_{x^2-y^2}$ wave superconductors. Phys. Rev. Lett. 72: 3871–3874.

[139] Barash YuS, Galaktionov AV, Zaikin AD (1995) Comment on "Superconducting pairing symmetry and Josephson tunneling". Phys. Rev. Lett. 75: 1676.

[140] Barash YuS, Galaktionov AV, Zaikin AD (1995) Charge transport in junctions between d-wave superconductors. Phys. Rev. B 52: 665–682.

[141] Xu JH, Shen JL, Miller JH Jr, Ting CS (1995) Xu et al. Reply. Phys. Rev. Lett. 75: 1677.

[142] Mineev VP, Samokhin KV (1999) Intoduction to Unconventional Superconductivity. Amsterdam: Gordon and Breach Science Publishers.

[143] Kashiwaya S, Tanaka Y (2000) Tunnelling effects on surface bound states in unconventional superconductors. Rep. Prog. Phys. 63: 1641–1724.

[144] Scalapino DJ (1995) The case for $d_{x^2-y^2}$ pairing in the cuprate superconductors. Phys. Rep. 250: 329–365.

[145] van Harlingen DJ (1995) Phase-sensitive tests of the symmetry of the pairing state in the high-temperature superconductors – evidence for $d_{x^2-y^2}$ symmetry. Rev. Mod. Phys. 67: 515–535.

[146] Leggett AJ (1996) Josephson experiments on the high-temperature superconductors. Phil. Mag. B 74: 509–522.

[147] Annett JF, Goldenfeld ND, Leggett AJ (1996) Experimental constraints on the pairing state of the cuprate superconductors: an emerging consensus. In: Ginsberg DM, editor. Physical Properties of High Temperature Superconductors V. River Ridge, NJ: World Scientific. pp. 375–461.

[148] Tsuei CC, Kirtley JR (2000) Pairing symmetry in cuprate superconductors. Rev. Mod. Phys. 72: 969–1016.

[149] Darminto AD, Smilde H-JH, Leca V, Blank DHA, Rogalla H, Hilgenkamp H (2005) Phase-sensitive order parameter symmetry test experiments utilizing $Nd_{2-x}Ce_xCuO_{4-y}/Nb$ zigzag junctions. Phys. Rev. Lett. 94: 167001.

[150] Kirtley JR, Tafuri F (2007) Tunneling measurements of the cuprate superconductors. In: Schrieffer JR, Brooks JS, editors. Handbook of High-Temperature Superconductivity. Theory and Experiment. New York: Springer Verlag. pp. 19–86.

[151] Tsuei CC, Kirtley JR (2008) Phase-sensitive tests of pairing symmetry in cuprate superconductors. In: Bennemann KH, Ketterson JB, editors. Superconductivity. Vol. 2: Novel Superconductors. Berlin: Springer Verlag. pp. 869–921.

[152] Kirtley JR (2011) Probing the order parameter symmetry in the cuprate high temperature superconductors by SQUID microscopy. C. R. Physique 12: 436–445.

[153] Sun AG, Gajewski DA, Maple MB, Dynes RC (1994) Observation of Josephson pair tunneling between a high-T_c cuprate ($YBa_2Cu_3O_{7-\delta}$) and a conventional superconductor (Pb). Phys. Rev. Lett. 72: 2267–2270.

[154] Kouznetsov KA, Sun AG, Chen B, Katz AS, Bahcall SR, Clarke J, Dynes RC, Gajewski DA, Han SH, Maple MB, Giapintzakis J, Kim JT, Ginsberg DM (1997) c-axis Josephson tunneling between $YBa_2Cu_3O_{7-\delta}$ and Pb: Direct evidence for mixed order parameter symmetry in a high-T_c superconductor. Phys. Rev. Lett. 79: 3050–3053.

[155] Mößle M, Kleiner R (1999) c-axis Josephson tunneling between $Bi_2Sr_2CaCu_2O_{8+x}$ and Pb. Phys. Rev. B 59: 4486–4496.

[156] Ponomarev YaG, Khi CS, Uk KK, Sudakova MV, Tchesnokov SN, Lorenz MA, Hein MA, Müller G, Piel H, Aminov BA, Krapf A, Kraak W (1999) Quasiparticle tunneling in

the c-direction in stacks of $Bi_2Sr_2CaCu_2O_{8+\delta}$ S–I–S junctions and the symmetry of the superconducting order parameter. Physica C 315: 85–90.

[157] Li Q, Tsay YN, Suenaga M, Klemm RA, Gu GD, Koshizuka N (1999) $Bi_2Sr_2CaCu_2O_{8+\delta}$ bicrystal c-axis twist josephson junctions: a new phase-sensitive test of order parameter symmetry. Phys. Rev. Lett. 83: 4160–4163.

[158] Komissinski PV, Il'ichev E, Ovsyannikov GA, Kovtonyuk SA, Grajcar M, Hlubina R, Ivanov Z, Tanaka Y, Yoshida N, Kashiwaya S (2002) Observation of the second harmonic in superconducting current-phase relation of $Nb/Au/(001)YBa_2Cu_3O_x$ heterojunctions. Europhys. Lett. 57: 585–591.

[159] Ovsyannikov GA, Komissinski PV, Il'ichev E, Kislinski YV, Ivanov ZG (2003) Josephson effect in Nb/Au/YBCO heterojunctions. IEEE Trans. Appl. Supercond. 13: 881–884.

[160] Smilde HJH, Golubov AA, Ariando, Rijnders G, Dekkers JM, Harkema S, Blank DHA, Rogalla H, Hilgenkamp H (2005) Admixtures to d-wave gap symmetry in untwinned $YBa_2Cu_3O_7$ superconducting films measured by angle-resolved electron tunneling. Phys. Rev. Lett. 95: 257001.

[161] Bussmann-Holder A (2012) Evidence for s+d wave pairing in copper oxide superconductors from an analysis of NMR and NQR data. J. Supercond. 25: 155–157.

[162] Klemm RA, Rieck CT, Scharnberg K (2000) Order-parameter symmetries in high-temperature superconductors. Phys. Rev. B 61: 5913–5916.

[163] Zhao G-m (2001) Experimental constraints on the physics of cuprates. Phil. Mag. B 81: 1335–1388.

[164] Brandow BH (2002) Arguments and evidence for a node-containing anisotropic s-wave gap form in the cuprate superconductors. Phys. Rev. B 65: 054503.

[165] Arnold GB, Klemm RA, Körner W, Scharnberg K (2003) Comment on "c-axis Josephson tunneling in $d_{x^2--y^2}$-wave superconductors". Phys. Rev. B 68: 226501.

[166] Brandow BH (2003) Strongly anisotropic s-wave gaps in exotic superconductors. Phil. Mag. 83: 2487–2519.

[167] Harshman DR, Kossler WJ, Wan X, Fiory AT, Greer AJ, Noakes DR, Stronach CE, Koster E, Dow JD (2004) Nodeless pairing state in single-crystal $YBa_2Cu_3O_7$. Phys. Rev. B 69: 174505.

[168] Klemm RA (2005) Bi2212 c-axis twist bicrystal and artificial and natural cross-whisker Josephson junctions: strong evidence for s-wave superconductivity and incoherent c-axis tunneling. J. Supercond. 18: 697–700.

[169] Zhao G-m (2007) Precise determination of the superconducting gap along the diagonal direction of $Bi_2Sr_2CaCu_2O_{8+y}$: Evidence for an extended s-wave gap symmetry. Phys. Rev. B 75: 140510.

[170] Zhao G-m (2009) Fine structure in the tunneling spectra of electron-doped cuprates: No coupling to the magnetic resonance mode. Phys. Rev. Lett. 103: 236403.

[171] Zhao G-m (2010) Nearly isotropic s-wave gap in the bulk of the optimally electron-doped superconductor $Nd_{1.85}Ce_{0.15}CuO_{4-y}$. Phys. Rev. B 82: 012506.

[172] Zhao G-M, Wang J (2010) Specific heat evidence for bulk s-wave gap symmetry of optimally electron-doped $Pr_{1.85}Ce_{0.15}CuO_{4-y}$ superconductors. J. Phys.: Condens. Matter 22: 352202.

[173] Zhao G-m (2011) The pairing mechanism of high-temperature superconductivity: experimental constraints. Phys. Scripta 83: 038302.

[174] Zhao G-m (2011) Reply to Comment on "The pairing mechanism of high-temperature superconductivity: experimental constraints". Phys. Scripta 83: 038304.

[175] Harshman DR, Fiory AT, Dow JD (2011) Concerning the superconducting gap symmetry in $YBa_2Cu_3O_{7-\delta}$, $YBa_2Cu_4O_8$, and $La_{2-x}Sr_xCuO_4$ determined from muon spin rotation in mixed states of crystals and powders. J. Phys.: Condens. Matter 23: 315702.

[176] Sadovskii MV (2001) Pseudogap in high-temperature superconductors. Usp. Fiz. Nauk 171: 539–564.

[177] Klemm RA (2004) Origin of the pseudogap in high temperature superconductors. In: Morawetz K, editor. Nonequilibrium Physics at Short Time Scales. Formation of Correlations. Berlin: Springer Verlag. pp. 381–400.

[178] Norman M, Pines D, Kallin C (2005) The pseudogap: friend or foe of high T_c? Adv. Phys. 54: 715–733.

[179] Deutscher G (2006) Superconducting gap and pseudogap. Fiz. Nizk. Temp. 32: 740–745.

[180] Li Y, Balédent V, Yu G, Barišić N, Hradil K, Mole RA, Sidis Y, Steffens P, Zhao X, Bourges P, Greven M (2010) Hidden magnetic excitation in the pseudogap phase of a high-T_c superconductor. Nature 468: 283–285.

[181] Lee PA, Rice TM, Anderson PW (1973) Fluctuation effects at a Peierls transition. Phys. Rev. Lett. 31: 462–465.

[182] Lebed AG, editor (2008) The Physics of Organic Superconductors and Conductors. Berlin: Springer Verlag.

[183] Eremin MV, Larionov IA, Varlamov S (1999) CDW scenario for pseudogap in normal state of bilayer cuprates. Physica B 259-261: 456–457.

[184] Gupta AK, Ng K-W (2002) Non-conservation of density of states in $Bi_2Sr_2CaCu_2O_y$: Coexistence of pseudogap and superconducting gap. Europhys. Lett. 58: 878–884.

[185] Pereg-Barnea T, Franz M (2005) Quasiparticle interference patterns as a test for the nature of the pseudogap phase in the cuprate superconductors. Int. J. Mod. Phys. B 19: 731–761.

[186] Li J-X, Wu C-Q, Lee D-H (2006) Checkerboard charge density wave and pseudogap of high-T_c cuprate. Phys. Rev. B 74: 184515.

[187] Borisenko SV, Kordyuk AA, Yaresko A, Zabolotnyy VB, Inosov DS, Schuster R, Büchner B, Weber R, Follath R, Patthey L, Berger H (2008) Pseudogap and charge density waves in two dimensions. Phys. Rev. Lett. 100: 196402.

[188] Kordyuk AA, Borisenko SV, Zabolotnyy VB, Schuster R, Inosov DS, Evtushinsky DV, Plyushchay AI, Follath R, Varykhalov A, Patthey L, Berger H (2009) Nonmonotonic pseudogap in high-T_c cuprates. Phys. Rev. B 79: 020504.

[189] Kondo T, Khasanov R, Takeuchi T, Schmalian J, Kaminski A (2009) Competition between the pseudogap and superconductivity in the high-T_c copper oxides. Nature 457: 296–300.

[190] Yuli O, Asulin I, Kalcheim Y, Koren G, Millo O (2009) Proximity-induced pseudogap: Evidence for preformed pairs. Phys. Rev. Lett. 103: 197003.

[191] Norman MR (2010) Fermi-surface reconstruction and the origin of high-temperature superconductivity. Physics 3: ID 86.

[192] Alexandrov AS, Beanland J (2010) Superconducting gap, normal state pseudogap, and tunneling spectra of bosonic and cuprate superconductors. Phys. Rev. Lett. 104: 026401.

[193] Dubroka A, Yu L, Munzar D, Kim KW, Rössle M, Malik VK, Lin CT, Keimer B, Wolf Th, Bernhard C (2010) Pseudogap and precursor superconductivity in underdoped cuprate high temperature superconductors: A far-infrared ellipsometry study. Eur. Phys. J. Spec. Topics 188: 73–88.

[194] Okada Y, Kuzuya Y, Kawaguchi T, Ikuta H (2010) Enhancement of superconducting fluctuation under the coexistence of a competing pseudogap state in $Bi_2Sr_{2-x}R_xCuO_y$. Phys. Rev. B 81: 214520.

[195] Kristoffel N, Rubin P (2011) Interband nodal-region pairing and the antinodal pseudogap in hole doped cuprates. In: Bonča J, Kruchinin S, editors. Physical Properties of Nanosystems. Dordrecht: Springer Verlag. pp. 141–152.

[196] Okada Y, Kawaguchi T, Ohkawa M, Ishizaka K, Takeuchi T, Shin S, Ikuta H (2011) Three energy scales characterizing the competing pseudogap state, the incoherent, and the coherent superconducting state in high-T_c cuprates. Phys. Rev. B 83: 104502.

[197] Greco A, Bejas M (2011) Short-ranged and short-lived charge-density-wave order and pseudogap features in underdoped cuprate superconductors. Phys. Rev. B 83: 212503.

[198] Nistor RA, Martyna GJ, Newns DM, Tsuei CC, Müser MH (2011) *Ab initio* theory of the pseudogap in cuprate superconductors driven by C4 symmetry breaking. Phys. Rev. B 83: 144503.

[199] Rice TM, Yang K-Y, Zhang FC (2012) A phenomenological theory of the anomalous pseudogap phase in underdoped cuprates. Rep. Prog. Phys. 75: 016502.

[200] Yoshida T, Hashimoto M, Vishik IM, Shen Z-X, Fujimori A (2012) Pseudogap, superconducting gap, and Fermi arc in high-T_c cuprates revealed by angle-resolved photoemission spectroscopy. J. Phys. Soc. Jpn. 81: 011006.

[201] Fujita K, Schmidt AR, Kim E-A, Lawler MJ, Lee DH, Davis JC, Eisaki H, Uchida S-i (2012) Spectroscopic imaging scanning tunneling microscopy studies of electronic structure in the superconducting and pseudogap phases of cuprate high-T_c superconductors. J. Phys. Soc. Jpn. 81: 011005.

[202] Rourke PMC, Mouzopoulou I, Xu X, Panagopoulos C, Wang Y, Vignolle B, Proust C, Kurganova EV, Zeitler U, Tanabe Y, Adachi T, Koike Y, Hussey NE (2011) Phase-fluctuating superconductivity in overdoped $La_{2-x}Sr_xCuO_4$. Nature Phys. 7: 455–458.

[203] Nakayama K, Sato T, Xu Y-M, Pan Z-H, Richard P, Ding H, Wen H-H, Kudo K, Sasaki T, Kobayashi N, Takahashi T (2011) Two pseudogaps with different energy scales at the antinode of the high-temperature $Bi_2Sr_2CuO_6$ superconductor using angle-resolved photoemission spectroscopy. Phys. Rev. B 83: 224509.

[204] Schmidt AR, Fujita K, Kim E-A, Lawler MJ, Eisaki H, Uchida S, Lee D-H, Davis JC (2011) Electronic structure of the cuprate superconducting and pseudogap phases from spectroscopic imaging STM. New J. Phys. 13: 065014.

[205] Zabolotnyy VB, Kordyuk AA, Evtushinsky D, Strocov VN, Patthey L, Schmitt T, Haug D, Lin CT, Hinkov V, Keimer B, Büchner B, Borisenko SV (2012) Pseudogap in the chain states of $YBa_2Cu_3O_{6.6}$. Phys. Rev. B 85: 064507.

[206] Orenstein J, Millis AJ (2000) Advances in the physics of high-temperature superconductivity. Science 288: 468–474.

[207] Johnston DC (2010) The puzzle of high temperature superconductivity in layered iron pnictides and chalcogenides. Adv. Phys. 59: 803–1061.

[208] Bourges P, Sidis Y (2011) Novel magnetic order in the pseudogap state of high-T_c copper oxides superconductors. C. R. Physique 12: 461–479.

[209] Li Y, Balédent V, Barišić N, Cho YC, Sidis Y, Yu G, Zhao X, Bourges P, Greven M (2011) Magnetic order in the pseudogap phase of $HgBa_2CuO_{4+\delta}$ studied by spin-polarized neutron diffraction. Phys. Rev. B 84: 224508.

[210] Gabovich AM, Voitenko AI (2009) Model for the coexistence of d-wave superconducting and charge-density-wave order parameters in high-temperature cuprate superconductors. Phys. Rev. B 80: 224501.

[211] Voitenko AI, Gabovich AM (2010) Charge density waves in d-wave superconductors. Fiz. Nizk. Temp. 36: 1300–1311.

[212] Einzel D, Schürrer I (1999) Weak coupling theory of clean $(d+s)$-wave superconductors. J. Low Temp. Phys. 117: 15–52.

[213] Ghosh A, Adhikari SK (2001) Mixing of superconducting $d_{x^2-y^2}$ state with s-wave states for different filling and temperature. Physica C 355: 77–86.

[214] Yeh N-C, Chen C-T, Hammerl G, Mannhart J, Schmehl A, Schneider CW, Schulz RR, Tajima S, Yoshida K, Garrigus D, Strasik M (2001) Evidence of doping-dependent pairing symmetry in cuprate superconductors. Phys. Rev. Lett. 87: 087003.

[215] Gorbonosov AE, Kulik IO (1967) Temperature dependence of the Josephson current in anisotropic superconductors. Fiz. Met. Metalloved. 23: 803–812.

[216] Ledvij M, Klemm RA (1995) Dependence of the Josephson coupling of unconventional superconductors on the properties of the tunneling barrier. Phys. Rev. B 51: 3269–3272.

[217] Kouznetsov K, Coffey L (1996) Theory of tunneling and photoemission spectroscopy for high-temperature superconductors. Phys. Rev. B 54: 3617–3621.

[218] Nie Y-m, Coffey L (1998) Elastic and spin-fluctuation-mediated inelastic Josephson tunneling between anisotropic superconductors. Phys. Rev. B 57: 3116–3122.

[219] Nie Y-m, Coffey L (1999) Elastic and inelastic quasiparticle tunneling between anisotropic superconductors. Phys. Rev. B 59: 11982–11989.

[220] Shukrinov YuM, Namiranian A, Najafi A (2001) Modeling of tunneling spectroscopy in high-T_c superconductors. Fiz. Nizk. Temp. 27: 15–23.

[221] Hanaguri T, Lupien C, Kohsaka Y, Lee D-H, Azuma M, Takano M, Takagi H, Davis JC (2004) A "checkerboard" electronic crystal state in lightly hole-doped $Ca_{2-x}Na_xCuO_2Cl_2$. Nature 430: 1001–1005.

[222] McElroy K, Lee D-H, Hoffman JE, Lang KM, Lee J, Hudson EW, Eisaki H, Uchida S, Davis JC (2005) Coincidence of checkerboard charge order and antinodal state decoherence in strongly underdoped superconducting $Bi_2Sr_2CaCu_2O_{8+\delta}$. Phys. Rev. Lett. 94: 197005.

[223] Ma J-H, Pan Z-H, Niestemski FC, Neupane M, Xu Y-M, Richard P, Nakayama K, Sato T, Takahashi T, Luo H-Q, Fang L, Wen H-H, Wang Z, Ding H, Madhavan V (2008) Coexistence of competing orders with two energy gaps in real and momentum space in the high temperature superconductor $Bi_2Sr_{2-x}La_xCuO_{6+\delta}$. Phys. Rev. Lett. 101: 207002.

[224] Shen KM, Ronning F, Lu DH, Baumberger F, Ingle NJC, Lee WS, Meevasana W, Kohsaka Y, Azuma M, Takano M, Takagi H, Shen Z-X (2005) Nodal quasiparticles and antinodal charge ordering in $Ca_{2-x}Na_xCuO_2Cl_2$. Science 307: 901–904.

[225] Bianconi A, Lusignoli M, Saini NL, Bordet P, Kvick A, Radaelli PG (1996) Stripe structure of the CuO_2 plane in $Bi_2Sr_2CaCu_2O_{8+y}$ by anomalous x-ray diffraction. Phys. Rev. B 54: 4310–4314.

[226] Fujita M, Goka H, Yamada K, Tranquada JM, Regnault LP (2004) Stripe order, depinning, and fluctuations in $La_{1.875}Ba_{0.125}CuO_4$ and $La_{1.875}Ba_{0.075}Sr_{0.050}CuO_4$. Phys. Rev. B 70: 104517.

[227] Kohsaka Y, Taylor C, Fujita K, Schmidt A, Lupien C, Hanaguri T, Azuma M, Takano M, Eisaki H, Takagi H, Uchida S, Davis JC (2007) An intrinsic bond-centered electronic glass with unidirectional domains in underdoped cuprates. Science 315: 1380–1385.

[228] Sharoni A, Leibovitch G, Kohen A, Beck R, Deutscher G, Koren G, Millo O (2003) Scanning tunneling spectroscopy of alpha-axis $YBa_2Cu_3O_{7-\delta}$ films: k-selectivity and the shape of the superconductor gap. Europhys. Lett. 62: 883–889.

[229] Bruder C, van Otterlo A, Zimanyi GT (1995) Tunnel junctions of unconventional superconductors. Phys. Rev. B 51: 12904–12907.

[230] Barash YuS, Burkhardt H, Rainer D (1996) Low-temperature anomaly in the Josephson critical current of junctions in d-wave superconductors. Phys. Rev. Lett. 77: 4070–4073.

[231] Bilbro G, McMillan WL (1976) Theoretical model of superconductivity and the martensitic transformation in A15 compounds. Phys. Rev. B 14: 1887–1892.

[232] Gabovich AM, Gerber AS, Shpigel AS (1987) Thermodynamics of superconductors with charge- and spin-density waves. Δ/T_c ratio and paramagnetic limit. Phys. Status Solidi B 141: 575–587.

[233] Gabovich AM, Li MS, Szymczak H, Voitenko AI (2003) Thermodynamics of superconductors with charge-density waves. J. Phys.: Condens. Matter 15: 2745–2753.

[234] Ekino T, Gabovich AM, Li MS, Pękała M, Szymczak H, Voitenko AI (2008) Temperature-dependent pseudogap-like features in tunnel spectra of high-T_c cuprates as a manifestation of charge-density waves. J. Phys.: Condens. Matter 20: 425218.

[235] Voitenko AI, Gabovich AM (2010) Charge-density waves in partially dielectrized superconductors with d-pairing. Fiz. Tverd. Tela 52: 20–27.

[236] Barash YuS, Svidzinskii AA (1997) Current–voltage characteristics of tunnel junctions between superconductors with anisotropic pairing. Zh. Éksp. Teor. Fiz. 111: 1120–1146.

[237] Mineev VP, Samokhin KV (1998) Intoduction into the Theory of Non-conventional Superconductors. Moscow: MFTI Publishing House. in Russian.

[238] Markiewicz RS (1997) A survey of the Van Hove scenario for high-T_c superconductivity with special emphasis on pseudogaps and striped phases. J. Phys. Chem. Sol. 58: 1179–1310.

[239] Khodel VA, Yakovenko VM, Zverev MV, Kang H (2004) Hot spots and transition from d-wave to another pairing symmetry in the electron-doped cuprate superconductors. Phys. Rev. B 69: 144501.

[240] Kordyuk AA, Zabolotnyy VB, Evtushinsky DV, Inosov DS, Kim TK, Büchner B, Borisenko SV (2010) An ARPES view on the high-T_c problem: Phonons vs. spin-fluctuations. Eur. Phys. J. Special Topics 188: 153–162.

[241] Annett JF (1995) Unconventional superconductivity. Contemp. Phys. 36: 423–437.

[242] Ambegaokar V, Baratoff A (1963) Tunneling between superconductors. Phys. Rev. Lett. 10: 486–489.

[243] Sommerfeld A, Bethe H (1933) Elektronentheorie der Metalle. Berlin: Springer Verlag.

[244] Harrison WA (1961) Tunneling from an independent particle point of view. Phys. Rev. 123: 85–89.

[245] Bardeen J (1961) Tunneling from a many-particle point of view. Phys. Rev. Lett. 6: 57–59.

[246] Xu JH, Shen JL, Miller JH Jr, Ting CS (1994) Superconducting pairing symmetry and Josephson tunneling. Phys. Rev. Lett. 73: 2492–2495.

[247] Kawayama I, Kanai M, Kawai T, Maruyama M, Fujimaki A, Hayakawa H (1999) Properties of c-axis Josephson tunneling between $Bi_2Sr_2CaCu_2O_8$ and Nb. Physica C 325: 49–55.

[248] Takeuchi I, Gim Y, Wellstood FC, Lobb CJ, Trajanovic Z, Venkatesan T (1999) Systematic study of anisotropic Josephson coupling between $YBa_2Cu_3O_{7-x}$ and PbIn using in-plane aligned a-axis films. Phys. Rev. B 59: 7205–7208.

[249] Anderson PW (1964) Special effects in superconductivity. In: Caianiello ER, editor. Lectures on the Many-Body Problem, Vol. 2. New York: Academic Press. pp. 113–135.

[250] Ekino T, Sezaki Y, Fujii H (1999) Features of the energy gap above T_c in $Bi_2Sr_2CaCu_2O_{8+\delta}$ as seen by break-junction tunneling. Phys. Rev. B 60: 6916–6922.

[251] Lee WS, Vishik IM, Tanaka K, Lu DH, Sasagawa T, Nagaosa N, Devereaux TP, Hussain Z, Shen Z-X (2007) Abrupt onset of a second energy gap at the superconducting transition of underdoped Bi2212. Nature 450: 81–84.

[252] Kurosawa T, Yoneyama T, Takano Y, Hagiwara M, Inoue R, Hagiwara N, Kurusu K, Takeyama K, Momono N, Oda M, Ido M (2010) Large pseudogap and nodal superconducting gap in $Bi_2Sr_{2-x}La_xCuO_{6+\delta}$ and $Bi_2Sr_2CaCu_2O_{8+\delta}$: Scanning tunneling microscopy and spectroscopy. Phys. Rev. B 81: 094519.

[253] Gürlich C, Goldobin E, Straub R, Doenitz D, Ariando, Smilde H-JH, Hilgenkamp H, Kleiner R, Koelle D (2009) Imaging of order parameter induced π phase shifts in cuprate superconductors by low-temperature scanning electron microscopy. Phys. Rev. Lett. 103: 067011.

[254] Yip S (1995) Josephson current-phase relationships with unconventional superconductors. Phys. Rev. B 52: 3087–3090.

[255] Gabovich AM, Moiseev DP (1986) Metalloxide superconductor $BaPb_{1-x}Bi_xO_3$: unusual properties and new applications. Usp. Fiz. Nauk 150: 599–623.

[256] Šmakov J, Martin I, Balatsky AV (2001) Josephson scanning tunneling microscopy. Phys. Rev. B 64: 212506.

[257] Löfwander T, Shumeiko VS, Wendin G (2001) Andreev bound states in high-T_c superconducting junctions. Supercond. Sci. Technol. 14: R53–R77.

[258] Kimura H, Barber RP Jr, Ono S, Ando Y, Dynes RC (2009) Josephson scanning tunneling microscopy: A local and direct probe of the superconducting order parameter. Phys. Rev. B 80: 144506.

[259] Gabovich AM (1993) Josephson and quasiparticle current in partially-dielectrized superconductors with spin density waves. Fiz. Nizk. Temp. 19: 641–654.

[260] Voitenko AI, Gabovich AM (1997) Josephson and one-particle currents between partially-dielectrized superconductors with charge-density waves. Fiz. Tverd. Tela 39: 991–999.

[261] Gabovich AM, Voitenko AI (1999) Nonstationary Josephson effect for superconductors with spin-density waves. Phys. Rev. B 60: 14897–14906.

[262] Gabovich AM, Voitenko AI (2000) Nonstationary Josephson tunneling involving superconductors with spin-density waves. Physica C 329: 198–230.

[263] Koren G, Levy N (2002) Experimental evidence for a small s-wave component in the order parameter of underdoped $YBa_2Cu_3O_{6+x}$. Europhys. Lett. 59: 121–127.

[264] Babcock SE, Vargas JL (1995) The nature of grain boundaries in the high-T_c superconductors. Annu. Rev. Mater. Sci. 25: 193–222.

[265] Hilgenkamp H, Mannhart J, Mayer B (1996) Implications of $d_{x^2-y^2}$ symmetry and faceting for the transport properties of grain boundaries in high-T_c superconductors. Phys. Rev. B 53: 14586–14593.

[266] Neils WK, Van Harlingen DJ, Oh S, Eckstein JN, Hammerl G, Mannhart J, Schmehl A, Schneider CW, Schulz RR (2002) Probing unconventional superconducting symmetries using Josephson interferometry. Physica C 368: 261–266.

[267] Aligia AA, Kampf AP, Mannhart J (2005) Quartet formation at (100)/(110) interfaces of d-wave superconductors. Phys. Rev. Lett. 94: 247004.

[268] Graser S, Hirschfeld PJ, Kopp T, Gutser R, Andersen BM, Mannhart J (2010) How grain boundaries limit supercurrents in high-temperature superconductors. Nature Phys. 6: 609–614.

[269] Aswathy PM, Anooja JB, Sarun PM, Syamaprasad U (2010) An overview on iron based superconductors. Supercond. Sci. Technol. 23: 073001.

[270] Seidel P (2011) Josephson effects in iron based superconductors. Supercond. Sci. Technol. 24: 043001.

[271] Johrendt D (2011) Structure-property relationships of iron arsenide superconductors. J. Mater. Chem. 21: 13726–13736.

[272] Wen J, Xu G, Gu G, Tranquada JM, Birgeneau RJ (2011) Interplay between magnetism and superconductivity in iron-chalcogenide superconductors: crystal growth and characterizations. Rep. Prog. Phys. 74: 124503.

[273] Hirschfeld PJ, Korshunov MM, Mazin II (2011) Gap symmetry and structure of Fe-based superconductors. Rep. Prog. Phys. 74: 124508.

[274] Richard P, Sato T, Nakayama K, Takahashi T, Ding H (2011) Fe-based superconductors: an angle-resolved photoemission spectroscopy perspective. Rep. Prog. Phys. 74: 124512.

[275] Hoffman JE (2011) Spectroscopic scanning tunneling microscopy insights into Fe-based superconductors. Rep. Prog. Phys. 74: 124513.

Chapter 5

Flux-Periodicity Crossover from hc/e in Normal Metallic to hc/2e in Superconducting Loops

Florian Loder, Arno P. Kampf and Thilo Kopp

Additional information is available at the end of the chapter

1. Introduction

One of the most important properties of superconductors is their perfectly diamagnetic response to an external magnetic field, the Meissner-Ochsenfeld effect. It is a pure quantum effect and therefore reveals the existence of a macroscopic quantum state with a pair condensate. A special manifestation of the diamagnetic response is observed for superconducting rings threaded by a magnetic flux: flux quantization and a periodic current response.

Persistent currents and periodic flux dependence are also known in normal metal rings and best known in form of the Aharonov-Bohm effect predicted theoretically in 1959 [4]. Since the wavefunction of an electron moving on a ring must be single valued, the phase of the wave function acquired upon moving once around the ring is a integer multiple of 2π. A magnetic flux threading the ring generates an additional phase difference $2\pi\varphi = (e/\hbar c) \oint_C d\mathbf{r} \cdot \mathbf{A}(\mathbf{r}) = (2\pi e/hc) \Phi$, where C is a closed path around the ring and $\mathbf{A}(\mathbf{r})$ the vector potential generating the magnetic flux Φ threading the ring. Here, e is the electron charge, c the velocity of light, and h is Planck's constant. Thus, the electron wave function is identical whenever φ has an integer value and therefore the system is periodic in the magnetic flux Φ with a periodicity of

$$\Phi_0 = \frac{hc}{e}, \tag{1}$$

the flux quantum in a normal metal ring. In particular, the persistent current $J(\varphi)$ induced by the magnetic flux is zero whenever $\varphi = \Phi/\Phi_0$ is an integer.

The periodic response of a superconducting ring to a magnetic flux is of similar origin as in a normal metal ring, though the phase winding of the condensate wavefunction has to be reconsidered. On account of the macroscopic phase coherence of the condensate, flux oscillations must be more stable in superconductors, and London predicted their existence in superconducting loops already ten years before the work of Aharonov and

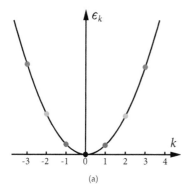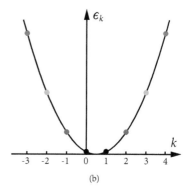

(a)   (b)

Figure 1. Scheme of the pairing of angular-momentum eigenstates in a one dimensional metal loop for (a) $\Phi = 0$ and (b) $\Phi = \Phi_0/2$, as used by Schrieffer in [43] to illustrate the origin of the $\Phi_0/2$ periodicity in superconductors. Paired are always electrons with equal energies, leading to center-of-mass angular momenta $q = 0$ in (a) and $q = 1$ in (b).

Bohm [35]. London expected that the magnetic flux threading a loop is quantized in multiples of Φ_0 because the interior of an ideal superconductor was known to be current free. Although London the pairing theory of superconductivity was not known yet, he anticipated the existence of electron pairs carrying the supercurrent and speculated that the flux quantum in a superconductor might be $\Phi_0/2$. This point of view became generally accepted with the publication of the 'Theory of Superconductivity' by Bardeen, Cooper, and Schrieffer (BCS) in 1957 [8]. Direct measurements of magnetic flux quanta $\Phi_0/2$ trapped in superconducting rings followed in 1961 by Doll and Näbauer [18] and by Deaver and Fairbank [17], corroborated later by the detection of flux lines of $\Phi_0/2$ in the mixed state of type II superconductors [1, 22].

It is tempting to explain the $\Phi_0/2$ flux periodicity of superconducting loops simply by the charge $2e$ of Cooper pairs carrying the supercurrent, but pairing of electrons alone is not sufficient for the $\Phi_0/2$ periodicity. The Cooper-pair wavefunction extends over the whole loop, as does the single-electron wavefunction, and it is not obvious whether the electrons forming the Cooper pair are tightly bound or circulate around the ring separately. A microscopic model on the basis of the BCS theory is therefore indispensable for the description of the flux periodicity of a superconducting ring. In this chapter, we analyze this problem in detail and focus on a previously neglected aspect: how do the Φ_0 periodic flux oscillations in a normal metal ring transform into the $\Phi_0/2$ periodic oscillations in a superconducting ring?

A theoretical description of the origin of the half-integer flux quanta was first found independently in 1961 by Byers and Yang [13], by Onsager [38], and by Brenig [10] on the basis of BCS theory. They realized that there are two distinct classes of superconducting wavefunctions that are not related by a gauge transformation. An intuitive picture illustrating these two types can be found in Schrieffer's book on superconductivity [43], using the energy spectrum of a one-dimensional metal ring. The first class of superconducting wavefunctions, which London had in mind in his considerations about flux quantization, is related to pairing of electrons with angular momenta $\hbar k$ and $-\hbar k$, which have equal energies in a metal loop without magnetic flux, as schematically shown in figure 1 (a). The Cooper pairs in this state have a center-of-mass angular momentum (pair momentum) $\hbar q = 0$. The pairing wavefunctions of the superconducting state for all flux values Φ, which are integer multiples of Φ_0 and correspond to even pair momenta $\hbar q = 2\hbar\Phi/\Phi_0$, are related to the wavefunction for $\Phi = 0$ by a gauge transformation. For a flux value $\Phi_0/2$, pairing occurs between the electron states with angular momenta $\hbar k$ and $\hbar(-k+1)$, which have equal energies in this case [figure 1 (b)]. This leads to pairs with momentum $\hbar q = \hbar$. The corresponding pairing wavefunction is again related by a gauge transformation to those for flux values Φ which are half-integer multiples of Φ_0 and correspond to the odd pair momenta $\hbar q = 2\hbar\Phi/\Phi_0$.

For the system to be $\Phi_0/2$ periodic, it is required that the free energies of the two types of pairing states are equal. Byers and Yang, Onsager as well as Brenig showed that this is in fact the case in the thermodynamic limit. The free energy consists then of a series of parabolae with minima at integer multiples of Φ_0 (corresponding to even pair momenta) and half integer multiples of Φ_0 (corresponding to odd pair momenta). If the arm of the ring is wider than the penetration depth λ, the flux is quantized and the groundstate is given by the minimum closest to the value of the external flux. However, in microscopic finite systems this degeneracy of the even and odd q minima is lifted, although their position is fixed by gauge invariance to multiples of $\Phi_0/2$. The restoration of the $\Phi_0/2$ periodicity in the limit of large rings was studied only much later [26, 31, 45, 52]. We study the revival of the $\Phi_0/2$ periodicity in sections 2.1 and 2.2 for a one-dimensional ring at zero temperature and investigate the effects of many channels and finite temperatures in section 2.3.

From the flux periodicity of the free energy, the same flux periodicity can be derived for all other thermodynamic quantities [44]. A clear and unambiguous observation of flux oscillations is possible in the flux dependence of the critical temperature T_c of small superconducting cylinders. Such experiments have been performed first by Little and Parks in 1962 [29, 30, 39]. They measured the resistance R of the cylinder at a fixed temperature T within the finite width of the superconducting transition and deduced the oscillation period of T_c from the variation of R. These experiments confirmed the $\Phi_0/2$ periodicity in conventional superconductors very accurately. At this stage the question of the flux periodicity in superconductors seemed to be settled and understood. The interest then shifted to the amplitude of the supercurrent and also the normal persistent current and their dependence on the ring size, the temperature, and disorder [11, 15, 21, 27, 47]. However, the influence of finite system sizes on the flux periodicity remained unaddressed. Earlier, certain experiments had already indicated some unexpected complications. E.g., Little and Parks pointed out in reference [29] that in tantalum cylinders they could not detect any flux oscillations in R at all. Even more peculiar were the oscillations observed in an indium cylinder where signs of an additional $\Phi_0/8$ periodicity were clearly visible [29]. This was surprising because indium is a perfectly conventional superconductor otherwise. These results remained unexplained and drew attention only years later, when flux oscillations of unconventional superconductors were studied.

In the meantime a new type of flux sensitive systems was advanced: superconducting quantum-interference devices (SQUIDs). The measurement of flux oscillations in SQUIDs is similar to the Little-Parks experiment. Here the flux dependence of the critical current J_c through a superconducting loop including one or two Josephson junctions is measured. This

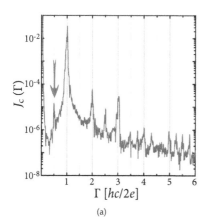
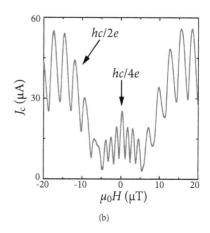

(a)   (b)

Figure 2. (a) Fourier transform $J_c(\Gamma)$ of the critical current $J_c(H)$ measured by Schneider *et al.* on a $24°$ grain boundary SQUID at $T = 77\,\text{K}$ as a function of the applied magnetic field where $\Phi_0/2 = 6.7\,\mu\text{T}$ [41]. The red arrow points out the Fourier peak corresponding to a Φ_0 periodic current contribution. (b) Critical current $J_c(H)$ over a $24°$ grain boundary SQUID at $T = 4.2\,\text{K}$, where $\Phi_0/2 = 2.7\,\mu\text{T}$. Clearly visible is the abrupt change of periodicity at $\mu_0 H \approx \pm 5\,\mu\text{T}$ [42].

has the advantage that flux oscillations can be observed at any temperature $T < T_c$, and they are most clearly visible in the critical current J_c. SQUIDs fabricated from conventional superconductors have been used in experiments and applications for five decades, and they proved to oscillate perfectly with the expected flux period $\Phi_0/2$. It was therefore a surprise that flux oscillations with different periodicities were found in 2003 by Lindström *et al.* [28] and Schneider *et al.* [41, 42] in SQUIDs fabricated from films of the high-T_c superconductor YBa$_2$Cu$_3$O$_y$ (YBCO) where the Josephson junctions arise from grain boundaries. Flux trapping experiments in loops showed that flux quantization in the cuprate class of high-T_c superconductors occurs in units of $\Phi_0/2$ [23], identically to what has been observed with conventional superconductors. In addition, Schneider *et al.* observed a variety of oscillation periods, depending on the geometry of the SQUID loop, the grain-boundary angle, the temperature, and the magnetic-field range of the SQUID.

Two distinct patterns of unconventional oscillations in YBCO SQUIDs have to be discerned. The first kind consists of oscillations which have a basic period of $\Phi_0/2$, overlaid by other periodicities, such that the Fourier transform $J_c(\Gamma)$ of $J_c(\Phi)$ contains peaks appear which do not correspond to the period $\Phi_0/2$ [41]. An example for such a measurement are shown in figure 2 (a). The peaks at integer values of Γ correspond to higher harmonics of $\Phi_0/2$, and their appearance is natural. However, there are clear peaks at $\Gamma = 1/2$ (red arrow) and $\Gamma = 5/2$, which correspond to Φ_0 periodicity and higher harmonics thereof. The origin of the Φ_0 periodicity in those experiments is so far not conclusively explained. There was, however, extensive research on the flux periodicity of unconventional (mostly d-wave) superconductors, which revealed that the periodicity of the normal state persists in the superconducting state if the energy gap symmetry allows for nodal states [6, 25, 32, 34, 51]. This effect derives directly from the analysis in this book chapter and is discussed in detail in reference [32].

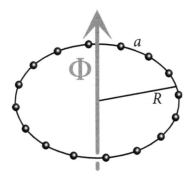

Figure 3. The simplest description of the many-particle state in a flux threaded loop, we use a tight-binding model on a discrete, one-dimensional ring with N lattice sites, lattice constant a and radius $R = Na/2\pi$. The magnetic flux Φ is confined to the interior of the ring and does not touch the ring itself.

The second kind of unconventional oscillations is more intriguing. In several different YBCO SQUIDs, the periodicity of sinusoidal oscillations changes abruptly with increasing magnetic flux. In the measurement shown in figure 2 (b), the period is $\Phi_0/4$ for small flux, and changes to $\Phi_0/2$ at a critical flux. As a possible explanation for the appearance of $\Phi_0/4$ periodicity, an unusually pronounced second harmonic in the critical current J_c of transparent Josephson junctions was proposed or, more fundamentally, an effect of interactions between Cooper pairs, leading to the formation of electron quartets [42]. The observation of similar abrupt changes to other fractional periodicities like $\Phi_0/6$ and $\Phi_0/8$ render this finding even more striking since it could indicate a transition into a new, non-BCS type of superconductivity. This concept, which we sketch briefly in the Conclusions, is a complex and promising topic for future research on unconventional superconductors.

2. The periodicity crossover

In this section we introduce the periodicity crossover and consider first the simplest model containing the relevant physics: a one dimensional ring consisting of N lattice sites and a lattice constant a (figure 3). The ring is threaded by a magnetic flux Φ focused through the center and not touching the ring itself. We use a tight-binding description with nearest-neighbor hopping parameter t, which sets the energy scale of the system. We start from the flux periodicity of the normal metal state of the ring, which varies for different numbers of electrons in the ring. On this basis we introduce a superconducting pairing interaction and investigate the flux periodicity of the groundstate upon increasing the interaction strength. For a ring with a finite width (an annulus) we investigate the flux dependence of the self-consistently calculated superconducting order parameter and study the temperature driven periodicity crossover when cooling the ring through the transition temperature T_c.

2.1. Normal state

The tight-binding Hamiltonian for an electronic system including a magnetic field is straightforwardly formulated using the annihilation and creation operators c_{is} and $c_{is}^\dagger$ for an electron with spin s on the lattice site i:

$$\mathcal{H}_0 = -t \sum_{\langle i,j \rangle, s} e^{i\varphi_{ij}} c_{is}^\dagger c_{js} - \mu \sum_{i,s} c_{is}^\dagger c_{is}. \tag{2}$$

Here $\langle i, j \rangle$ denotes all nearest-neighbor pairs i and j, $s =\uparrow, \downarrow$. The magnetic field $\mathbf{B} = \nabla \times \mathbf{A}$ enters into the Hamiltonian (2) through the Peierls phase factor $\varphi_{ij} = (e/hc) \int_i^j d\mathbf{l} \cdot \mathbf{A}$. The chemical potential μ controls the number of electrons in the ring. The flux periodicity is easiest to discuss for a particle-hole symmetric situation with $\mu = 0$, for which the Fermi energy is $E_F = 0$. We will later address the changes introduced through an arbitrary μ.

We assume that the N lattice sites are equally spaced along a ring with circumference $2\pi R = Na$ (figure 3). It follows that the Peierls phase factor for a magnetic field focused through the center of the ring simplifies to $\varphi_{ij} = 2\pi\varphi/N$, where $\varphi = \Phi/\Phi_0$ is the dimensionless magnetic flux. The Hamiltonian (2) is then written in momentum space as:

$$\mathcal{H}_0 = \sum_{k,s} \epsilon_k(\varphi) c_{ks}^\dagger c_{ks} \tag{3}$$

where $c_{ks}^\dagger$ creates an electron with angular momentum $\hbar k$. The energy dispersion is

$$\epsilon_k(\varphi) = -2t \cos\left(\frac{k - \varphi}{R/a}\right) - \mu. \tag{4}$$

The eigenenergies depend on the flux only in the combination $k - \varphi$, as is shown in figure 4 for three different cases: (a) $N/4$ is an integer, (b) $N/4$ is a half integer, and (c) N is an odd number. The φ dependent shift in $\epsilon_k(\varphi)$ is known as the Doppler shift since it is proportional to the velocity of the corresponding electron. In all three cases, the spectrum has obviously the periodicity 1 with respect to φ. However, the flux values, for which an energy level crosses E_F, are different. This number dependence, sometimes referred to as the "parity effect", is characteristic for discrete systems and not restricted to one dimension. It was discussed in detail in the context of the persistent current in metallic loops [15, 27, 47] and also in metallic nano clusters [36]; it is also essential for the discussion of superconducting rings.

Physical quantities of the normal metal ring can be expressed through the thermal average $n_s(k)$ of the number of electrons with angular momentum k and spin s: $n_s(k) = \langle c_{ks}^\dagger c_{ks} \rangle = f(\epsilon_k(\varphi))$, with the Fermi distribution function $f(\epsilon) = 1/(1 + e^{\epsilon/k_B T})$ for the temperature T. The groundstate is given by the minimum of the total energy E of the system

$$E(\varphi) = \langle \mathcal{H}_0 \rangle = \sum_{k,s} \epsilon_k(\varphi) n_s(k), \tag{5}$$

which is a piecewise quadratic function of the magnetic flux. The momentum distribution function $n_s(k)$ also depends on the magnetic flux only in the combination $k - \varphi$. The sum over k in equation (5) directly renders the Φ_0 flux periodicity of $E(\varphi)$. However, the position of the minima of $E(\varphi)$ depends on the highest occupied energy level and therefore also shows a parity effect (see figure 5).

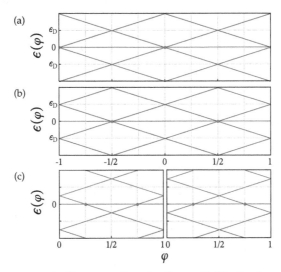

Figure 4. Energy spectrum of a discrete one-dimensional ring with N lattice sites, $\mu = 0$ and (a) $N/4$ is an integer, (b) $N/4$ a half integer, and (c) N an odd number. In (a) and (b), levels cross $E_F = 0$ at integer (a) or half-integer (b) values of φ. For odd N, two different spectra are possible [$N = 4n + 1$ (left) and $N = 4n - 1$ (right)], and for both, two levels cross E_F within one flux period (red points). ϵ_D denotes the maximum value of the Doppler shift.

The energy $E(\varphi)$ is maximal for those values of φ where an energy level reaches E_F (red points) and has minima in between. If $N/4$ is an integer, then the minima of $E(\varphi)$ are at half-integer values of φ (figure 5 (a), light blue curve), whereas if $N/4$ is a half-integer, the minima are at integer values of φ (figure 5 (a), dark blue curve). If N is odd, two different (but physically equivalent) spectra for $N = 4n \pm 1$ are possible, and for both, two levels cross E_F in one flux period. This results in a superposition of the two previous cases and there are minima of $E(\varphi)$ for both integer and half-integer values of φ; $E(\varphi)$ is therefore $\Phi_0/2$ periodic.

The normal persistent current $J(\varphi) = -(e/hc)\,\partial E(\varphi)/\partial\varphi$ (see equation 12 below) jumps whenever an energy level crosses E_F, because the population of left and right circulating states changes abruptly [figure 5 (b)]. The occupied state closest to E_F contributes dominantly to the current, because all other contributions tend to almost cancel in pairs. The Doppler shift decreases with the ring radius like $1/R$ [c.f. equation (4)] and so does the persistent current.

2.2. Superconducting state: Emergence of a new periodicity

The theory of flux threaded superconducting loops was first derived by Byers and Yang [13], Brenig [10], and Onsager [38] on the basis of the BCS theory. They showed the thermodynamic equivalence of the two superconducting states discussed above in the thermodynamic limit. However, in a strict thermodynamic limit the persistent (super-) current vanishes, and therefore a more precise statement is necessary with respect to the $\Phi_0/2$ periodicity of the supercurrent. Here we analyze the crossover from Φ_0 periodicity in the normal metal loop

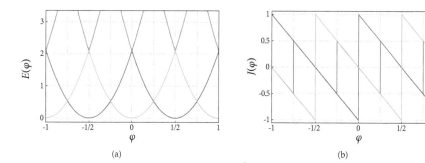

Figure 5. (a) Total energy $E(\varphi)$ of a ring with N sites as a function of the magnetic flux φ. If $N/4$ is an integer, then the minima are at half-integer values of φ (dark blue). If $N/4$ is a half-integer, the parabolae are shifted by $1/2$ (light blue). The gray lines above the crossing points of the parabolae correspond to possible excited states. (b) Persistent current $J(\varphi)$ corresponding to the systems described in (a). The purple curve shows the current obtained for odd N.

to the $\Phi_0/2$ periodicity in the superconducting loop upon turning on the pairing interaction. The discussion of this crossover enables precise statements about the periodicity.

For a one-dimensional superconducting loop (or any loop thinner than the penetration depth λ), finite currents flow throughout the superconductor. The magnetic flux is consequently not quantized, only the fluxoid $\Phi' = \Phi + (\Lambda/c) \oint d\mathbf{r} \cdot \mathbf{J}(\mathbf{r})$ is, which was introduced by F. London [35]. The flux Φ is the total flux threading the loop, including the current induced flux, and $\Lambda = 4\pi\lambda^2/c^2$. In the absence of flux quantization, φ is a continuous variable also in a superconducting system with a characteristic periodicity in φ.

In this section we focus on the emergence of a new periodicity when a superconducting order parameter arises. We therefore include an attractive on-site interaction of the general form [1]

$$\mathcal{H} = \mathcal{H}_0 - \frac{V}{2N^2} \sum_{k,k'} \sum_q c_{k\uparrow}^\dagger c_{-k+q\downarrow}^\dagger c_{-k'+q\downarrow} c_{k'\uparrow}, \tag{6}$$

where $V > 0$ is the interaction strength. In BCS theory it is *assumed* that electron pairs have zero center-of-mass (angular) momentum, i.e., the pairs are condensed in a macroscopic quantum state with $q = 0$, similar to a Bose-Einstein condensate of bosonic particles. In the case of a flux threaded ring, Byers and Yang [13], Brenig [10], and Onsager [38] showed that q is generally finite and has to be chosen to minimize the kinetic energy of the Cooper pairs in the presence of a magnetic flux. Nevertheless it is still assumed that pairing occurs only for one specific angular momentum q. For conventional superconductors, this assumption is generally true, although for superconductors with gap nodes, the situation may be different, as was shown for d-wave pairing symmetry in reference [33]. In this section, we use the

---

[1] In the literature the symmetric Hamiltonian $\tilde{\mathcal{H}} = \mathcal{H}_0 + (V/2N^2) \sum_{k,k'} \sum_q c_{k+q/2\uparrow}^\dagger c_{-k+q/2\downarrow}^\dagger c_{-k'+q/2\downarrow} c_{k'+q/2\uparrow}$ is often used [37]. $\tilde{\mathcal{H}}$ is naturally hc/e periodic in φ, but it is not well defined, although it yields the same physical quantities as $\mathcal{H}$. The introduction of half-integer angular momenta in $\tilde{\mathcal{H}}$ leads to two different limits $\Delta \to 0$ for even or odd q, corresponding to the two spectra for $N/2$ even or odd. Therefore the symmetric $\tilde{\mathcal{H}}$ is unsuitable for the discussion of the flux periodicity.

assumption of condensation into a state with one selected angular momentum q for all pairs, which allows us to write $\mathcal{H}$ in the decoupled form

$$\mathcal{H} = \mathcal{H}_0 + \sum_k \left[\Delta_q^*(\varphi) c_{-k+q\downarrow} c_{k\uparrow} + \Delta_q(\varphi) c_{k\uparrow}^\dagger c_{-k+q\downarrow}^\dagger \right] + \frac{\Delta_q^2(\varphi)}{V}, \tag{7}$$

where the order parameter is defined as $\Delta_q(\varphi) = (V/2) \sum_k \langle c_{k\uparrow} c_{k\downarrow} \rangle$. The mean-field Hamiltonian (7) is diagonalized with the standard Bogoliubov transformation

$$c_{k\uparrow} = u(k) a_{k+} + v(k) a_{k-}^\dagger, \qquad c_{-k+q\downarrow} = u(k) a_{k-}^\dagger - v(k) a_{k+} \tag{8}$$

with the coherence factors

$$u^2(k) = \left(1 + \frac{\epsilon(k, \varphi)}{E(k, \varphi)} \right) \quad \text{and} \quad v^2(k) = \left(1 - \frac{\epsilon(k, \varphi)}{E(k, \varphi)} \right), \tag{9}$$

which depend on φ and q through $E(k, \varphi) = \sqrt{\Delta_q^2 + \epsilon^2(k, \varphi)}$ and $\epsilon(k, \varphi) = [\epsilon_k(\varphi) + \epsilon_{-k+q}(\varphi)]/2$. The energy spectrum splits into the two branches

$$E_\pm(k, \varphi) = \frac{\epsilon_k(\varphi) - \epsilon_{-k+q}(\varphi)}{2} \pm \sqrt{\Delta_q^2 + \epsilon^2(k, \varphi)}, \tag{10}$$

where the Doppler shift term arises from the different energies of the two paired states with momenta k and $-k + q$. The order parameter $\Delta_q(\varphi)$ is determined self-consistently from

$$\frac{1}{N} \sum_k \frac{f(E_-(k, \varphi)) - f(E_+(k, \varphi))}{2\sqrt{\Delta_q(\varphi)^2 + \epsilon^2(k, \varphi)}} = \frac{1}{V}. \tag{11}$$

For the discussion of the periodicity of this system, we first disregard the self-consistency condition for the order parameter and set $\Delta_q(\varphi) \equiv \Delta$ to be constant. Importantly, while $\mathcal{H}_0$ is strictly Φ_0 periodic, $\mathcal{H}$ is *not* periodic in φ if $\Delta > 0$. The question of periodicity is therefore: *which* periodicity is restored by minimizing $E(\varphi) = \langle \mathcal{H} \rangle$ with respect to q and how is this achieved?

Figure 6 shows $E(\varphi)$ for two different values of Δ for $q = 0$ and $q = 1$. For small Δ, $E(\varphi)$ is still a series of parabolae with minima at integer values of φ, but the degeneracy of the minima is lifted [figure 6 (a) and (b)]. For even q, the energy minimum at $\varphi = q/2$ is lowered relative to the other minima, whereas for odd q, one new minimum emerges at $\varphi = q/2$, which is absent in the normal state. If Δ exceeds a certain threshold Δ_c, this new odd q minimum becomes deeper than the neighboring ones [figure 6 (d)]. We have thus identified the second class of states with minima in $E(\varphi)$ at half-integer flux values anticipated above and we find that the even and odd q minima become equal if Δ becomes large compared to Δ_c, a ring size dependent value which we will determine below. It is to be understood that the energies $E(\varphi)$ in figure 6 are not periodic in φ because the q-values are fixed, either to $q = 0$ in (a, c) or to $q = 1$ in (b, d). In loops thicker than the penetration depth, screening currents drive the system always into an energy minimum. In this case, the flux is then quantized in units of $\Phi_0/2$.

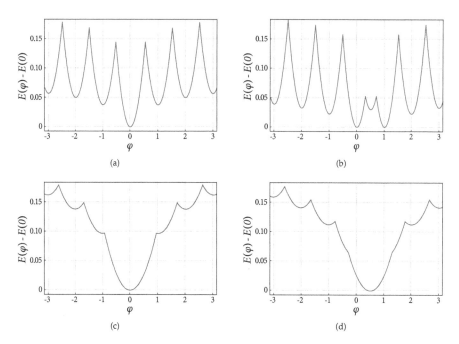

Figure 6. Energy $E(\varphi)$ in the superconducting state with $q = 0$ (a, c) and $q = 1$ (b, d) for $N = 50$. The upper panels (a, b) show the "small gap" case with $\Delta = 0.05\,t$ and the lower panels (c, d) the "large gap" case with $\Delta = 0.2\,t$.

Let us for the moment assume that the flux value, at which the energy minimizing q changes from one integer to the next, is well approximated by the half way between two minima: $q = \text{floor}(2\varphi + 1/2)$ (floor(x) is the largest integer smaller than x, e.g., $\varphi = 0 \rightarrow q = 0$; $\varphi = 1/4 \rightarrow q = 1$; $\varphi = 3/4 \rightarrow q = 2$). Small deviations from these values will be discussed for the self-consistent solution in section 2.3. The energy spectrum is then Φ_0 periodic, but discontinuous at the flux values where q changes, as shown in figure 7. Clearly distinguishable are now the "small gap" (a) and the "large gap" (b) regime: Δ_c represents the maximum of the flux-induced shift of the energy levels close to E_F, before q changes. If $\Delta < \Delta_c$, the energy gap closes at certain values of φ, whereas if $\Delta > \Delta_c$, an energy gap persists for all φ.

Although the spectra are Φ_0 periodic both in figure 7 (a) and (b), the closing of the gap in the "small gap" regime has significant effects on the periodicity of physical quantities like $E(\varphi)$. Even more prominent is the periodicity crossover for the persistent current in the ring. The supercurrent is given by $J(\varphi) = J_+(\varphi) + J_-(\varphi) = (e/h)\partial E(\varphi)/\partial \varphi$, where

$$J_\pm(\varphi) = \frac{e}{hc} \sum_k \frac{\partial \epsilon_k(\varphi)}{\partial k} n_\pm(k) \qquad (12)$$

with $n_+(k) = u^2(k)f(E_+(k, \varphi))$ and $n_-(k) = v^2(k)f(E_-(k, \varphi))$. $J_+(\varphi)$ and $J_-(\varphi)$, as well as $J(\varphi)$, are plotted in figure 8. The contribution $J_-(\varphi)$ forms a $\Phi_0/2$ periodic saw-tooth pattern,

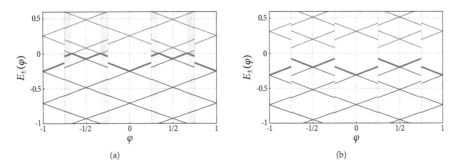

Figure 7. Eigenenergies $E_\pm(k,\varphi)$ (10) as a function of flux φ for $N = 50$ and a fixed order parameter: (a) "large gap" regime with $\Delta = 0.2\,t$, (b) "small gap" regime with $\Delta = 0.05\,t$. Blue lines: occupied states, grey lines: unoccupied states. The bold line marks the highest occupied state for all φ. In the blue shaded areas in (a), the energy gap has closed due to the Doppler shift.

both in the normal and in the superconducting state. The Φ_0 periodic part in the normal state is contained exclusively in $J_+(\varphi)$. A flux window where $E_+(k,\varphi)$ is partially occupied exists in each q sector when the energy gap has closed [shaded blue areas in figure 7 (a)]. These windows decrease for increasing Δ until $J_+(\varphi)$ vanishes for $\Delta = \Delta_c$. In the "large gap" regime, the supercurrent is carried entirely by $J_-(\varphi)$ and is therefore $\Phi_0/2$ periodic and essentially independent of Δ. The discontinuities in $J_-(\varphi)$ are not caused by energy levels crossing E_F, but by the reconstruction of the condensate when the pair momentum q changes to the next integer at the flux values $\varphi = (2n-1)/4$. Figure 9 shows the periodicity crossover of the persistent current in four different steps from the Φ_0 periodic normal current to the $\Phi_0/2$ periodic supercurrent. A very similar type of crossover was discussed earlier for loops consisting of a normal and a superconducting part. The flux periodicity of such a system changes from Φ_0 to $\Phi_0/2$ with an increasing ratio of superconducting to normal conducting parts [12, 14].

Further insight into the current periodicity is obtained by analyzing Δ_c. Close to E_F, the maximum energy shift is $\epsilon_D = at/2R$, and the condition for a direct energy gap (or $E_+(k,\varphi) > 0$ for all k, φ) and a $\Phi_0/2$-periodic current pattern is therefore $\Delta > \Delta_c = \epsilon_D$. The corresponding critical ring radius is $R_c = at/2\Delta$. It is instructive to compare R_c with the BCS coherence length $\xi_0 = \hbar v_F/\pi\Delta$, where v_F is the Fermi velocity and Δ the BCS order parameter at $T = 0$. On the lattice we identify $v_F = \hbar k_F/m$ with $k_F = \pi/2a$ and $m = \hbar^2/2a^2 t$, and obtain $\xi_0 = at/\Delta$ and thus $2R_c = \xi_0$. This signifies that the current response of a superconducting ring with a diameter smaller than the coherence length, is generally Φ_0 periodic [34]. In these rings the Cooper-pair wavefunction is delocalized around the ring.

We have hereby identified the basic mechanism underlying the crossover from Φ_0 periodicity in the normal state to $\Phi_0/2$ periodicity in the superconducting state. It is the crossing of E_F of energy levels as a function of the magnetic flux that leads to kinks in the energy and to discontinuities in the supercurrent (or the persistent current in the normal state). If the superconducting gap is large enough to prevent all energy levels from crossing the Fermi energy, the kinks and jumps occur only where the pair momentum q of the groundstate

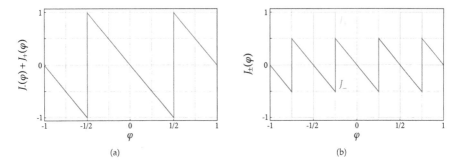

Figure 8. (a) The Φ_0 periodic persistent current $J(\varphi)$ (in units of t/Φ_0) in the normal state. (b) The contribution $J_-(\varphi)$ (dark blue) is $\Phi_0/2$ periodic and identical in the normal and the superconducting state. The Φ_0 periodicity in the "small gap" regime is entirely due to $J_-(\varphi)$ (light blue), which vanishes in the "large gap" regime.

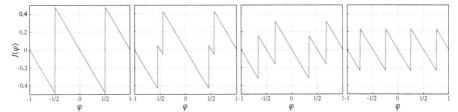

Figure 9. Crossover from the Φ_0-periodic normal persistent current to the $\Phi_0/2$-periodic supercurrent in a ring with $N = 26$ at $T = 0$. $J(\varphi)$ is in units of t/Φ_0. For this ring size $\Delta_c \approx 0.24\,t$. The discontinuities occur where the φ-derivative of the highest occupied state energy changes sign. From left to right: $\Delta = 0,\ 0.08\,t,\ 0.16\,t,\ 0.24\,t$.

changes. The latter is true, if the ring diameter is larger than the coherence length ξ_0 of the superconductor.

To conclude the discussion of the supercurrent we mention an issue raised by Little and Parks [39]. A simple theoretical model to predict the amplitude of the oscillations of T_c is the following: For all non-integer or non-half-integer values of φ, there is a persistent current $J(\varphi)$ circulating in the cylinder. The kinetic energy $E_{kin}(\varphi)$ associated with this current is proportional to $J^2(\varphi)$, as is the energy $E(\varphi)$ in figure 5 (a). It is therefore suggestive to subtract $E_{kin}(\varphi)$ from the condensation energy of the superconducting state and deduce the oscillations of T_c from those of $E(\varphi)$. This was done in a first approach by Little and Parks [30], by Tinkham [44], and by Douglass [19] within a Ginzburg-Landau ansatz, yet it was later shown to be incorrect by Parks and Little in a subsequent article [39]. They wrote that "the microscopic theory [i.e., the BCS theory] shows that it is *not the kinetic energy of the pairs* which raises the free energy of the superconducting phase ..., but rather it is due to *the difference in the energy* of the two members of the pairs", i.e., $\epsilon_k(\varphi) - \epsilon_{-k+q}(\varphi)$. It is remarkable that the results of Tinkham and Douglass are nevertheless identical to the microscopic result [40]. The notion whether it is the kinetic energy of the screening current that causes the oscillations, or rather an internal cost in condensation energy in the presence of a finite flux, is important insofar as it provides an explanation for an intriguing problem: In the same way as the pairing

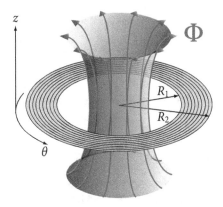

Figure 10. Flux threaded annulus with inner radius R_1 and outer radius R_2. For a magnetic flux threading the interior of the annulus, the radial part of the Bogoliubov - de Gennes equations is solved numerically with a discretized radial coordinate.

of electrons leads to a reduction of the fundamental flux period Φ_0 to $\Phi_0/2$, the pairing of pairs to quartets would lead to the quarter-period $\Phi_0/4$. Then the saw-tooth pattern of the supercurrent becomes $\Phi_0/4$ periodic and the maximum current is only half the value for unpaired Cooper pairs. If the oscillation in $E(\varphi)$ were due to the kinetic energy of the pairs, then the formation of quartets and the $\Phi_0/4$ periodicity would be energetically favorable. The fact that it is $\Phi_0/2$ periodic illustrates the remark by Parks and Little.

2.3. Multi channels and self consistency

Although the one-dimensional ring discussed above comprises all the qualitative features of the periodicity crossover at $T = 0$ upon entering the superconducting state, some additional issues need to be considered. First, the spectrum of a one dimensional ring is special insofar as only two energy levels exist that cross the Fermi energy in one flux period. This situation is ideal to investigate persistent currents, because there are maximally two jumps in one period. In an extended loop all radial channels contribute at the Fermi energy and have to be taken into account. Second, the self-consistency condition of the superconducting order parameter leads to corrections of the results obtained above, and third, the periodicity crossover upon entering the superconducting state by cooling through the transition temperature T_c is somewhat different from the $T = 0$ crossover. These points are addressed in this section.

Here we extend the ring to an annulus with an inner radius R_1 and an outer radius R_2, as shown in figure 10. For such an annulus, we choose a continuum approach on the basis of the Bogoliubov - de Gennes (BdG) equations, for which no complications arise from the parity effect. For spin singlet pairing the BdG equations are [16]

$$
\begin{aligned}
E_n u_{\mathbf{n}}(\mathbf{r}) &= \left[\frac{1}{2m}\left(i\hbar\boldsymbol{\nabla} + \frac{e}{c}\mathbf{A}(\mathbf{r})\right)^2 - \mu\right] u_{\mathbf{n}}(\mathbf{r}) + \Delta(\mathbf{r})\, v_{\mathbf{n}}(\mathbf{r}), \\
E_n v_{\mathbf{n}}(\mathbf{r}) &= -\left[\frac{1}{2m}\left(i\hbar\boldsymbol{\nabla} - \frac{e}{c}\mathbf{A}(\mathbf{r})\right)^2 - \mu\right] v_{\mathbf{n}}(\mathbf{r}) + \Delta^*(\mathbf{r}) u_{\mathbf{n}}(\mathbf{r}),
\end{aligned}
\tag{13}
$$

with the self-consistency condition (gap equation) for the order parameter $\Delta(\mathbf{r})$:

$$\Delta(\mathbf{r}) = V \sum_{\mathbf{n}} u_{\mathbf{n}}(\mathbf{r}) v_{\mathbf{n}}^*(\mathbf{r}) \tanh\left(\frac{E_{\mathbf{n}}}{2k_{\mathrm{B}}T}\right), \tag{14}$$

where V is the local pairing potential. For an annulus of finite width we separate the angular part of the quasi-particle wavefunctions $u_{\mathbf{n}}(\mathbf{r})$, $v_{\mathbf{n}}(\mathbf{r})$ using polar coordinates $\mathbf{r} = (r, \theta)$ with the ansatz

$$u_{\mathbf{n}}(r, \theta) = u_{\mathbf{n}}(r) e^{\frac{i}{2}(k+q)\theta},$$
$$v_{\mathbf{n}}(r, \theta) = v_{\mathbf{n}}(r) e^{\frac{i}{2}(k-q)\theta}, \tag{15}$$

where k and q are either both even or both odd integers. Thus $\hbar k$ is the angular momentum as for the one dimensional ring and $\mathbf{n} = (k, \rho)$ with a radial quantum number ρ. The order parameter factorizes into $\Delta(r, \theta) = \Delta(r) e^{iq\theta}$ where the radial component

$$\Delta(r) = V_0 \sum_{\mathbf{n}} u_{\mathbf{n}}(r) v_{\mathbf{n}}^*(r) \tanh\left(\frac{E_{\mathbf{n}}}{2k_{\mathrm{B}}T}\right) \tag{16}$$

is real. For a magnetic flux φ threading the interior of the annulus we choose the vector potential $\mathbf{A}(r, \theta) = \mathbf{e}_\theta \, \varphi/(2\pi r)$, where $\mathbf{e}_\theta$ is the azimuthal unit vector. With

$$\left(-i\boldsymbol{\nabla} \pm \frac{\varphi}{r}\mathbf{e}_\theta\right)^2 = -\frac{1}{r}\partial_r(r\partial_r) + \frac{1}{r^2}(-i\partial_\theta \pm \varphi)^2 \tag{17}$$

the BdG equations therefore reduce to radial differential equations for $u_{\mathbf{n}}(r)$ and $v_{\mathbf{n}}(r)$:

$$E_{\mathbf{n}} u_{\mathbf{n}}(r) = -\left[\frac{\hbar^2}{2m}\frac{\partial_r}{r}(r\partial_r) - \frac{\hbar^2 l_u^2}{2mr^2} + \mu\right] u_{\mathbf{n}}(r) + \Delta(r) v_{\mathbf{n}}(r),$$
$$E_{\mathbf{n}} v_{\mathbf{n}}(r) = \left[\frac{\hbar^2}{2m}\frac{\partial_r}{r}(r\partial_r) - \frac{\hbar^2 l_v^2}{2mr^2} + \mu\right] v_{\mathbf{n}}(r) + \Delta(r) u_{\mathbf{n}}(r), \tag{18}$$

with the canonical angular momenta

$$\hbar l_u = \frac{\hbar}{2}(k + q - 2\varphi), \tag{19}$$

$$\hbar l_v = \frac{\hbar}{2}(k - q + 2\varphi). \tag{20}$$

For integer and half-integer flux values, equations (18) can be solved analytically whereas for an arbitrary magnetic flux a numerical solution is required (for details, see reference [31]). Within this procedure, the radial coordinate is discretized into $N_\perp$ values r_n separated by the distance $a_\perp = (R_2 - R_1)/N_\perp$. The number q plays the same role as in the previous section. Here we choose q for each value of the flux to minimize the total energy of the system. The flux for which q changes to the next integer can therefore deviate from the values $(2n - 1)/4$, at which we fixed the jump to the next q for the one-dimensional model.

In the normal state ($\Delta = 0$), the number of eigenstates sufficiently close to E_F which may cross E_F as a function of φ is controlled by the average charge density n and is approximately

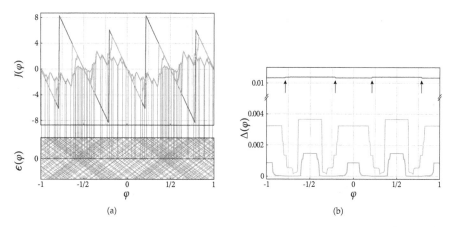

(a)   (b)

Figure 11. Self-consistent calculations for a discretized annulus with an inner radius $R_1 = 100a_\perp$ and an outer radius $R_2 = 150a_\perp$. The (super-) current (a) $J(\varphi)$ jumps whenever an energy level crosses E_F. The energy levels $\epsilon(\varphi)$ of the normal state are indicated by the grey lines. (b) displays the self-consistent order parameter Δ as a function of φ. The lines correspond to the pairing interaction $V = 0$ (orange), $V = 0.28\,t$ (green), $V = 0.32\,t$ (light blue), and $V = 0.38\,t$ (dark blue). The black arrows mark the positions of the q-jump for $V = 0.38\,t$. Here the energy units are $t = \hbar^2/2m_e a_\perp$.

$(R_2 - R_1)/a_\perp$ for $n = 1$. For each crossing, a jump appears in the current as a function of φ, as shown in figure 11 (a). The persistent current is therefore proportional to the level spacing at E_F for each flux value, i.e., it is small for a large density of states and large for a small density of states and vanishes in the limit of a continuous density of states (c.f. reference [32]). The amplitude of the normal persistent current is thus a measure for the difference of the energy spectrum at integer and half integer flux values. It is maximal for the one dimensional case, but might be very small in real metal loops (c.f. measurements of the Aharonov-Bohm effect in metal rings [4]).

Upon entering the superconducting state, an energy gap develops around E_F preventing energy levels from crossing E_F. Thus the persistent supercurrent arises as in the one dimensional ring independent of the density of states. The large jumps in the supercurrent appear at the value of φ where the energies of the even-q and odd-q states become degenerate and q switches to the next integer. In the flux regimes with no crossings of E_F the supercurrent is linear and the total energy quadratic in φ. For the largest value shown ($\Delta = 0.006\,t$), there is a direct gap for all values of φ. Even for this large Δ, the current and the energy are not precisely $\Phi_0/2$-periodic because of the energy difference of the even and odd q states in finite systems [34, 45]. The offset of the q-jump is only relevant for values of the pairing interaction V for which Δ is finite for all φ. In figure 11 (b), the offset is clearly visible for the largest two values of V (marked with black arrows). Its sign depends on the shape of the annulus and the value of V— the offset changes sign for increasing V (cf. reference [45]).

The introduction of self-consistency for the order parameter does not fundamentally change these basic observations [figure 11 (b)]. One finds that $\Delta(R_1) < \Delta(R_2)$, but if $(R_2 - R_1)/R_1 \lesssim 1$, the difference is small. In the following, we denote the average of $\Delta(r)$ by Δ. The crossover

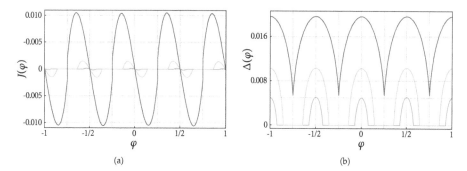

Figure 12. The order parameter $\Delta(\varphi)$ and the persistent current $J(\varphi)$ for the temperature driven transition from the normal to the superconducting state in an annulus with inner radius $R_1 = 30a_\perp$ and outer radius $R_2 = 36a_\perp$. The pairing interaction is $V_0 = 0.7\,t$, with a critical temperature of $k_B T_c \approx 0.0523\,t$ for zero flux. For these parameters $\Delta(T = 0) \approx 0.1\,t$. The lines (from top to bottom) correspond to the temperatures $k_B T = 0.0513\,t$ (dark blue), $k_B T = 0.0520\,t$ (light blue), and $k_B T = 0.0522\,t$ (green). Notice that Δ is slightly different for the flux values $\varphi = 0$ and $\varphi = \pm 1/2$.

is then controlled by the pairing interaction strength V, for which we chose such values as to reproduce the crossover from the normal state to a state with direct energy gap for all flux values. The order parameter Δ is now also a function of φ. If $\Delta(\varphi = 0) \lesssim 0.006\,t$ [c.f. figure 11 (b)], the gap closes with φ, and Δ decreases whenever a state crosses E_F. Unlike in a one dimensional ring, Δ does not drop to zero at the closing of the energy gap, but decreases stepwise. This is because in two or three dimensions, Δ is stabilized beyond the depairing velocity by contributions to the condensation energy from pairs with relative momenta perpendicular to the direction of the current flow; the closing of the indirect energy gap does not destroy superconductivity [7, 50].

Experimentally more relevant is to control the crossover through temperature. With the pairing interaction V sufficiently strong to produce a $T = 0$ energy gap much larger than the maximum Doppler shift, the crossover regime is reached for temperatures slightly below T_c. For the annulus of figure 10, the crossover proceeds within approximately one percent of T_c. The crossover regime becomes narrower for larger rings, proportional to the decrease of the Doppler shift. In the limit of a quasi one-dimensional ring of radius R we can be more precise: If we define the crossover temperature T^* by $\Delta(T^*) = \Delta_c$ and assuming $\Delta_c \ll \Delta$, we can use the Ginzburg-Landau form of the order parameter

$$\frac{\Delta(T)}{\Delta(0)} \approx 1.75 \sqrt{1 - \frac{T}{T_c}} \tag{21}$$

and obtain

$$\frac{T_c - T^*}{T_c} \approx \frac{\Delta_c^2}{3.1 \Delta^2(0)} = \frac{t^2}{12.4 \Delta^2(0)(R/a)^2} = \frac{E_F^2}{3.1 (k_B T_c)^2 (R/a)^2}, \tag{22}$$

where we used the relations $\Delta_c = at/2R$ and $E_F = 2t$ for a discretized one-dimensional ring at half filling, and the BCS relation $\Delta^2(0) \approx 3.1(k_B T_c)^2$. For a ring with a radius of 2500 lattice constants ($\approx 10\,\mu m$) and $\Delta(0) = 0.01\,t$ ($\approx 3\,meV$) one finds the ratio $(T_c - T^*)/T_c \approx 1.3 \times 10^{-4}$. This is in reasonable quantitative agreement with the experimental results of Little and Parks [30, 39], discussed also by Tinkham in reference [44]. Their prediction was similar to equation (22), up to a factor in which they include a finite mean free path. But they did not included the difference introduced through even and odd q states. This difference was considered in calculations of T_c by Bogachek et al. [9] in the single-channel limit and found to be exponentially small. A detailed study of the normal- to superconducting phase boundary was also done by Wei and Goldbart in reference [49] in which they considered the Φ_0 periodic contributions. In equation (22) the value of $\Delta(0)$ is in fact different for even and odd q. Although quantitative predictions of $T_c - T^*$ of the theory presented here might be too large as compared to the experiment, it serves as an upper limit because it describes the maximum possible persistent current. Inhomogeneities and scattering processes in real systems further reduce the difference of the energy spectra in the even- and odd-q flux sectors and thereby reduce $T_c - T^*$.

For temperatures close to T_c, the difference of the eigenenergies of even and odd q states is less important than at $T = 0$. Thus the deviation from the $\Phi_0/2$ periodicity of the current and of the order parameter is smaller. Furthermore, persistent currents in the normal state are exponentially small compared to the supercurrents below T_c. Their respective Φ_0 periodic behavior is therefore essentially invisible for the flux values where $\Delta = 0$. In figure 12, the difference between $\Delta(\varphi = 0)$ and $\Delta(\varphi = 1/2)$ is still visible, but the corresponding differences in the current are too small. Only for a superconductor with very small T_c, we expect the periodicity crossover to be visible.

Although we found within the framework of the BCS theory, that the crossover to a $\Phi_0/2$ periodic supercurrent takes place slightly below T_c, detailed studies by Ambegaokar and Eckern [5] and by von Oppen and Riedel [48] including superconducting fluctuations reveals that the crossover might actually take place above T_c. This fluctuation driven crossover is broader than the BCS crossover with a similarly, exponentially suppressed Φ_0 periodic normal current contribution. For a superconductor with a T_c small enough to observe a normal persistent current above T_c, Eckern and Schwab suggested that the crossover regime, where both Φ_0 and $\Phi_0/2$ periodic current contributions are present, should be observable at a temperature $T \approx 2T_c$ [20, 21].

The discussion of the periodicity crossover in a multi-channel loop also gives insight into the flux periodicity of loops of unconventional superconductors with gap nodes like a d-wave superconductor. In nodal superconductors the density of states is finite arbitrarily close to E_F. Therefore some energy levels cross E_F as a function of the flux, regardless of the size of the order parameter and consequently, the "small gap" situation extends to arbitrarily large loops [6, 25, 32, 34]. Of course, the number of energy levels crossing E_F decreases with increasing ring size and thus also the Φ_0 periodic contribution to the supercurrent. The dependence of this Φ_0 periodic contribution on the ring size depends on the order parameter symmetry. The careful study in reference [32] revealed that for d-wave superconductors, the relation between the Φ_0 and $\Phi_0/2$ periodic current contributions is proportional to $1/R_1$. It was estimated that for a ring of a cuprate superconductor with a circumference of $\sim 1\,\mu m$, this ratio is about 1% and should be observable experimentally.

3. Conclusion

We analyzed the crossover from the Φ_0 periodic persistent currents as a function of magnetic flux in a metallic loop to the $\Phi_0/2$ periodic supercurrent in the groundstate of the loop. We considered conventional s-wave pairing in a one-dimensional as well as in a multi-channel annulus. Although a one-dimensional superconducting ring is a rather idealized system, it proves valuable for discussing the physics of this crossover, which includes the emergence of a new minimum in the free energy for odd center-of-mass angular momenta $\hbar q$ of the Cooper pairs and the restoration of the flux periodicity of the free energy. The physical concepts, which we illustrated in a simplified form in section 2.2, remain thereby valid even in the more complex context of the self consistent calculations on the annulus.

In the superconducting state, a distinguished minimum in the free energy develops at $\varphi = q/2$. Choosing the proper value for q at each flux value leads to a series of minima at integer and half-integer flux values which, however, differ in energy for finite systems. In rings with a radius smaller than half the superconducting coherence length, the two electrons forming a Cooper pair are not forced to circulate the ring as a pair, and the supercurrent shows a Φ_0 periodicity. Only if the order parameter Δ is larger than the maximal Doppler shift ϵ_D, the supercurrent is $\Phi_0/2$ periodic. This is equivalent to the condition that the maximum flux induced current is smaller than the critical current J_c. Assuming that the relations obtained from the one-dimensional model remain valid on a ring with finite thickness $R_2 - R_1 \ll R_1$, as indeed suggested by the multi-channel model, the critical radius to observe $\Phi_0/2$ periodicity, $R_c = at/2\Delta$, would be of the order of $1\,\mu m$ for aluminum rings. Within the temperature controlled crossover upon cooling through T_c, Φ_0 periodicity might by difficult to observe since the differences in the energy spectra for integer and half-integer flux values are exponentially suppressed by temperature. Φ_0 periodicity is therefore only observable if a normal persistent current would be observable at the same temperature if superconductivity was absent.

In the introduction we referred to experiments where flux oscillations with "fractional periodicities", i.e., fractions of $\Phi_0/2$, were observed. Among various suggested origins, there is one particularly elegant approach based on a standard two-electron interaction. Consider the order parameter $\Delta_q(\varphi)$ for electron pairs with center-of-mass angular momentum $\hbar q$. In real space, q describes the phase winding of the order parameter $\Delta(\theta, \varphi) = \Delta(\varphi)e^{iq\theta}$, where θ is the angular coordinate in the ring and $\Delta(\varphi)$ is real. To ensure that $\Delta(\theta, \varphi)$ is a single valued and continuous function, q must be an integer number. If, however, $\Delta(\theta, \varphi)$ is zero somewhere on the ring, it can change sign. Such a sign changing order parameter is modeled as

$$\tilde{\Delta}(\theta, \varphi) = \frac{1}{2}\left[\Delta_0(\varphi) + \Delta_q(\varphi)e^{iq\theta}\right] \xrightarrow{\Delta_q = \Delta_0} \Delta(\varphi)e^{-iq\theta/2}\cos\left(\frac{q}{2}\theta\right), \tag{23}$$

which displays a phase-winding number $q/2$ if $\Delta_q = \Delta_0$, and consequently a vanishing supercurrent at the fractional flux value $q\Phi_0/4$. In momentum space, $\tilde{\Delta}(\theta, \varphi)$ is represented by the two-component order parameter $\{\Delta_0(\varphi), \Delta_q(\varphi)\}$. Such a superconducting state is typically referred to as a "pair-density wave" (PDW) state [3], since the real-space order parameter is periodically modulated and therefore q can no longer be interpreted as an angular momentum. Agterberg and Tsunetsugu showed within a Ginzburg-Landau approach, that a PDW superconductor indeed allows for vortices carrying a $\Phi_0/4$ flux quantum [2, 3].

Based on a microscopic model, it was shown in reference [33] that a PDW state cannot result from an on-site pairing interaction. However, the PDW state can be a stable alternative for unconventional superconductors with gap nodes, specifically for d-wave superconductivity as realized in the high-T_c cuprate superconductors. More work is, however, needed to analyze under which conditions the PDW state indeed develops an energy minimum at multiples $\Phi_0/4$, and these minima become degenerate in the limit of large loops. The notion of the absence of coexistence of Cooper pairs with different center-of-mass momenta for conventional superconductors (as, e.g., in the PDW state) justifies the reduction of the sum over q in the Hamiltonian (6) in section 2.2 to one specific q in order to derive the periodicity crossover in superconducting rings.

Here we mention a further system where a similar mechanism as described above may lead to $\Phi_0/4$ flux periodicity: Sr_2RuO_4. Experimental evidence exists that Sr_2RuO_4 is a spin triplet p-wave superconductor. A triplet superconductor can be represented by the two-component order parameter $\{\Delta_{q_1}^{\uparrow\uparrow}(\varphi), \Delta_{q_2}^{\downarrow\downarrow}(\varphi)\}$, with the center-of-mass momenta q_1 and q_2 for the $s_z = 1$ and the $s_z = -1$ condensates. This realizes a similar situation as for the PDW state where $\Phi_0/4$ periodicity is possible [46]. Indeed Jang et al. observed recently that the flux through a microscopic Sr_2RuO_4 ring is quantized in units of $\Phi_0/4$ [24].

Acknowledgements

The authors acknowledge discussions with Yuri Barash, Ulrich Eckern, Jochen Mannhart, and Christof Schneider. This work was supported by the Deutsche Forschungsgemeinschaft through TRR 80.

Author details

Loder Florian
Experimental Physics VI & Theoretical Physics III, Center for Electronic Correlations and Magnetism, Institute of Physics, University of Augsburg, 86135 Augsburg, Germany

Kampf Arno P.
Theoretical Physics III, Center for Electronic Correlations and Magnetism, Institute of Physics, University of Augsburg, 86135 Augsburg, Germany

Kopp Thilo
Experimental Physics VI, Center for Electronic Correlations and Magnetism, Institute of Physics, University of Augsburg, 86135 Augsburg, Germany

4. References

[1] Abrikosov, A. A. [1957]. On the magnetic properties of superconductors of the second group, *Soviet Physics – JETP* 5: 1174.

[2] Agterberg, D. F., Sigrist, M. & Tsunetsugu, H. [2009]. Order parameter and vortices in the superconducting q phase of $CeCoIn_5$, *Phys. Rev. Lett.* 102: 207004.

[3] Agterberg, D. F. & Tsunetsugu, H. [2008]. Dislocations and vortices in pair-density-wave superconductors, *Nature Phys.* 4: 639.

[4] Aharonov, Y. & Bohm, D. [1959]. Significance of electromagnetic potentials in the quantum theory, *Phys. Rev.* 115: 485.

[5] Ambegaokar, V. & Eckern, U. [1991]. Diamagnetic response of mesoscopic superconducting rings above T_c, *Phys. Rev. B* 44: 10358.

[6] Barash, Y. S. [2008]. Low-energy subgap states and the magnetic flux periodicity in d-wave superconducting rings, *Phys. Rev. Lett.* 100: 177003.

[7] Bardeen, J. [1962]. Critical fields and currents in superconductors, *Rev. Mod. Phys.* 34: 667.

[8] Bardeen, J., Cooper, L. N. & Schrieffer, J. R. [1957]. Theory of superconductivity, *Phys. Rev.* 108: 1175.

[9] Bogachek, E. N., Gogadze, G. A. & Kulik, I. O. [1975]. Doubling of the period of flux quantization in hollow superconducting cylinders due to quantum effects in the normal state, *Phys. Stat. Sol. (b)* 67: 287.

[10] Brenig, W. [1961]. Remark concerning quantized magnetic flux in superconductors, *Phys. Rev. Lett.* 7: 337.

[11] Büttiker, M., Imry, Y. & Landauer, R. [1983]. Josephson behavior in small normal one-dimensional rings, *Phys. Lett. A* 96: 365.

[12] Büttiker, M. & Klapwijk, T. M. [1986]. Flux sensitivity of a piecewise normal and superconducting metal loop, *Phys. Rev. B* 33: 5114.

[13] Byers, N. & Yang, C. N. [1961]. Theoretical considerations concerning quantized magnetic flux in superconducting cylinders, *Phys. Rev. Lett.* 7: 46.

[14] Cayssol, J., Kontos, T. & Montambaux, G. [2003]. Isolated hybrid normal/superconducting ring in a magnetic flux: From persistent current to josephson current, *Phys. Rev. B* 67: 184508.

[15] Cheung, H., Gefen, Y., Riedel, E. K. & Shih, W. [1988]. Persistent currents in small one-dimensional metal rings, *Phys. Rev. B* 37: 6050.

[16] de Gennes, P. G. [1966]. *Superconductivity of Metals and Alloys*, Addison Wesley Publishing Company, chapter 5.

[17] Deaver, B. S. & Fairbank, W. M. [1961]. Experimental evidence for quantized flux in superconducting cylinders, *Phys. Rev. Lett.* 7: 43.

[18] Doll, R. & Näbauer, M. [1961]. Experimental proof of magnetic flux quantization in a superconducting ring, *Phys. Rev. Lett.* 7: 51.

[19] Douglass, D. H. [1963]. Properties of a thin hollow superconducting cylinder, *Rev. Rev.* 132: 513.

[20] Eckern, U. [1994]. Superconductivity in restricted geometries, *Physica B* 203: 448.

[21] Eckern, U. & Schwab, P. [1995]. Normal persistent currents, *Adv. in Physics* 44: 387.

[22] Essmann, U. & Träuble, H. [1967]. The direct observation of individual flux lines in type ii superconductors, *Phys. Lett. A* 24: 526.

[23] Gough, C. H., Colclough, M. S., Forgan, E. M., Jordan, R. g. & Keene, M. [1987]. Flux quantization in a high-T_c superconductor, *Nature* 326: 855.

[24] Jang, J., Ferguson, D. G., Vakaryuk, V., Budakian, R., Chung, S. B., Goldbart, P. M. & Maeno, Y. [2011]. Observation of half-height magnetization steps in Sr_2RuO_4, *Science* 331: 186.

[25] Juričić, V., Herbut, I. F. & Tešanović, Z. [2008]. Restoration of the magnetic hc/e-periodicity in unconventional superconductors, *Phys. Rev. Lett.* 100: 187006.

[26] Khavkine, I., Kee, H.-Y. & Maki, K. [2004]. Supercurrent in nodal superconductors, *Phys. Rev. B.* 70: 184521.

[27] Landauer, R. & Büttiker, M. [1985]. Resistance of small metallic loops, *Phys. Rev. Lett.* 54: 2049.

[28] Lindström, T., Charlebois, S. A., Tzalenchuk, A. Y., Ivanov, Z., Amin, M. H. S. & Zagoskin, A. M. [2003]. Dynamical effects of an unconventional current-phase relation in YBCO dc SQUIDs, *Phys. Rev. Lett.* 90: 117002.

[29] Little, W. A. [1964]. Long-range correlations in superconductivity, *Rev. Mod. Phys.* 36: 264.

[30] Little, W. A. & Parks, R. D. [1962]. Observation of quantum periodicity in the transition temperature of a superconducting cylinder, *Phys. Rev. Lett.* 9: 9.

[31] Loder, F., Kampf, A. P. & Kopp, T. [2008]. Crossover from hc/e to $hc/2e$ current oscillations in rings of s-wave superconductors, *Phys. Rev. B* 78: 174526.

[32] Loder, F., Kampf, A. P. & Kopp, T. [2009]. Flux periodicities in loops of nodal superconductors, *New J. Phys.* 11: 075005.

[33] Loder, F., Kampf, A. P. & Kopp, T. [2010]. Superconducting state with a finite-momentum pairing mechanism in zero external magnetic field, *Phys. Rev. B.* 81: 020511(R).

[34] Loder, F., Kampf, A. P., Kopp, T., Mannhart, J., Schneider, C. & Barash, Y. [2008]. Magnetic SSux periodicity of h/e in superconducting loops, *Nature Phys.* 4: 112.

[35] London, F. [1950]. *Superfluids*, John Wiley & Sons, New York.

[36] Mineev, V. P. & Samokhin, K. V. [1999a]. *Introduction to Unconventional Superconductivity*, Gordon and Breach science publishers, chapter 8.

[37] Mineev, V. P. & Samokhin, K. V. [1999b]. *Introduction to Unconventional Superconductivity*, Gordon and Breach science publishers, chapter 17.

[38] Onsager, L. [1961]. Magnetic flux through a superconducting ring, *Phys. Rev. Lett.* 7: 50.

[39] Parks, R. D. & Little, W. A. [1964]. Fluxoid quantization in a multiply-connected superconductor, *Phys. Rev.* 133: A97.

[40] Peshkin, M. [1963]. Fluxoid quantization, pair symmetry, and the gap energy in the current carrying bardeen-cooper-schrieffer state, *Phys. Rev.* 132: 14.

[41] Schneider, C. [2007]. Conference on *"Superconductivity and Magnetism in the Perovskites and Other Novel Materials"*, Tel Aviv, unpublished.

[42] Schneider, C., Hammerl, G., Logvenov, G., Kopp, T., Kirtley, J. R., Hirschfeld, P. & Mannhart, J. [2004]. Half-$h/2e$ critical current-oscillations of squids, *Europhys. Lett.* 68: 86.

[43] Schrieffer, J. R. [1964]. *Theory of Superconductivity*, Addison Wesley Publishing Company, chapter 8.

[44] Tinkham, M. [1963]. Effect of fluxoid quantization on transitions of superconducting films, *Phys. Rev.* 129: 2413.

[45] Vakaryuk, V. [2008]. Universal mechanism for breaking the $hc/2e$ periodicity of flux-induced oscillations in small superconducting rings, *Phys. Rev. Lett.* 101: 167002.

[46] Vakaryuk, V. & Vinokur, V. [2011]. Effect of half-quantum vortices on magnetoresistance of perforated superconducting films, *Phys. Rev. Lett.* 107: 037003.

[47] von Oppen, F. & Riedel, E. K. [1991]. Average persistent current in a mesoscopic ring, *Phys. Rev. Lett.* 66: 84.

[48] von Oppen, F. & Riedel, E. K. [1992]. Flux-periodic persistent current in mesoscopic superconducting rings close to T_c, *Phys. Rev. B* 46: 3203.

[49] Wei, T.-C. & Goldbart, P. M. [2008]. Emergence of h/e-period oscillations in the critical temperature of small superconducting rings threaded by magnetic flux, *Phys. Rev. B* 77: 224512.

[50] Zagoskin, A. M. [1998]. *Quantum Theory of Many-Body Systems*, Springer, chapter 4.

[51] Zha, G.-Q., Milošević, M. V., Zhou, S.-P. & Peeters, F. M. [2009]. Magnetic flux periodicity in mesoscopic d-wave symmetric and asymmetric superconducting loops, *Phys. Rev. B* 80: 144501.

[52] Zhu, J. & Wang, Z. D. [1994]. Supercurrent determined from the aharonov-bohm effect in mesoscopic superconducting rings, *Phys. Rev. B* 50: 7207.

Chapter 6

Composite Structures of d-Wave and s-Wave Superconductors (d-Dot): Analysis Using Two-Component Ginzburg-Landau Equations

Masaru Kato, Takekazu Ishida, Tomio Koyama and Masahiko Machida

Additional information is available at the end of the chapter

1. Introduction

Superconductivity is a macroscopic quantum phenomenon [1]. Therefore it shows quite interesting properties because of its quantum nature. Such properties are described by a macroscopic complex wave function of the superconductivity. Especially, a phase of the macroscopic wave function play an important role in these properties. For example, superconducting devices, such as, superconducting charge, flux and phase qubits, superconducting single flux quantum device, and intrinsic Josephson junction Terahertz emitter of high Tc cuprate superconductors, use such quantum nature of superconductivity. They have attracted much attention recently.

In conventional superconductors, there is only single phase ϕ of the superconducting order parameter or the macroscopic wave function $\Delta = |\Delta| e^{i\phi}$. This phase causes interference effect, such as, Josephson effect, and quantization of vortices in the superconductor. But unconventional and anisotropic superconductivity have different phase that comes from internal degree of freedom of the superconducting order parameter. Because in the superconductors, electrons are paired and if their paring symmetry is an s-wave, as in the conventional superconductors, the order parameter is just a single complex number. But if the symmetry is other one such as p-wave or d-wave, then the order parameter has an internal phase [2,3]. For example, d-wave superconductors, especially $d_{x^2-y^2}$-wave superconductors have a symmetry that is shown in Fig. 1. This symmetry is internal and it appears in momentum space that means the wave function of the Cooper pair moving along the x-axis has + sign and that moving along y-axis has − sign. This is also another phase of superconducting order parameter and it affects the interference phenomena in the superconductors.

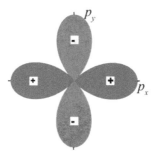

Figure 1. The symmetry of $d_{x^2-y^2}$-wave superconductivity. The color shows the sign of wave function (order parameter) and the shape shows the amplitude of the wave function (order parameter).

The high-Tc cuprate superconductors are typical example of d-wave superconductors. Since the discovery of the high-Tc superconductors, the pairing symmetry, as well as its origin, has been controversial, but phase-sensitive experiment is crucial for determining the symmetry of the Cooper pairs. One such phase-sensitive experiment is a corner junction experiment [4,5]. In this experiment, a square-shaped high-Tc superconductor is connected with a conventional s-wave superconductor with two Josephson junctions A and B, around a corner and each junction is perpendicular to either of the x or the y axes of the high-Tc superconductor, as shown in Fig. 2.

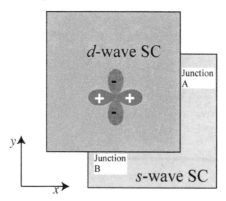

Figure 2. Configuration of corner junction between d- and s-wave superconductors. There are two junctions, A and B.

In such geometry if the Cooper pair tunnels through junction A has phase 0, then the Cooper pair tunnels through junction B has phase π. The critical current between the s- and the d-wave superconductors through these two junctions is zero under zero external magnetic field because each supercurrent cancels with each other. This is in apparent contrast with a corner junction between both s-wave superconductors for which the critical current is largest under zero external field. Therefore, from this experiment, the symmetry of the Cooper pairs in high-Tc superconductors was determined as d-wave, especially $d_{x^2-y^2}$.

In addition, this experiment showed that the critical current become smax when the total magnetic flux through the corner junction was $\Phi_0/2$. Here, $\Phi_0 = hc/2e$ is the flux quantum. Because the phase difference π between two Josephson currents through junctions A and B, is compensated by additional phase,

$$\chi = \int_C \frac{2e}{c\hbar} A \cdot ds = \int_s \frac{2e}{c\hbar} curl A \cdot dS = 2\pi \frac{2e}{ch}\Phi = 2\pi\frac{\Phi}{\Phi_0} \tag{1}$$

where C is a contour around the corner junction and S is the area surrounded by the C and Φ is the total magnetic flux through the junctions. When total magnetic flux $\Phi = \Phi_0/2$, this phase becomes $\chi = \pi$. With this phase, total phase difference becomes 0 or 2π. Therefore two Josephson currents now are reinforced with each other.

This result shows, the stable superconducting state under zero external current and zero external field becomes nontrivial. For such stable state, the free energy of whole system, especially the superconducting condensation energies of both d- and s-wave superconductors should be low. Therefore the order parameter should become continuous across both of junctions. And then the phase of order parameter cannot be uniform because of the phase difference π between two junctions. This phase difference π should be compensated by changing the phase of the order parameter spatially in both s- and d-wave superconducting regions because of single valuedness of the order parameter. This spatial variation of the order parameter causes the supercurrent around the corner junction and then spontaneous magnetic flux is created at the corner without an external field. Because the associated phase change of this supercurrent is not 2π but only π, the spontaneous magnetic flux is not singly quantized magnetic flux Φ_0, but half-flux quantum magnetic flux ($\Phi_0/2$).

An experiment showing this property was done by Higenkamp et al., who made a zigzag junction between conventional s-wave superconductor Nb and high-Tc d-wave superconductor YBCO, which consists of successive corner junctions [6]. And they observed spontaneous magnetic fluxes at every corner under a zero external field, using scanning SQUID microscope. The spontaneous magnetic fluxes aligned antiferromagnetically, because of the attractive vortex-anti-vortex interaction. They also made small ring with two junctions between Nb and YBCO, which is called π-rings, and controlled the spontaneous half-quantum magnetic fluxes [7].

When spontaneous magnetic flux appears in the zero external magnetic field, this state does not have the time reversal symmetry, which the original system has. Therefore there are always two degenerate stable states. In these states, supercurrent directions are opposite and henceforth directions of spontaneous magnetic fluxes are opposite. This spontaneously appeared magnetic flux is useful and can be used as a spin or a bit by it self. Ioffe et al. [8] proposed a quantum bit using this half-quantum flux. Using the spontaneous magnetic flux as an Ising spin system, Kirtley et al. made a frustrated triangular lattice of π-rings [7]. In their systems, the π-rings were isolated and interacted with each other purely by the electromagnetic force.

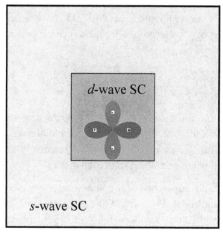

Figure 3. Schematic diagram of a d-dot.

In contrast to the previous approach for using the spontaneous half-flux quantum, we considered nano-sized d-wave superconductor embedded in an s-wave matrix, as shown in Fig. 3 [9-11]. We want to consider the whole d-dot system as a single element, not the individual half-quantum fluxes. As in the single half-quantum system, our d-dot has two degenerate states if the spontaneous magnetic fluxes appear, because the state with spontaneous magnetic fluxes under zero external field also breaks time-reversal symmetry. This property is independent from the shape, and the d-dot in any shape always has two degenerate stable states. Therefore, the d-dot as a whole can be considered as a single element with two level states and it might be used as a spin or a bit also as a qubit. It has better properties than those of single flux quantum element, which will be shown in following sections.

In the following, we first show a phenomenological superconducting theory, which describes the spontaneous magnetic fluxes in these composite structures, especially in d-dot systems and then we discuss the basic properties of this d-dot, based on this phenomenological theory. Also, we discuss the difference between a single half-quantum flux system and our d-dot system.

2. Model: two components Ginzburg-Landau free energy

In order to discuss the basic properties of the d-dot, especially to describe the appearance of spontaneous magnetic fluxes, we use the phenomenological Ginzburg-Landau (GL) theory. However, for anisotropic superconductors, such as the $d_{x^2-y^2}$-wave high-Tc cuprate superconductors, their anisotropy cannot be treated by the simple GL theory. This is because using only up to quartic term and quadratic term of gradients of the single order parameters in free energy, there are no anisotropic terms. Anisotropy of the vortices in d-wave superconductors within the phenomenological theory was treated by Ren et al., who used a two-component GL theory [12-15]. Here, two components mean the two components of the order parameter of with s and d symmetries. They derived the two-component GL free energy from the Gor'kov equations.

Multi-components GL equations were used for exotic superconductors, e.g. heavy fermion superconductors[16-18] and Sr_2RuO_4[19,20]. Also, recently, two components GL equations are studied for two-band or two-gap superconductors, such as MgB_2 [21,22].

The model by Ren.et al. especially emphasize the anisotropy of d-wave superconductivity. Therefore, we use the following two-component Ginzburg-Landau (GL) free energy for d-wave superconductors;

$$
F_d\left(\Delta_s, \Delta_d, \mathbf{A}\right) = \int_\Omega \left(\frac{3\alpha}{8}\lambda_d\left[|\Delta_d|^2 - \frac{4\ln\left(T_{cd}/T\right)}{3\alpha}\right]^2 + \alpha\lambda_d\left[|\Delta_s|^4 + 2\frac{\alpha_s}{\alpha}|\Delta_s|^2\right] \right.
$$
$$
+ \frac{1}{4}\alpha\lambda_d v_F^2 \left[\left|\Pi\Delta_d^*\right|^2 + 2|\Pi\Delta_s|^2 + \left(\Pi_x^*\Delta_s\Pi_x\Delta_d^* - \Pi_y^*\Delta_s\Pi_y\Delta_d^* + \text{H.c.}\right)\right]
$$
$$
\left. + \alpha\lambda_d\left[2|\Delta_s|^2|\Delta_d|^2 + \frac{1}{2}\left(\Delta_s^{*2}\Delta_d^2 + \Delta_s^2\Delta_d^{*2}\right)\right] + \frac{1}{8\pi}\left[|\mathbf{h}-\mathbf{H}|^2 + \left(\text{div}\,\mathbf{A}\right)^2 \right]\right) d\Omega
$$
(2)

where $\Pi = \frac{\hbar}{i}\nabla - \frac{2e}{c}\mathbf{A}$ is a generalized momentum operator that is gauge invariant and $\alpha = \frac{7\zeta(3)}{8(\pi T)^2}$. Δ_d and Δ_s are the d-wave and the s-wave components of the order parameter, respectively. $\lambda_d = V_d N(0)$ is the strengths of the coupling constants for the d-wave interaction channel and $\alpha_s = \frac{1+\dfrac{V_s}{V_d}}{\lambda_d}$. Here, $N(0)$ is the density of states of electrons at the Fermi energy and V_d and V_s are interaction constants between electrons for d- and s-wave channels, respectively. We assume attractive and repulsive interactions for the d- and the s-wave channels, respectively. T_{cd} is the transition temperature of the d-wave superconductivity under zero-external field. $\mathbf{H}$ is an external field and $\mathbf{h} = \text{curl}\,\mathbf{A}$. We take the London gauge i.e. $\nabla\cdot\mathbf{A} = 0$ and $\mathbf{A}\cdot\mathbf{n} = 0$ at the surface of superconductor. The term $(\nabla\cdot\mathbf{A})^2$ in the integrand of Eq. 2 is added for fixing the gauge.

Also, for s-wave superconductors, we use the following two-component GL equation with attractive and repulsive interactions for the s-wave and the d-wave channels, respectively:

$$F_s\left(\Delta_s,\Delta_d,\mathbf{A}\right)=\int_\Omega\left(\frac{\alpha}{2}\lambda_s\left[\left|\Delta_s\right|^2-\frac{\ln\left(T_{cs}/T\right)}{\alpha}\right]^2+\frac{\alpha\lambda_s}{2}\left[\frac{3}{8}\left|\Delta_d\right|^4+\frac{\alpha_d}{\alpha}\left|\Delta_d\right|^2\right]\right.$$

$$+\frac{1}{8}\alpha\lambda_s v_F^2\left[\left|\Pi\Delta_d^*\right|^2+2\left|\Pi\Delta_s\right|^2+\left(\Pi_x^*\Delta_s\Pi_x\Delta_d^*-\Pi_y^*\Delta_s\Pi_y\Delta_d^*+\text{H.c.}\right)\right] \tag{3}$$

$$+\frac{\alpha\lambda_s}{2}\left[2\left|\Delta_s\right|^2\left|\Delta_d\right|^2+\frac{1}{2}\left(\Delta_s^{*2}\Delta_d^2+\Delta_s^2\Delta_d^{*2}\right)\right]+\frac{1}{8\pi}\left[\left|\mathbf{h}-\mathbf{H}\right|^2+\left(\nabla\cdot\mathbf{A}\right)^2\right]\right)d\Omega$$

Here $\lambda_s=V_sN(0)$ is the strengths of the coupling constants for the s-wave interaction

channeland $\alpha_d=\dfrac{2\dfrac{V_s}{V_d}-1}{\lambda_s}$.

In these free energies, the anisotropy of the d-wave superconductivity appears in the coupling terms of the gradient of both components of order parameters, $\left(\Pi_x^*\Delta_s\Pi_x\Delta_d^*-\Pi_y^*\Delta_s\Pi_y\Delta_d^*+\text{H.c.}\right)$, where the two terms that contain the gradients along the x and the y directions have different signs.

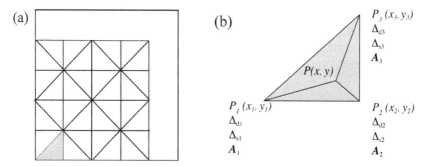

Figure 4. (a) Triangle elements of superconductors. (b) Nodes of a triangle element and the value of physical quantities at the nodes.

By minimizingthe sum of these free energies, we obtain the Ginzburg-Landau (GL) equations. For this purpose, we use the finite-element method [23-25], because we want to investigate variously shaped d-dots. Hereafter we consider two-dimensional system and ignore the variation of physical quantities along the direction perpendicular to the two dimensional system. In the finite-element method for two-dimensional system, we divide the superconductor region into small triangular elements (see Fig. 4 (a)) and then expand the order parameters and the vector potential in each element by using the area coordinate, which is defined as,

$$N_i^e\left(x,y\right)=\begin{cases}\dfrac{1}{2S_e}\left(a_i+b_ix+c_iy\right)&\text{inside of element}\\[2mm]0&\text{outside of element}\end{cases}\quad(i=1,2,3) \tag{4}$$

S_e is an area of e-th element and the coefficients are defined as,

$$a_i = x_j^e y_k^e - x_k^e y_j^e \tag{5}$$

$$b_i = y_j^e - y_k^e \tag{6}$$

$$c_i = x_k^e - x_j^e \tag{7}$$

where (i, j, k) is a cyclic permutation of $(1,2,3)$ and (x_i^e, y_i^e) is the coordinate of the i-th nodes of the e-th element (see Fig. 4(b)).These area coordinates are localized functions as shown in Fig. 5.Also the area coordinate $N_i^e(x,y)$ represents normalized area of a triangular of which nodes are P, P_j and P_k in Fig. 5 (b), where (i, j, k) is a cyclic permutation of $(1,2,3)$. And then they have following properties;

$$\sum_{i=1}^{3} N_i^e(x,y) = 1 \tag{8}$$

$$N_i^e\left(x_j^e, y_j^e\right) = \delta_{ij} \tag{9}$$

(a)

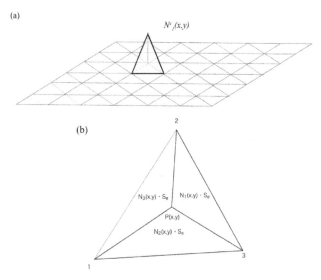

(b)

Figure 5. (a) A bird's-eye view of an area coordinate $N_i^e(x,y)$. (b) Area coordinates represent normalized area of triangular elements.

Using these properties, the order parameters are expanded as

$$\Delta_d(x,y) = \sum_{e}\sum_{i=1}^{3} \Delta_{di}^e N_i^e(x,y) \tag{10}$$

$$\Delta_s(x,y) = \sum_e \sum_{i=1}^3 \Delta_{si}^e N_i^e(x,y) \tag{11}$$

and also the magnetic vector potential is expanded as,

$$\mathbf{A}(x,y) = \sum_e \sum_{i=1}^3 \mathbf{A}_i^e N_i^e(x,y) \tag{12}$$

where Δ_{di}^e and Δ_{si}^e are the value of the d-wave and the s-wave order parameters at the i-th vertex of e-th element, respectively (see Fig. 4 (b)). Also, $\mathbf{A}_i^e$ is the value of the vector potential at the i-th vertex of the e-th element (see Fig. 4 (b)). Inserting these expansions into the free energies Eqs. 2 and 3, the free energies are expressed by those values, Δ_{di}^e, Δ_{si}^e and $\mathbf{A}_i^e$. Then, minimizing the free energies, we get following GL equations. For the d-wave superconducting order parameter,

$$\sum_j \left[P_{ij}^{dd}(\{A\}) + P_{ij}^{dd2R}(\{\Delta\}) \right] \mathrm{Re}\,\Delta_{dj} + \sum_j \left[Q_{ij}^{dd}(\{A\}) + Q_{ij}^{dd2}(\{\Delta\}) \right] \mathrm{Im}\,\Delta_{dj}$$
$$+ \sum_j P_{ij}^{ds}(\{A\})\mathrm{Re}\,\Delta_{sj} + \sum_j Q_{ij}^{ds}(\{A\})\mathrm{Im}\,\Delta_{sj} = V_i^{dR}(\{\Delta\}) \tag{13}$$

$$\sum_j \left[-Q_{ij}^{dd}(\{A\}) + Q_{ij}^{dd2}(\{\Delta\}) \right] \mathrm{Re}\,\Delta_{dj} + \sum_j \left[P_{ij}^{dd}(\{A\}) + P_{ij}^{dd2I}(\{\Delta\}) \right] \mathrm{Im}\,\Delta_{dj}$$
$$+ \sum_j P_{ij}^{ds}(\{A\})\mathrm{Im}\,\Delta_{sj} - \sum_j Q_{ij}^{ds}(\{A\})\mathrm{Re}\,\Delta_{sj} = V_i^{dI} \tag{14}$$

And for the s-wave superconducting order parameter,

$$\sum_j \left[P_{ij}^{ss}(\{A\}) + P_{ij}^{ss2R}(\{\Delta\}) \right] \mathrm{Re}\,\Delta_{sj} + \sum_j \left[Q_{ij}^{ss}(\{A\}) + Q_{ij}^{ss2}(\{\Delta\}) \right] \mathrm{Im}\,\Delta_{sj}$$
$$+ \sum_j P_{ij}^{ds}(\{A\})\mathrm{Re}\,\Delta_{dj} + \sum_j Q_{ij}^{ds}(\{A\})\mathrm{Im}\,\Delta_{dj} = V_i^{sR}(\{\Delta\}) \tag{15}$$

$$\sum_j \left[-Q_{ij}^{ss}(\{A\}) + Q_{ij}^{ss2}(\{\Delta\}) \right] \mathrm{Re}\,\Delta_{sj} + \sum_j \left[P_{ij}^{ss}(\{A\}) + P_{ij}^{ss2I}(\{\Delta\}) \right] \mathrm{Im}\,\Delta_{sj}$$
$$+ \sum_j P_{ij}^{ds}(\{A\})\mathrm{Im}\,\Delta_{dj} - \sum_j Q_{ij}^{ds}(\{A\})\mathrm{Re}\,\Delta_{dj} = V_i^{sI}(\{\Delta\}) \tag{16}$$

Coefficients for d-wave region and s-wave region are different and are given in Appendix.

Also for the vector potential following equations are obtained as follows,

$$\sum_j \left[R_{ij}^e(\{\Delta_d,\Delta_s\}) + R_{ij}^{2e}(\{\Delta_d,\Delta_s\}) \right] A_{jx} + \sum_j S_{ij}^e A_{jy} = T_i^{ex} - T_i^{2ex} - U_i^{ey} \tag{17}$$

$$\sum_j \left[R_{ij}^e(\{\Delta_d,\Delta_s\}) - R_{ij}^{2e}(\{\Delta_d,\Delta_s\}) \right] A_{jy} - \sum_j S_{ij}^e A_{jx} = T_i^{ey} + T_i^{2ey} - U_i^{ex} \tag{18}$$

Here coefficients are also given in Appendix. At the boundary of d- and s-wave superconductors, because the wave function is continuous, following boundary conditions are applied:

$$\frac{\Delta_{1s}}{V_{1s}} = \frac{\Delta_{2s}}{V_{2s}} \tag{19}$$

$$\frac{\Delta_{1d}}{V_{1d}} = \frac{\Delta_{2d}}{V_{2d}} \tag{20}$$

In this model, d- and s-wave components of the order parameter interfere anisotropically with each other in the both of d-wave and s-wave superconductors. And the boundaries between both superconductors are assumed clean. But we cannot take into account the roughness of the boundary.

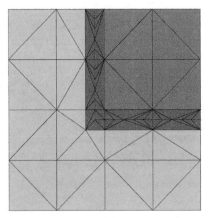

Figure 6. Mesh partition used in the finite element method. Red, blue and grey regions are d-wave and s-wave superconducting and junction regions, respectively.

For treating roughness of the junctions, we consider second model. In contrast to the first model, in the second model, the s-wave superconductor is connected to the d-wave superconductor through a thin metal or an insulator layers. In the second model, we consider only the d-wave (s-wave) component of the order parameter in the d-wave (s-wave) superconducting region, but in the thin metal layer, we take both components. So, free energies are given as,

$$F_d(\Delta_d, \mathbf{A}) = \int_\Omega \left(\frac{3\alpha}{8} \lambda_d \left[|\Delta_d|^2 - \frac{4\ln T_{cd}/T}{3\alpha} \right]^2 + \frac{1}{4}\alpha\lambda_d v_F^2 \left|\Pi\Delta_d^*\right|^2 + \frac{1}{8\pi}|\mathbf{h} - \mathbf{H}|^2 + \frac{1}{8\pi}(\operatorname{div}\mathbf{A})^2 \right) d\Omega \tag{21}$$

$$F_s(\Delta_s, \mathbf{A}) = \int_\Omega \left(\frac{\alpha}{2} \lambda_s \left[|\Delta_s| - \frac{\ln T_{cs}/T}{\alpha} \right]^2 + \frac{1}{4}\alpha\lambda_s v_F^2 \left|\Pi\Delta_s^*\right|^2 + \frac{1}{8\pi}|\mathbf{h} - \mathbf{H}|^2 + \frac{1}{8\pi}(\operatorname{div}\mathbf{A})^2 \right) d\Omega \tag{22}$$

$$F_M\left(\Delta_s,\Delta_d,\mathbf{A}\right)=\int_\Omega\left(|\lambda_d|\frac{3b}{2}\left[|\Delta_d|^2+\frac{a_d}{3b}\right]^2+b\left[|\Delta_s|^2+\frac{a_s}{2b}\right]^2|\lambda_d|+b|\lambda_d|\left[2|\Delta_s|^2|\Delta_d|^2+\frac{1}{2}\left(\Delta_s^{*2}\Delta_d^2+\Delta_s^2\Delta_d^{*2}\right)\right.\right.$$

$$\left.\left.+\frac{1}{4}v_F^2\left\{|\Pi\Delta_d^*|^2+2|\Pi\Delta_s^*|^2+\left(\Pi_x^*\Delta_s\Pi_x\Delta_d^*-\Pi_y^*\Delta_s\Pi_y\Delta_d^*+\mathrm{H.c.}\right)\right\}\right]+\frac{1}{8\pi}|\mathbf{h}-\mathbf{H}|^2+\frac{1}{8\pi}(\mathrm{div}\,\mathbf{A})^2\right)d\Omega$$

(23)

Here $F_d\left(\Delta_d,\mathbf{A}\right)$, $F_s\left(\Delta_s,\mathbf{A}\right)$ and $F_M\left(\Delta_s,\Delta_d,\mathbf{A}\right)$ are the free energies for d-wave, s-wave superconducting and junction regions respectively. A typical division of superconducting region to d- and s-wave superconducting and junction regions is given in Fig. 6. In the junction region, order parameters and the vector potential vary rapidly and therefore smaller mesh sizes are taken. At the boundary, we assume that each component of the order parameter is continuous.

3. Appearance of the half-quantum magnetic flux in d-dot

Using the two models in Section 2, the spontaneous magnetic flux can be described. In Fig.7, the s-wave and d-wave order parameters and magnetic field distribution are shown for a corner junction between s-wave and d-wave superconductors under zero external magnetic field.

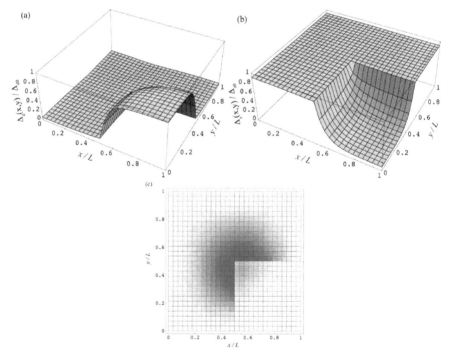

Figure 7. Appearance of spontaneous half-quantum magnetic flux at a corner junction. (a) Amplitude of s-wave component of order parameter. (b) Amplitude of s-wave component of order parameter. (c) Magnetic field distribution at zero external field.

In this calculation, the physical parameters, such as the GL parameters and the transition temperatures, for Nb and YBCO in the s- and the d-wave superconducting regions are used, and system size is set as $L = 25\xi_{d0}$. Here, ξ_{d0} is the coherence length of the d-wave superconductors at zero temperature. In Fig.7 (c), right-lower part is the d-wave superconductor and other part is s-wave superconductor. Both components of the order parameter exist in both of s- and d-wave superconductors. Especially, s-wave component penetrates into the d-wave superconducting region and interferes with d-wave component. Such interference causes the spontaneous magnetic field, which is perpendicular to the plane. In Fig. 7 (c), blue color means magnetic field H_z is positive. Total magnetic flux around the corner is approximately $\Phi_0/2$. So a half-quantum magnetic flux appears spontaneously. In this figure, the size of d-wave superconductor is $10\xi_{d0} \times 10\xi_{d0}$. The coherence length of high-Tc superconductors is rather small ($\xi_{d0} = 2 \sim 4$ nm for YBCO) and the size of d-wave region is 40 nm, which is rather small compare to the experiments by Hilgencamp et al. [6]. But for such nano-sized d-wave superconductors, half-quantum magnetic fluxes appear, as this result shows. For larger d-wave region, the spontaneous magnetic flux is expected to appear easily.

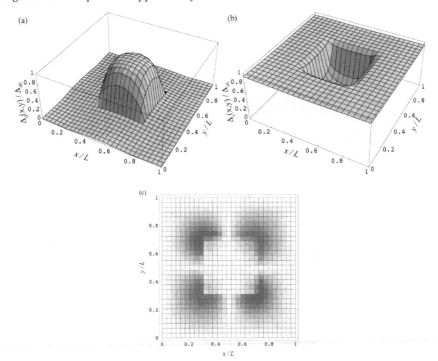

Figure 8. Superconducting order parameter and magnetic field structures for a square d-wave superconductor in an s-wave superconductor. (a) Amplitude of d-wave component of the order parameter. (b) Amplitude of s-wave component of the order parameter. (c) Magnetic field distribution. Blue means $H_z < 0$.

How such spontaneous magnetic flux appears, when a square d-wave superconductor is embedded in an s-wave superconductor? In Fig.8, the typical magnetic flux structure of such square d-dot in which edge of square is parallel to the x- or y-directions, under zero magnetic field is shown. In this figure, red color means $H_z < 0$. Magnetic fluxes appear at four corners of the square d-dot, and they order antiferromagnetically. Such antiferromagnetic order is already obtained by the experiment for zigzag junctions by Hilgencamp et al. [6]. The reason of this antiferromagnetic order can be understood by considering the interaction between vortices. Usual theory of vortex interaction tells us that parallel vortices repel each other and antiparallel vortices attract each other because of the current around the vortices and Lorentz force by this current. For the d-dot in Fig.8, the spontaneous current flows form top-center of the d-wave superconductors, turns right and left and flows into left and right sides of d-wave superconductor. So this current naturally creates upward flux at the left-top corner and downward flux at the right-top corner. Fluxes at the lower two corners can be explained similarly.

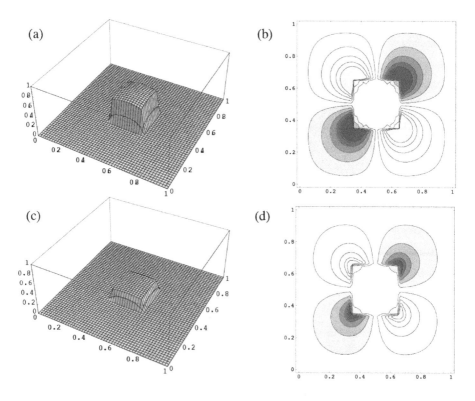

Figure 9. The d-wave superconducting order parameter amplitudes ((a) and (c)) and magnetic field distribution ((b) and (c)) for smaller d-dot with d=1.7 ξ. ((a) and (b)) with d=1.3 ξ. ((c) and (d)).

The second model of d-dots (Eqs.21-23) also describes these spontaneous half-quantum magnetic fluxes. Using this model, we argue the size-dependence of these magnetic fluxes. Question is: The d-dot in Fig.8 have four half-quantum magnetic fluxes, but if the size of d-dot becomes small, then what happens? In Fig.9, the typical size dependences of the order parameter structure and the magnetic fluxes are shown. In these figures, for smaller d-dots the amplitudes of d-wave order parameter and magnetic field distributions are shown. When the size of the d-dot becomes small, then the amplitude of d-wave component of the order parameter becomes small, and the spontaneous magnetic field becomes small. This is because the spontaneous current, which flows around each corner, does not decrease much at the center of the edge of the square when the size of the d-dot is comparable to the coherence length. Then, the total magnetic flux becomes less than the half-flux quantum, although the fluxoid is still a half-flux quantum.

The temperature dependences of amplitude of order parameters and the spontaneous magnetic fluxes also show similar tendency, because in this case, the coherence length increases with increasing temperature.

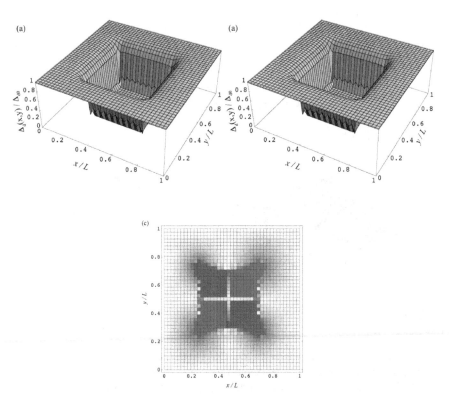

Figure 10. Order parameter ((a) d-wave and (b) s-wave) and magnetic field (c) distribution for an s-dot.

These spontaneous magnetic fluxes do not appears for the geometry for a square d-wave superconductor embedded in an s-wave superconducting matrix, but they appear for a square s-wave superconductor embedded in d-wave superconducting matrix and various shaped d-wave superconductors embedded in the s-wave superconducting matrix. The s-wave superconducting dot case is called "s-dot". For an s-dot, order parameter structures and field distribution are shown in Fig. 10. There appear spontaneous magnetic fluxes around the corners antiferromagnetically. These magnetic fluxes also are explained similarly by the phase anisotropy of d-wave superconductivity. But the distribution of magnetic field is opposite to the d-dot case, where magnetic field appears mainly outside of inner superconducting region, but in this case, magnetic field appears mainly inside of inner superconducting region. The s-dot is also useful but in the following we focus on the d-dot case, which is more easily fabricated, we think.

Next we discuss the shape dependence of the d-dot. Even if the shape of the d-wave superconducting region is different from the square that is parallel to the crystal axis or x- and y-axis, the spontaneous magnetic field is also expected. In Fig. 11, distributions of the order parameters and magnetic field for an equilateral triangle plate are shown. For this case the spontaneous magnetic fluxes appear along the upper edges connected to the top corner. The spontaneous current flows the top corner and return to the intermediate points of upper edges. Also spontaneous magnetic fluxes appear around the lower corners. These spontaneous magnetic field along the edge also appear for rotated or diagonal squares, as shown in Fig. 12. In these figures, the spontaneous currents across the junction mainly flow along x- or y- directions. And direction of junction between d- and s-wave superconductors is important for the appearance of magnetic flux.

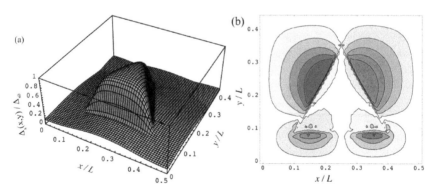

Figure 11. Spatial distribution of d-wave component of the order parameter (a) and magnetic field for a triangular d-dot.

Also we note that not only square d-dots that is parallel to the x- and y-axis but also arbitrary shaped d-dots that show spontaneous magnetic field, such as in Figs. 11 and 12, have doubly degenerate stable states. And the shape of d-dot controls 15 the magnetic field distribution.

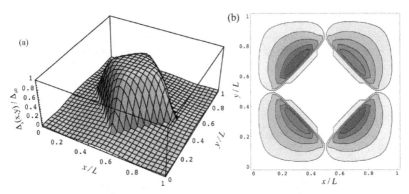

Figure 12. Spatial distribution of d-wave component of the order parameter (a) and magnetic field for a rotated square d-dot.

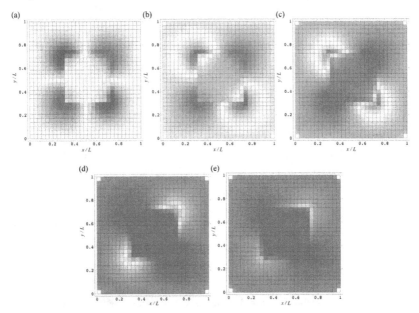

Figure 13. Spatial distribution of magnetic field for a square d-dot. (a) $H = 0$, (b) $H = 0.02\Phi_0/2\pi L^2$, (c) $H = 0.05\Phi_0/2\pi L^2$, (d) $H = 0.1\Phi_0/2\pi L^2$, and (e) $H = 0.2\Phi_0/2\pi L^2$.

These doubly degenerate states have good properties for applications. First they remain under weak external magnetic field. In Fig. 13, external field dependence of spontaneous magnetic field distributions for a square d-dot is shown. The spontaneous magnetic flux parallel (anti-parallel) to the external field becomes large (small), respectively. Although the degeneracy from broken time reversal symmetry is lifted under the external magnetic field, the state with π/2 rotated magnetic field distributions are equally stable. These doubly degenerate states come from the broken four-fold symmetry of the square shape. This property depends on the shape of d-dots. For asymmetric shaped d-dots, one of the doubly degenerate states becomes more stable than another state. This means that we can control these degenerate states using the magnetic field for asymmetric d-dots.

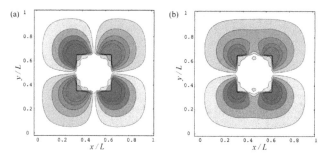

Figure 14. Magnetic field distributions of (a) the most stable state (udud) and (b) an excited state (udud).

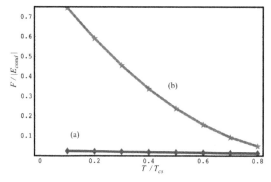

Figure 15. Free energies of (a) the most stable state (udud) and (b) an excited state (uudd). E_{cond} is the condensation energy of the superconductor and T_{cs} is the critical temperature of the s-wave superconductor.

Second property is the stability of the degenerate stable states. For the square d-dots, most stable states show antiferromagnetic order of spontaneous magnetic fluxes, and we call these state udud (up down up down) (Fig. 4 (a)). There are other states, which have higher free energy. In Fig. 14 (b), one of such states is shown. In this state, spontaneous magnetic fluxes do not show antiferromagnetic order, but parallel magnetic fluxes align at the upper or lower edges. We call this state uudd (up up down down). The free energies of the udud and uudd states are shown in Fig. 15. Well below the critical temperature of the s-wave superconductor T_{cs}, free energy difference between udud (a) and uudd (b) states becomes comparable to the condensation energy of the superconductor. Therefore we can treat them as two-level systems.

4. Interaction between d-dots

As shown in previous section, the d-dots have double degenerate stable states. So we can use them as bits or 1/2 spins. In order to use them as artificial spins, the d-dots will be placed periodically or randomly. Then the interaction between them is important for these spin systems. For using the d-dots as computational bits, they are also placed to transform the information. Therefore interaction between d-dots also important for these applications.

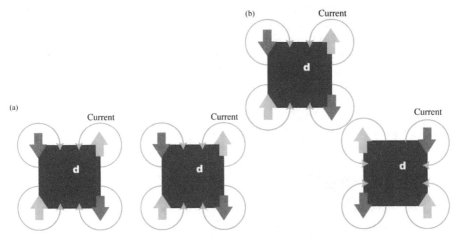

Figure 16. Pairs of d-dots in a parallel (a) or a diagonal (b) positions.

How the d-dots interact with each other? Interaction between d-dots basically comes from the interaction between spontaneous magnetic fluxes or vortices. If the spontaneous vortices are independent from each other, that is, if there is no current flow between the vortices, then via a purely electromagnetic interaction, they interact. This is the case of the π-ring system of Kirtley et al. [4]. If the vortices are interacting in the same superconductors, there is a supercurrent flow around the vortices and ordinary vortices interact with each other through this current. The current distribution around a singly quantized vortex is given by the first order Bessel function. And therefore the interaction force to vortex 1 from vortex 2 is given as,

$$\mathbf{f}_{12} = \frac{(\boldsymbol{\Phi}_0)_1 (\boldsymbol{\Phi}_0)_2}{8\pi^2 \lambda^3} K_1\left(\frac{r_{12}}{\lambda}\right)\hat{\mathbf{r}}_{12} \tag{24}$$

Where r_{12} and $\hat{\mathbf{r}}_{12}$ are distance between two vortices and a unit vector from the vortex 2 to the vortex 1, respectively, and K_1 is the first order modified Bessel function. Directions of vortices are expressed by $(\boldsymbol{\Phi}_0)_i$ and if two vortices are parallel (anti-parallel) then interaction is repulsive (attractive), respectively.

For d-dots, there is an s-wave region between d-wave islands, and the spontaneous currents around the corners affect each other as usual supercurrent around singly quantized vortices, mentioned above. Then, an interaction between d-dots arises through the spontaneous currents in the s-wave region. If two square d-dots are in a line (Fig. 16 (a)), nearest vortex pair should be antiparallel (e.g. right-upper flux in the left d-dot and left upper vortex in the right d-dot should be antiparallel). And then we can expect that two d-dots will have a same spontaneous magnetic flux distribution, as shown in Fig. 16 (a). Therefore when we regard a d-dot as a spin, they interact ferromagnetically. In contrast to this configuration, if two square d-dots are placed diagonally (Fig. 16 (b)), then we expect an antiferromagnetic interaction by the same argument. In Fig. 17, stable states for parallel two d-dots are shown. For short distance ((e)), the nearest magnetic fluxes disappear and the flux distributions of two d-dots are same. Increasing the distance, the nearest magnetic fluxes appear and becomes gradually large ((f)-(h)) and the states of the d-dots are still same.Therefore ferromagnetic states are always stable, as we expected, and this is independent from the distance.

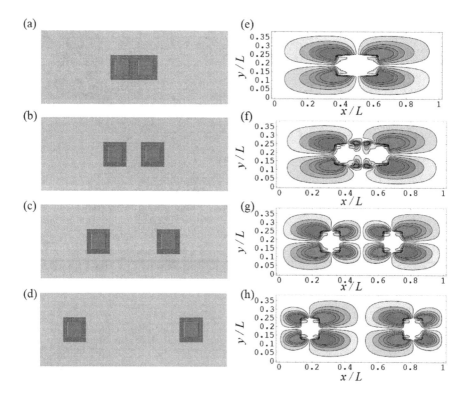

Figure 17. Stable states for parallel two d-dots. (a)-(d) : configurations of d- and s-wave superconductors. (e)-(h): Stable magnetic field distributions.

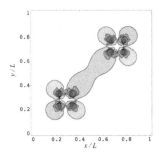

Figure 18. Stable magnetic field distribution around a pair of d-dots, which are placed diagonally. The states of these d-dots are same and therefore they interact ferromagnetically.

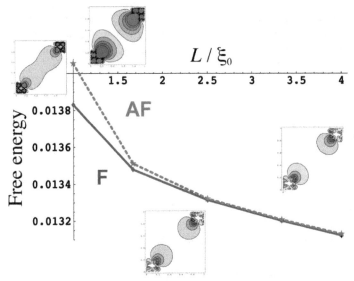

Figure 19. Free energies for Ferromagnetic (F) and Antiferromagnetic states in diagonally placed two d-dots.

However, for diagonally placed two d-dots, the magnetic field distribution is not so simple. In Fig. 18, the stable state is shown for short distance between two diagonally placed d-dots. Unlike to the expectation from the usual vortex-vortex interaction, ferromagnetic state becomes stable. This is because two adjacent half-quantum fluxes are connected and form broad single quantum flux and this can occur when two fluxes are parallel. Therefore the ferromagnetic state becomes stable. It seems that the energy of this configuration becomes lower when two half-quantum fluxes are closer to each other. This property does not appear for ordinary vortices and we call this a fusion of half-quantum vortices. In Fig. 19, distance dependence of the free energies for the ferromagnetic and antiferromagnetic states are plotted. When distance between two d-dots is short, the ferromagnetic state has much lower free energy than that of the antiferromagnetic state. For longer distance, free energies of these states are almost same, because the spontaneous magnetic fluxes become almost independent as shown in the insets of Fig. 19.

These interactions between d-dots can be used for changing the state of the d-dots [26]. Using these interactions, the d-dots have a potential applicability to the superconducting devices[27].

5. Conclusions

We showed that the anisotropic pairing of superconductivity causesinteresting phenomena when two different superconductors are combined. Spontaneous half-quantized magnetic flux around the junction between d-wave and s-wave superconductors is one of such phenomena. It is simulated by the two-components Ginzburg-Landau equations, which can treat the anisotropy of d-wave superconductivity. This half-quantum magnetic state for the d-wave superconductors embedded in the s-wave superconductor has good properties for applications to superconducting devices, especially for classical bits or qubits.

Appendix

In this appendix, the coefficients in Eqs. 13-18 aregiven. For d-wave superconducting regions, they are defined as,

$$P_{ij}^{dd}\left(\{\mathbf{A}\}\right)\equiv\frac{3}{4}\xi_d^2\left(\Delta_d^0\right)^2\left(\sum_{\alpha=x,y}K_{ij}^{\alpha\alpha}+\sum_{\substack{i_1i_2\\\alpha=x,y}}I_{i_1i_2ij}A_{i_1\alpha}A_{i_2\alpha}\right)-\frac{3}{4}\left(\Delta_d^0\right)^2I_{ij} \tag{25}$$

$$P_{ij}^{dd2R}\left(\{\Delta\}\right)\equiv\left(\frac{\lambda_s}{\lambda_d}\right)^2\sum_{i_1i_2}I_{i_1i_2ij}\left(3\,\mathrm{Re}\Delta_{si_1}\,\mathrm{Re}\Delta_{si_2}+\mathrm{Im}\Delta_{si_1}\,\mathrm{Im}\Delta_{si_2}\right)$$
$$+\frac{3}{4}\sum_{i_1i_2}I_{i_1i_2ij}\left(3\,\mathrm{Re}\Delta_{di_1}\,\mathrm{Re}\Delta_{di_2}+\mathrm{Im}\Delta_{di_1}\,\mathrm{Im}\Delta_{di_2}\right) \tag{26}$$

$$P_{ij}^{dd2I}\left(\{\Delta\}\right)\equiv\left(\frac{\lambda_s}{\lambda_d}\right)^2\sum_{i_1i_2}I_{i_1i_2ij}\left(\mathrm{Re}\Delta_{si_1}\,\mathrm{Re}\Delta_{si_2}+3\,\mathrm{Im}\Delta_{si_1}\,\mathrm{Im}\Delta_{si_2}\right)$$
$$+\frac{3}{4}\sum_{i_1i_2}I_{i_1i_2ij}\left(\mathrm{Re}\Delta_{di_1}\,\mathrm{Re}\Delta_{di_2}+3\,\mathrm{Im}\Delta_{di_1}\,\mathrm{Im}\Delta_{di_2}\right) \tag{27}$$

$$Q_{ij}^{dd}\left(\{\mathbf{A}\}\right)\equiv-\frac{3}{4}\xi_d^2\left(\Delta_d^0\right)^2\sum_{\substack{i_1\\\alpha=x,y}}\left(J_{i_1j}^\alpha-J_{j_1i}^\alpha\right)A_{i_1}^\alpha \tag{28}$$

$$Q_{ij}^{dd2}\left(\{\Delta\}\right)\equiv\left(\frac{\lambda_s}{\lambda_d}\right)^2\sum_{i_1i_2}I_{i_1i_2ij}2\,\mathrm{Im}\Delta_{si_1}\,\mathrm{Re}\Delta_{si_2}+\frac{3}{4}\sum_{i_1i_2}I_{i_1i_2ij}2\,\mathrm{Im}\Delta_{di_1}\,\mathrm{Re}\Delta_{di_2} \tag{29}$$

$$P_{ij}^{ds}\left(\{\mathbf{A}\}\right) \equiv \frac{3}{4}\xi_d^2\left(\Delta_d^0\right)^2\frac{\lambda_s}{\lambda_d}\left(K_{ij}^{xx} - K_{ij}^{yy} + \sum_{\substack{i_1 i_2 \\ \alpha=x,y}} I_{i_1 i_2 ij}\left(A_{i_1 x}A_{i_2 x} - A_{i_1 y}A_{i_2 y}\right)\right) \tag{30}$$

$$Q_{ij}^{ds}\left(\{\mathbf{A}\}\right) \equiv \frac{3}{4}\xi_d^2\left(\Delta_d^0\right)^2\frac{\lambda_s}{\lambda_d}\sum_{i_1}\left\{\left(J_{i i_1 j}^x - J_{j i_1 i}^x\right)A_{i_1}^x - \left(J_{i i_1 j}^y - J_{j i_1 i}^y\right)A_{i_1}^y\right\} \tag{31}$$

$$V_i^{dR}\left(\{\Delta\}\right) \equiv \frac{3}{2}\sum_{i_1 i_2 i_3} I_{i_1 i_2 i_3 i}\Delta_{i_1}\Delta_{i_2}^*\,\mathrm{Re}\Delta_{i_3} \tag{32}$$

$$V_i^{dI}\left(\{\Delta\}\right) \equiv \frac{3}{2}\sum_{i_1 i_2 i_3} I_{i_1 i_2 i_3 i}\Delta_{i_1}\Delta_{i_2}^*\,\mathrm{Im}\Delta_{i_3} \tag{33}$$

$$P_{ij}^{ss}\left(\{\mathbf{A}\}\right) \equiv \frac{3}{2}\xi_d^2\left(\Delta_d^0\right)^2\left(\frac{\lambda_s}{\lambda_d}\right)^2\left(\sum_{\alpha=x,y} K_{ij}^{\alpha\alpha} + \sum_{\substack{i_1 i_2 \\ \alpha=x,y}} I_{i_1 i_2 ij}A_{i_1 \alpha}A_{i_2 \alpha}\right) + 2\left(\frac{\lambda_s}{\lambda_d}\right)^4\left(\Delta_s^0\right)^2 I_{ij} \tag{34}$$

$$P_{ij}^{ss2R}\left(\{\Delta\}\right) \equiv \left(\frac{\lambda_s}{\lambda_d}\right)^2\sum_{i_1 i_2} I_{i_1 i_2 ij}\left(3\mathrm{Re}\Delta_{d i_1}\,\mathrm{Re}\Delta_{d i_2} + \mathrm{Im}\Delta_{d i_1}\,\mathrm{Im}\Delta_{d i_2}\right)$$
$$+2\left(\frac{\lambda_s}{\lambda_d}\right)^4\sum_{i_1 i_2} I_{i_1 i_2 ij}\left(3\mathrm{Re}\Delta_{s i_1}\,\mathrm{Re}\Delta_{s i_2} + \mathrm{Im}\Delta_{s i_1}\,\mathrm{Im}\Delta_{s i_2}\right) \tag{35}$$

$$P_{ij}^{ss2I}\left(\{\Delta\}\right) \equiv \left(\frac{\lambda_s}{\lambda_d}\right)^2\sum_{i_1 i_2} I_{i_1 i_2 ij}\left(\mathrm{Re}\Delta_{d i_1}\,\mathrm{Re}\Delta_{d i_2} + 3\mathrm{Im}\Delta_{d i_1}\,\mathrm{Im}\Delta_{d i_2}\right)$$
$$+2\left(\frac{\lambda_s}{\lambda_d}\right)^4\sum_{i_1 i_2} I_{i_1 i_2 ij}\left(\mathrm{Re}\Delta_{s i_1}\,\mathrm{Re}\Delta_{s i_2} + 3\mathrm{Im}\Delta_{s i_1}\,\mathrm{Im}\Delta_{s i_2}\right) \tag{36}$$

$$Q_{ij}^{ss}\left(\{\mathbf{A}\}\right) \equiv -\frac{3}{2}\xi_d^2\left(\Delta_d^0\right)^2\left(\frac{\lambda_s}{\lambda_d}\right)^2\sum_{\substack{i_1 \\ \alpha=x,y}}\left(J_{i i_1 j}^\alpha - J_{j i_1 i}^\alpha\right)A_{i_1}^\alpha \tag{37}$$

$$Q_{ij}^{ss2}\left(\{\Delta\}\right) \equiv \left(\frac{\lambda_s}{\lambda_d}\right)^2\sum_{i_1 i_2} I_{i_1 i_2 ij}\,2\mathrm{Im}\Delta_{d i_1}\,\mathrm{Re}\Delta_{d i_2} + 2\left(\frac{\lambda_s}{\lambda_d}\right)^4\sum_{i_1 i_2} I_{i_1 i_2 ij}\,2\mathrm{Im}\Delta_{s i_1}\,\mathrm{Re}\Delta_{s i_2} \tag{38}$$

$$V_i^{sR}\left(\{\Delta\}\right) \equiv 4\left(\frac{\lambda_s}{\lambda_d}\right)^4\sum_{i_1 i_2 i_3} I_{i_1 i_2 i_3 i}\Delta_{s i_1}\Delta_{s i_2}^*\,\mathrm{Re}\Delta_{s i_3} \tag{39}$$

$$V_i^{sI}\left(\{\Delta\}\right) \equiv 4\left(\frac{\lambda_s}{\lambda_d}\right)^4\sum_{i_1 i_2 i_3} I_{i_1 i_2 i_3 i}\Delta_{s i_1}\Delta_{s i_2}^*\,\mathrm{Im}\Delta_{s i_3} \tag{40}$$

For s-wave superconducting regions, they are defined similarly.

The coefficients in Eqs. 17 and 18 in the d-wave superconducting region are given as,

$$R_{ij}(\{\mathbf{A}\}) \equiv \frac{3}{2} \kappa_d^2 \xi_d^2 \left(\Delta_d^0\right)^2 \left(\sum_{\alpha=x,y} K_{ij}^{\alpha\alpha} \right) + \frac{3}{4} \sum_{i_1 i_2} I_{i_1 i_2 ij} \left[2\,\mathrm{Re}\left(\Delta_{d_{i_1}} \Delta_{d_{i_2}}^*\right) + 4\left(\frac{\lambda_s}{\lambda_d}\right)^2 \mathrm{Re}\left(\Delta_{s_{i_1}} \Delta_{s_{i_2}}^*\right) \right], \qquad (41)$$

$$R_{ij}^2(\{\mathbf{A}\}) \equiv 3\sum_{i_1 i_2} I_{i_1 i_2 ij} \left[\left(\frac{\lambda_s}{\lambda_d}\right) \mathrm{Re}\left(\Delta_{d_{i_1}}^* \Delta_{s_{i_2}}\right) \right], \qquad (42)$$

$$S_{ij} \equiv \frac{3}{2} \kappa_d^2 \xi^2 \left(\Delta_d^0\right)^2 \left(K_{ij}^{xy} - K_{ij}^{yx} \right), \qquad (43)$$

$$T_i^\alpha \equiv \frac{3}{2} \sum_{i_1 i_2} J_{i_1 i_2 i}^\alpha \left[\mathrm{Im}\left(\Delta_{d_{i_1}}^* \Delta_{d_{i_2}}\right) + 2\left(\frac{\lambda_s}{\lambda_d}\right)^2 \mathrm{Re}\left(\Delta_{s_{i_1}}^* \Delta_{s_{i_2}}\right) \right] (\alpha = x, y), \qquad (44)$$

$$T_i^{2\alpha} \equiv \frac{3}{2} \sum_{i_1 i_2} J_{i_1 i_2 i}^\alpha \left(\frac{\lambda_s}{\lambda_d}\right) \left[\mathrm{Im}\left(\Delta_{d_{i_1}}^* \Delta_{s_{i_2}}\right) + \mathrm{Im}\left(\Delta_{s_{i_1}}^* \Delta_{d_{i_2}}\right) \right] (\alpha = x, y), \qquad (45)$$

$$U_i^\alpha \equiv \frac{3}{4} \kappa_d^2 \xi_d^2 \left(\Delta_d^0\right)^2 \frac{2\pi}{\Phi_0} H J_i^\alpha \ (\alpha = x, y). \qquad (46)$$

Author details

Masaru Kato
Department of Mathematical Sciences, Osaka Prefecture University1-1, Gakuencho, Sakai, Osaka, Japan

Takekazu Ishida
Department of Physics and Electronics, Osaka Prefecture University, 1-1, Gakuencho, Sakai, Osaka, Japan

Tomio Koyama
Institute of Material Sciences, Tohoku University, Sendai, Japan

Masahiko Machida
CCSE, Japan Atomic Energy Agency, Tokyo, Japan

Acknowledgement

One of the authors (M. K.) thanks his former and present graduate students, M. Ako, M. Hirayama, S. Nakajima, H. Suematsu, T.Minamino, S. Tomita, Y. Niwa, and D. Fujibayashi for useful discussions. This work is partly supported by "The Faculty Innovation Research Project" of Osaka Prefecture University and it was supported by the CREST-JST project.

6. References

[1] Tinkham M. Introduction to Superconductivity. New York: McGrawa-Hill, inc; 1996.

[2] Sigrist M, Ueda K. Rev. Mod. Phys. 1991; 63(2): 239-311.

[3] Tsuneto T. Superconductivity and Superfluidity. Cambridge: Cambridge University Press; 1998.

[4] Tsuei C C, Kirtley J R. Pairing symmetry in cuprate superconductors. Rev. Mod. Phys. 2000;72(4): 969-1016.

[5] Van Harlingen D J. Phase-sensitive tests of the symmetry of the pairing state in the high-temperature superconductors—Evidence for dx2-y2 symmetry.Rev. Mod. Phys. 1995;67(2): 515-535.

[6] Hilgenkamp A, Ariando, Smilde H-J H, Blank D H A, Rijnders G, Rogalla H, Kirtley J R, Tsuei C C. Ordering and manipulation of the magnetic moments in large-scale superconducting -loop arrays. Nature 2003; 422, 50 -53.
     http://www.nature.com/nature/journal/v422/n6927/full/nature01442.html

[7] Kirtley J R, Tsuei C C, Ariando, Smilde H-J H, Hilgenkamp H. Antiferromagnetic ordering in arrays of superconducting π-rings. Phys. Rev. B 2005; 72 (21): 214521 (1-11).

[8] IoffeL B, Geshkenbein V B, Feigel'man M V, Fauchère A L, Blatter G. Environmentally decoupled sds -wave Josephson junctions for quantum computing. Nature 1999; 398: 679-681.

[9] KatoM, AkoM, MachidaM,Koyama T, IshidaT.Ginzburg–Landau calculations of d-wave superconducting dot in s-wave superconducting matrix. Physica C 2004; 412-414: 352-357.

[10] Kato M, Ako M, Machida M, Koyama T, Ishida T. Structure of magnetic flux in nano-scaled superconductors. J. Mag. Mag. Mat. 2004; 272-276, 171-172.

[11] Ako M, Ishida T, Machida M, Koyama T, Kato M. Vortex state of nano-scaled superconducting complex structures (d-dot).Physica C 2004; 412-414: 544-547.

[12] Ren Y, Xu J-H, Ting C S. Ginzburg-Landau Equations and Vortex Structure of a dx2-y2 Superconductor. Phys. Rev. Lett. 1995;74: 3680-3683.

[13] Xu J H, Ren Y, Ting C S. Ginzburg-Landau equations for a d-wave superconductor with applications to vortex structure and surface problems. Phys. Rev. B 1995; 52:7663-7674.

[14] XuJ H, RenY, Ting C S.Structures of single vortex and vortex lattice in a d-wave superconductor. Phys. Rev. B 1996; 53:R2991-R2994.

[15] Li Q, Wang Z D, Wang Q H. Vortex structure for a d+is-wave superconductor. Phys. Rev. B; 1999; 59:613-618.

[16] Garg A. Ginzhurg-Landau Theory of the Phase Diagram of Superconducting UPt3. Phys. Rev. Lett. 1992; 69: 676-679.

[17] Garg A, Chen D-C. Two-order-parameter theory of the phase diagram of superconducting UPt3. Phys. Rev. B 1994;49: 479-493.

[18] Martisovits V, Zaránd G, Cox D L. Theory of "Ferrisuperconductivity" in U1-xThxBe13. Phys. Rev. Lett. 2000; 84: 5872-5875.

[19] Wang Q H, Wang Z D. Vortex flow in a two-component unconventional superconductor. Phys. Rev. B 1998; 57: 10307-10310.

[20] Agterberg D F. Square vortex lattices for two-component superconducting order parameters. Phys. Rev. 1998; 58; 14484-14489.

[21] Babaev E, Speight M. Semi-Meissner state and neither type-I nor type-II superconductivity in multicomponent superconductors. Phys. Rev. B 2005; 72; 180502(R)1-4.

[22] Dao V H, Chibotaru L F, Nishio T, Moshchalkov V V. Giant vortices, rings of vortices, and reentrantbehavior in type-1.5 superconductors. Phys. Rev. B 2011; 83: 020503 (R) 1-4.

[23] Du Q, Gunzburger M D, Peterson J S. Analysis and approximation of the Ginzburg Landau model of superconductivity. SIAM Rev. 1992;34:54-81.

[24] Du Q, Gunzburger M D, Peterson J S.Solving the Ginzburg-Landau equations by finite-element methods. Phys. Rev. B1992; 46: 9027-9034.

[25] Wang Z D, Wang Q H. Vortex state and dynamics of a d-wave superconductor: Finite-element analysis; Phys. Rev. B 1997; 55: 11756-11765.

[26] Nakajima S, Kato M, Koyama T, Machida M, Ishida T,Nori F. Simulation of logic gate using d-dot's.PhysicaC 2008; 468: 769-772.

[27] Koyama T, Machida M, Kato M, Ishida T. Quantum theory for the Josephson phase dynamics in a d-dot. Physica C 2005; 426-431: 1561-1565.

A Description of the Transport Critical Current Behavior of Polycrystalline Superconductors Under the Applied Magnetic Field

C.A.C. Passos, M. S. Bolzan, M.T.D. Orlando, H. Belich Jr,
J.L. Passamai Jr., J. A. Ferreira and E. V. L. de Mello

Additional information is available at the end of the chapter

1. Introduction

Since their discovery in 1986 the high-T_c superconductors (HTSC) have been employed in several applications.The expectation with the discover of new devices sparked the beginning of an intense research to understand the parameters which control the physical properties of these materials. With the goal to the practical applications, the critical current density (J_c) is one of the crucial parameters that must be optimized for HTSC [1]. Thus the aim of this chapter is to describe he transport critical current behavior of polycrystalline superconductors under the applied magnetic field.

According to Gabovich and Mosieev [2], there is a dependence of the superconducting properties on the macrostructure of ceramic. They studied the $BaPb_{1-x}Bi_xO_3$ metal oxide superconductor properties which are a consequence of the granularity of the ceramic macrostructure and the existence of weak Josephson links between the grains. In this case, the superconductivity depends strongly on the presence of grain boundaries and on the properties of the electronic states at the grain boundaries. This determines the kinetic characteristics of the material. For instance, the temperature dependence of the electrical conductivity of oxide superconductor is related to complex Josephson medium.

Nowadays it is well known that the J_c in polycrystalline superconductors is determined by two factors: the first is related to the defects within the grains (intragrain regions) such as point defects, dislocations, stacking faults, cracks, film thickness, and others [3, 4]. When polycrystalline samples are submitted to magnetic field, the intragranular critical current can be limited by the thermally activated flux flow at high magnetic fields. Secondly the critical current depends on the grain connectivity, that is, intergrain regions. Rosenblatt *et al.* [5] developed an idea to discuss the key concept of granularity and its implications for localization in the normal state and paracoherence in the superconducting state. For arrays formed by niobium grains imbedded in epoxy resin [6] the coherent penetration depth or screening current are influenced by the intergrain regions. In fact, the main obstacles to intergranular critical current flow are weak superconductivity regions between the grains [7], called weak links (WLs) [8]. Ceramic superconductor samples present a random network for the supercurrent path, with the critical current being limited by the weakest links in each path. This Josephson-type mechanism of conduction is responsible to the dependence of the critical current density on the magnetic field $J_c(H)$, as noted in several experimental studies [9–11]. On the other hand, the intragranular critical current is limited by an activated flux flow at high temperature and a high magnetic field [9, 12, 13].

Considering these factors, Altshuler *et al.* [14] and Muller and Matthews [15] introduced the possibility of calculating the $J_c(H)$ characteristic under any magnetic history following the proposal of Peterson and Ekin [16]. Basically the model considers that the transport properties of the junctions are determined by an "effective field" resulting from superposition of a external applied field and the field associated with magnetization of the superconducting grains.

Another theoretical approach to the $J_c(H)$ dependence in a junction took into account the effect of the magnetic field within the grains. This study has revealed that the usual Fraunhofer-like expression for $J_c(H)$ [17, 18] should be written as $J_c(H)$ $\propto$ $sin(bH^{1/2})/(bH^{1/2})$, which we call the modified Fraunhofer-like expression [19]. Mezzetti *et al.* [20] and González *et al.* [21] also proposed models to describe $J_c(H)$ behavior taking into account the latter expression. In both studies the authors concluded that a Gamma-type WL distribution controls the transport critical current density.

González *et al.* considered two different regimes [21]: for low applied magnetic field, a linear decrease in J_c with the field was observed, whereas for high fields $J_c(H) \propto (1/B)^{0.5}$ dependence was found. Here we have decided to follow the same approach and extend the analytical results to all applied magnetic fields.

Usually polycrystalline ceramics samples contain grains of several sizes and the junction length changes from grain to grain. In addition, the granular samples may exhibit electrical, magnetic or other properties which are distinct from those of the material into the grains [5]. The average $J_c(H)$ is obtained by integrating $J_c(H)$ for each junction and taking into account a distribution of junction lengths in the sample. It was demonstrated that the WL width follows a Gamma-type distribution [22]. This function yields positive unilateral values and is always used to represent positive physical quantities. Furthermore, this Gamma distribution is the classical distribution used to describe the microstructure of granular samples [23] and satisfactorily reproduces the grain radius distribution in high-T_c ceramic superconductors [24].

2. Basic properties

A superconductor exhibits two interesting properties: the first one is the electrical resistance of the material abruptly drops to zero at critical temperature T_c. The superconductor is able to carry electrical current without resistance. This phenomenon is related to the perfect diamagnetism. The second feature of superconductivity is also known as Meissner effect. In this case a superconductor expels an external applied magnetic field into its interior.

This struggle between superconductivity and magnetic field penetration select two important behaviors. If a superconductor does not permit any applied magnetic flux, it is known as Type I superconductor. In this case, if the superconducting state is put in the presence of a too high magnetic field, the superconductivity is destroyed when the magnetic field magnitude exceeds the critical value H_c. Other superconductor category is the Type II material in which the magnetic properties are more complex. For this material the superconductor switches from the Meissner state to a state of partial magnetic flux penetration. The penetration of magnetic flux starts at a lower field H_{c1} to reach at an upper a higher field H_{c2}.

In addition to the two limiting parameters T_c and H_c, the superconductivity is also broken down when the material carries an electrical current density that exceeds the critical current density J_c. In the Ginzburg-Landau theory, the superconducting critical current density can be written as

$$J_c = \left(\frac{2}{3}\right)^{2/3} \frac{H_c}{\lambda}. \tag{1}$$

The current density given by Eq. (1) is sometimes called the *Ginzburg-Landau depairing current density*.

Once into the superconductor state, it is possible to cross the superconductor surface changing only the current. In this case, even for $T < T_c$ and $H < H_c$ with the material reaching its normal state, and with loss of its superconductor properties.

3. Josephson-type mechanism

Following the discovery of the electron tunneling (barrier penetration) in semiconductor, Giaever [25] showed that electron can tunnel between two superconductors. Subsequently, Josephson predicted that the Cooper pairs should be able to tunnel through the insulator from one superconductor to the other even zero voltage difference such the supercurrent is given by [26]

$$J = J_c \sin(\theta_1 - \theta_2) \tag{2}$$

where J_c is the maximum current in which the junction can support, and θ_i ($i = 1, 2$) is the phase of wave function in *ith* superconductor at the tunnel junctions. This effect takes in account dc current flux in absence of applied electric and magnetic fields, called as the *dc Josephson effect*.

If a constant nonzero voltage V is maintained across the Josephson junction (barrier or weak link), an ac supercurrent will flow through the barrier produced by the single electrons tunneling. The frequency of the ac supercurrent is $\nu = 2eV/\hbar$. The oscillating current of Cooper pairs is known as the *ac Josephson effect*. These Josephson effects play a special role in superconducting applications.

It was mentioned that the behavior of a superconductor is sensitive to a magnetic field, so that the Josephson junction is also dependent. Therefore another mode of pair tunneling is a tunneling current with an oscillatory dependence on the applied magnetic flux $\sin(\pi\Phi/\Phi_0)$, where Φ_0 is the quantum of magnetic flux. This phenomenon is known as *macroscopic quantum interference effect*.

3.1. Basic equations of Josephson effect

As mentioned, the Josephson effect can occur between two superconductors weakly connected. Some types of linkage are possible such as sketched in Figure (1). These configurations depends on the application types in which the weak link can be:

1. an insulating, corresponding to a SIS junction, in which case the insulate layer can be in order of 10 - 20 Å;

2. a normal metal, corresponding to a SNS junction of typical dimensions 102 - 104 Å;

3. a very fine superconducting point presses on a flat superconductor;

4. a narrow constriction (microbridge) of typical dimensions like the coherence length 1 μm.

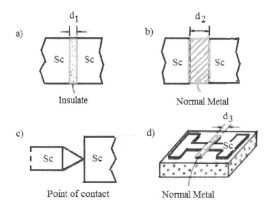

Figure 1. Four types of Josephson junctions: (a) SIS with d_1 = 10 - 20 Å, (b) SNS where d_2 = 102 - 104 Å, (c) Point of contact, and (d) microbridge with $d_3 \approx 1\mu$m [27].

Consider that two superconductors are separated from each other by an insulating layer. The junction is of thickness d normal to the y-axis with cross-sectional dimensions a and c along x and z, respectively. A voltage is applied between the superconductors and the junction is thick enough so that one assumes the potential to be zero in the middle of the barrier. Figure (2) displays Josephson junctions corresponding to a SIS junction.

In Feynman approach Ψ_1 and Ψ_2 are the quantum mechanical wavefunction of the superconducting state in the left and the right superconductor, respectively. This system is determined by coupled time-dependent Schroedinger equations:

$$i\hbar\frac{\partial\Psi_1}{\partial t} = -2eV_1\Psi_1 + K\Psi_2$$

$$i\hbar\frac{\partial\Psi_2}{\partial t} = -2eV_2\Psi_1 + K\Psi_1, \tag{3}$$

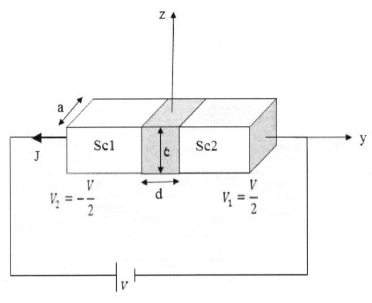

Figure 2. Application of current through the Josephson junctions [28].

where K is a coupling constant for the wave functions across the barrier. The functions Ψ_1 and Ψ_1 may be written as

$$\Psi_1 = (n_{s1})^{1/2} \exp(i\theta_1)$$

$$\Psi_2 = (n_{s2})^{1/2} \exp(i\theta_2), \tag{4}$$

where n_{s1} and n_{s2} are their superelectrons densities. And substituting equation (4) into equation (3) it obtains

$$\hbar\frac{\partial n_{s1}}{\partial t} = 2K\sqrt{n_{s1}n_{s2}}\sin(\theta_2 - \theta_1)$$

$$\hbar\frac{\partial n_{s2}}{\partial t} = 2K\sqrt{n_{s1}n_{s2}}\sin(\theta_2 - \theta_1) \tag{5}$$

Taking the time derivative of the Cooper pair density as being the supercurrent density can be written as

$$J = e\frac{\partial}{\partial t}(n_{s1} - n_{s2}),$$

such that the result is

$$J = J_c \sin(\delta),$$

where $J_c = \frac{4eK(n_{s1}n_{s2})^{1/2}}{\hbar}$ and $\delta = \theta_1 - \theta_2$. We interpret these results as describing a charge transport, that is, a phase difference δ between either side of a superconductor junction causes a dc current tunnel for this simple case.

The behavior of J_c can also be analyzed the Ambegaokar and Baratoff theory [29]. Ambegaokar and Baratoff generalized the Josephson tunnel theory and derived the tunnelling supercurrent on the basis of the BCS theory for a s-wave homogeneous superconductor. In this approach, the temperature dependence of critical current is given with the following expression:

$$J_c = \frac{\pi}{2eR_N S}\Delta(T) tanh\left[\frac{\Delta(T)}{2k_B T}\right]$$ (6)

when T near T_c, $\Delta(T) \simeq 1.74\Delta_0(1 - T/T_c)^{1/2}$ is the superconducting gap parameter from the BCS theory. R_N is the normal-state resistance of the junction, S is the cross section of a junction, and e and k_B are electron charge and Boltzmann constant, respectively. For temperature relatively close to T_c, we can suppose the condition $\Delta(T) \ll k_B T$ and the $tanh[\Delta(T)/2k_B T] \approx \Delta(T)/2k_B T$. Taking this into account, Eq. (8) is transformed into [21]

$$J_c \approx \frac{\pi}{4eR_N S}\Delta_0^2\left[1 - \frac{T}{T_c}\right].$$ (7)

And in limiting $T \to 0$,

$$J_c \approx \frac{\pi}{4eR_N S}\frac{\Delta_0}{e}.$$ (8)

To calculate the Josephson coupling energy for cuprate superconductors that have a d-wave order parameter with nodes, we recall the work of Bruder and co-workers [30]. They have found that the tunneling current behaves in a similar fashion of s-wave superconductors junction and the leading behavior is determined by tunneling from a gap node in one side of a junction into the effective gap in the other side. Consequently, as a first approximation to the Josephson coupling energy EJ , we describe the theory of s-wave granular superconductors [29] to an average order parameter Δ in the grains.

In 1974, Rosenblatt [31] also analysed the tunnelling supercurrent through Josephson barriers, but in bulk granular superconductors (BGS). He proposed that the superconducting order parameter of an assembly of superconducting grains in the absence of applied current can be represented by a set of vectors in the complex plane $\Delta_\alpha = |\Delta| \exp(i\phi_\alpha)$, where ϕ_α is the superconducting phase in αth grain. He showed the arrays of Josephson junctions become superconducting in two stages. At the bulk transition temperature T_0, the magnitude of the order parameter of each grain becomes nonzero [32]. He considered two neighboring grains along an axis with complex superconducting order parameters Δ_1 and Δ_2. Therefore, the Josephson junction can be modelled by [33]

$$H_t = -\sum_{\langle ij \rangle} J_{ij}S_i^+ S_j^-$$ (9)

where H_t is pair tunnelling Hamilton, $S_{i;j}^\pm$ is destruction and creation operators,

$$J_{ij} = \frac{R_c}{2R_{ij}}\Delta(T) tanh\left[\frac{\Delta(T)}{2k_B T}\right],$$ (10)

the Josephson coupling energy between grains i and j, $R_c = \pi\hbar/2e^2$ and R_{ij} is normal state resistance of the junctions between grains i and j.

Until now it was discussed Josephson junctions independent of magnetic field. However the Josephson contacts exhibit macroscopy quantum effects under magnetic field. In order to examine the effect of applying a magnetic field into the junctions, considering the Josephson junctions as sketched in Figure (3) with a magnetic field $B_0\vec{k}$ applied along the vertical z direction,

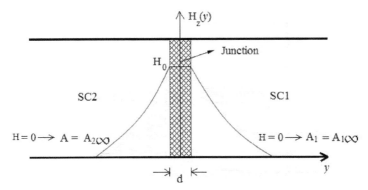

Figure 3. Behavior of the magnetic field in a Josephson junction. A_1 is potential in the superconductor 1 and A_2 is potential in the superconductor 2 [28].

It is assumed because of symmetry, the magnetic field H(y) has no x- or z-direction dependence, but it varies in y-direction insofar the field penetrates into superconductor.

$$\vec{H} = H_z(y)\vec{k}.$$

It is known that magnetic field is derived from potential vector $\vec{H} = \nabla \times \vec{A}$, such that

$$\vec{A} = A_x(y)\vec{i}$$

with $|y| = \frac{d}{2}$. Inside the barrier the material is not superconducting and $H_z = H_0$. Now it must choose an integration contour as shown in Figure (4)

Consider the equation that relates the gradient of the phase of the wave function of the superconducting state with the magnetic vector potential integration of a closed path where the current is zero.

$$\oint \vec{\nabla}\Theta(r)d\vec{l} = \frac{2\pi}{\Phi_0} \oint \vec{A} \cdot d\vec{l}. \tag{11}$$

For the integration path ABCD

$$\int_A^B \vec{\nabla}\Theta(r)d\vec{l} + \int_B^C \vec{\nabla}\Theta(r)d\vec{l} + \int_C^D \vec{\nabla}\Theta(r)d\vec{l} = \frac{2\pi}{\Phi_0}\left(\int_A^B \vec{A} \cdot d\vec{l} + \int_B^C \vec{A} \cdot d\vec{l} + \int_C^D \vec{A} \cdot d\vec{l}\right). \tag{12}$$

The integrals in AB and CD are zero due to orthogonality of the vectors $\vec{A} \cdot d\vec{l}$.

$$\Theta_1 - \Theta_{10} = \frac{2\pi}{\Phi_0}A_{1\infty}(x - x_0). \tag{13}$$

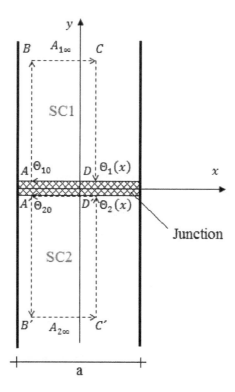

Figure 4. Path of integration around the Josephson junction. One superconductor SC1 is to the above of the insulating barrier and the other superconductor SC2. It was considered a weak-link tunnel short junction. [28]

To $\Theta_2(x)$ the integration result is similar, but it must be to consider the A'B'C'D' path

$$\Theta_2 - \Theta_{20} = -\frac{2\pi}{\Phi_0} A_{2\infty}(x - x_0). \tag{14}$$

Since the aim is to obtain the phase difference $\delta(x)$,

$$\delta(x) = \Theta_2(x) - \Theta_1(x) = \delta_0 + \frac{2\pi}{\Phi_0}(A_{1\infty} + A_{2\infty})x, \tag{15}$$

where $\delta_0 = \Theta_{20} - \Theta_{10}$ is phase difference at x_0. The total magnetic flux is given by

$$\Phi = \int \vec{A} \cdot d\vec{l} = \int \vec{H} \cdot dS\vec{n},$$

and then

$$\Phi = a(A_{1\infty} + A_{2\infty}).$$

Thus the phase difference is given by

$$\delta(x) = \delta_0 + \frac{2\pi\Phi}{\Phi_0}\left(\frac{x}{a}\right). \tag{16}$$

Inserting equation (16) into equation (8) yields after integration over area $S = ac$ cross section of the barrier, the tunneling current is

$$I = J_c \int \sin(\delta_0)dxdz = cJ_c \int_{-a/2}^{+a/2} \sin(\delta_0 + \frac{2\pi\Phi}{\Phi_0})dx. \tag{17}$$

Taking $u = \delta_0 + \frac{2\pi\Phi}{\Phi_0}$, the tunneling current becomes

$$I = acJ_c \frac{\Phi_0}{\pi\Phi}\sin(\delta_0)\sin(\frac{\pi\Phi}{\Phi_0}) \tag{18}$$

When is this current maximum? The answer is for phase difference $\delta_0 = \frac{\pi}{2}$. Hence,

$$I_{max} = I_c \left|\frac{\sin(\pi\Phi/\Phi_0)}{(\pi\Phi/\Phi_0)}\right|, \tag{19}$$

where $I_c = acJ_c$ is the critical current. This is named the *Josephson junction diffraction equation* and shows a Fraunhofer-like dependence of the magnetic field as is displayed in Figure (5)

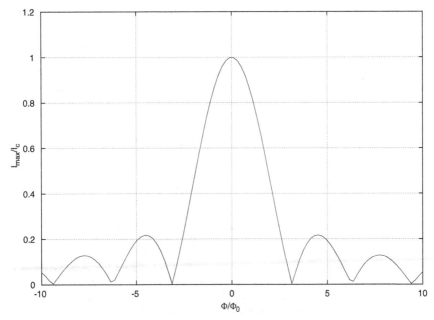

Figure 5. Josephson Fraunhofer diffraction pattern dependence of magnetic field [34].

4. Critical current model

It is well known which ceramic superconductor samples present a random network for the supercurrent path, with the critical current being limited by the weakest links in each path. Moreover, magneto-optical studies have demonstrated that the magnetic field first penetrates grains associated with these regions, even for very low values of H. Consequently, it would be interesting to estimate the influence of the magnetic field on the overall J_c of a sample taking into account the previous remarks. There are some general hypotheses about transport properties in polycrystalline ceramic superconductors on application of a magnetic field. (i) The electric current percolates through the material and heating begins to occur in WLs and in channels between them. This means that the critical current measured in the laboratory is an intergranular current. (ii) The junction widths among grains are less than the Josephson length, and the magnetic field penetrates uniformly into the junctions. (iii) The sample temperature during transport measurement must be close to the critical temperature. Under these conditions, the junction widths are less than the bulk coherence length and the Cooper-pairs current is given by Josephson tunneling. (iv) Near the critical temperature the magnetic field first penetrates WLs and, at practically the same time, the grains.

Normally polycrystalline ceramics samples contain grains of several sizes and the junction length changes from grain to grain. The average $J_c(H)$ is obtained by integrating $J_c(H)$ for each junction and taking into account a distribution of junction lengths in the sample. This function yields positive unilateral values and is always used to represent positive physical quantities. Furthermore, this Gamma distribution is the classical distribution used to describe the microstructure of granular samples [23] and reproduces the grain radius distribution in high-T_c ceramic superconductors [24].

Following the previous discussion, we can describe $J_c(H)$ as a statistical average of the critical current density through a grain boundary. In the same way as Mezzetti *et al.* [20] and González *et al.* [21], we consider that the weak-link width fits a Gamma-type distribution [35]. For a magnetic field higher or lower than the first critical field, the usual Fraunhofer diffraction pattern or the modified pattern is used to describe $J_c(H)$ for each grain boundary. Thus,

$$J_c(H) = J_{c0} \int_{-\infty}^{+\infty} P(u) \left| \frac{sin(\pi u / u_0)}{\pi u / u_0} \right| du \tag{20}$$

$$P(u) = \begin{cases} \frac{u^{m-1}e^{(-u/\eta)}}{\eta^m \Gamma(m)} & u \geq 0 \\ 0 & u < 0, \end{cases}$$

where $\Gamma(m)$ is the Gamma function which is widely tabulated [22, 36]:

$$\Gamma(m) = \int_0^\infty w^{m-1}e^{-w} \, dw$$

when m is a real number. Or $\Gamma(m) = (m-1)!$ if m is a positive integer. The parameters m and η, both positive integer, determine the distribution form and scale (width and height), respectively [21]. The variable u represents the WL length. The quantity u_0 is defined as $u_0 = \phi_0 / \Lambda_0 H$, where ϕ_0 is the quantum flux and Λ_0 is the effective thickness of the WL.

Following the same González *et al.* [21] procedure, we have:

$$J_c(\alpha) = J_{c0} \frac{\alpha^m (-1)^m}{(m-1)!} \frac{\partial^{m-2}}{\partial \alpha^{m-2}} \left[\frac{coth(\alpha/2)}{\alpha^2 + \pi^2} \right], \tag{21}$$

where the variable α is defined as $\alpha = u_0/\eta = \phi_0/(\eta \Lambda_0 H)$. To obtain a simpler expression for the transport critical current density, we develop Eq. (21) in the ranges $0 < \alpha < \pi/2$ and $\alpha \geq \pi/2$ [37].

4.1. Expression for $J_c(H)$ for $0 < \alpha < \pi/2$

The function $F(z) = [coth(z/2)] = (z^2 + \pi^2)$ has singular points at $z = \pm i\pi$, but is analytical at all remaining points on the disc $|z| = \pi$. Thus, we expanded $F(z)$ for the disc $|z| < \pi$.

The hyperbolic cotangent has the expansion [36]

$$z \, coth(z/2) = 2 \left[\sum_{n=0}^{\infty} \frac{b_{2n}}{2n!} z^{2n} \right] \qquad |z| < \pi, \tag{22}$$

where b_{2n} are the Bernoulli numbers ($b_0 = 1, b_2 = 1/6, b_4 = -1/30, b_6 = 1/42, \ldots$) given by [36]:

$$b_{2n} = [(-1)^{n-1} 2(2n)!] / [(2\pi)^{2n}] \zeta(2n),$$

where $\zeta(2n)$ is the Zeta Riemann function. The function $[1/(z^2 + \pi^2)]$ is represented by the Taylor series around zero:

$$\frac{1}{z^2 + \pi^2} = \frac{1}{\pi^2} \sum_{j=0}^{\infty} (-1)^j \frac{z^{2j}}{\pi^{2j}}. \tag{23}$$

Now we can compute the Cauchy product of the series (22) and (23) to obtain:

$$\frac{z \, coth(z/2)}{z^2 + \pi^2} = \frac{2}{\pi^2} \left[1 + \sum_{n=1}^{\infty} \left(\sum_{j=0}^{\infty} \frac{(-1)^{n+j} b_{2j}}{\pi^{2n-2j}(2j)!} \right) z^{2n} \right]. \tag{24}$$

It is convenient to define

$$\beta_n = \left[\sum_{j=0}^{\infty} \frac{(-1)^{n+j} b_{2j}}{\pi^{2n-2j}(2j)!} \right] = \frac{2}{\pi^{2n}} \left[\sum_{j=0}^{\infty} \frac{(-1)^{n-1}}{(2)^{2j}} \zeta(2j) \right]. \tag{25}$$

Thus, we can rewrite Eq. (24) as:

$$\frac{z \, coth(z/2)}{z^2 + \pi^2} = \frac{2}{\pi^2} \left[\frac{1}{z} + \sum_{n=1}^{\infty} \beta_n z^{2n-1} \right] \qquad z \neq 0. \tag{26}$$

The critical current density $J_c(\alpha)$ is calculated by taking $z = \alpha$ in Eq. (26) and differentiating it $(m-2)$ times, term by term. This yields:

$$J_c(\alpha) = \frac{2J_{c0}}{\pi^2(m-1)} \alpha \left[1 + (-1)^m \sum_{n_0}^{\infty} \binom{2n-1}{m-2} \beta_n \alpha^{2n} \right]$$

$$(0 < \alpha < \pi/2), \tag{27}$$

where n_0 is the lower integer and $n_0 \geq (m-1)/2$. This expression (27) for J_c is valid for the range $0 < \alpha < \pi$. However, for more efficient calculation, we suggest that it is only used for the range $0 < \alpha < \pi/2$.

Finally, we can express J_c as a function of H by substituting the definition of α, $\alpha = u_0/\eta = \phi_0/(\eta \Lambda_0 H)$ in Eq. (27). Thus [37],

$$
J_c(H) = \frac{2 J_{c0}}{\pi^2 (m-1)} 1.02 \sqrt{\frac{H_0^*}{H}} \Big[1 +
$$

$$
(-1)^m \sum_{n_0}^{\infty} (1.02)^{2n} (m)^{2n} \binom{2n-1}{m-2} \beta_n \left(\frac{H_0^*}{H} \right)^{2n} \Big], \tag{28}
$$

where $H_0^* = \phi_0(\langle m \eta \rangle \Lambda_0) = \phi_0/\bar{u}\Lambda_0 = \phi_0/A$ is the effective magnetic field characteristic for each polycrystalline superconductor, and A is the area perpendicular to the magnetic field direction. It is important to emphasize that the first term of Eq. (28) was determined by González et al. [21] for high magnetic fields ($\alpha \ll 1$).

The estimate errors for the series in Eq. (28) were calculated as [37]:

$$
E_N = \frac{5\pi^2}{12} \Big[\frac{4}{3} \left(\frac{2N-1}{m-2} \right) + \left(1 + \frac{(-1)^m}{3^{m-1}} \right) \Big] (1/2)^{2N}, \tag{29}
$$

where E_N is defined as the N-order error of the series in Eq. (28).

4.2. Expression for $J_c(H)$ for $\alpha \geq \pi/2$

The hyperbolic cotangent in Eq. (21) can also be written as:

$$
\coth(z/2) = 1 + \left(\frac{2e^{-z}}{1 - e^{-z}} \right) = 1 + 2 \sum_{k=1}^{\infty} e^{-kz}
$$

$$
(\text{Re } z > 0), \tag{30}
$$

since the series in (30) is a Dirichlet type and is convergent at all semi-planes $\text{Re} z > 0$. Dividing Eq. (30) by $(z^2 + \pi^2)$, we obtain:

$$
\frac{\coth(z/2)}{z^2 + \pi^2} = \frac{1}{z^2 + \pi^2} + \frac{2}{z^2 + \pi^2} \sum_{k=1}^{\infty} e^{-kz}
$$

$$
(\text{Re } z > 0). \tag{31}
$$

The expression for the critical current density for $\alpha \geq \pi/2$ is obtained by differentiation of Eq. (31) in Eq. (21) $(m-2)$ times. Thus,

$$
J_c(z) = \frac{J_{c0} z^m (-1)^m}{(m-1)!} \Big[D^{m-2} \left(\frac{1}{z^2 + \pi^2} \right) + D^{m-2} \left(\frac{2}{z^2 + \pi^2} \sum_{k=1}^{\infty} e^{-kz} \right) \Big],
$$

$$
J_c(z) = \frac{J_{c0} z^m (-1)^m}{(m-1)!} \Big[f_{m-2}(z) + R(z) \Big], \tag{32}
$$

A Description of the Transport Critical Current Behavior of Polycrystalline Superconductors Under the
Applied Magnetic Field

181

where

$$f_{m-2}(z) = D^{m-2}\left(1/(z^2 + \pi^2)\right)$$

and

$$R(z) = D^{m-2}[2/(z^2 + \pi^2 \sum_{k=1}^{\infty} e^{-kz})].$$

For $z = \alpha$

$$R(\alpha) = 2\sum_{k=1}^{\infty}\left[\sum_{p=0}^{m-2}(-1)^{m-p}\binom{m-2}{p}\times\right.$$

$$\left. f_p(\alpha)k^{m-p-2}\right]e^{-k\alpha} \tag{33}$$

and $f_p(\alpha)$ can be written as

$$f_p(p) = \frac{(-1)^p p!}{(\alpha^2 + \pi^2)^{p+1}}\left[\frac{(p+1)!}{p!}\alpha^p - \frac{(p+1)!}{(p-2)!3!}\pi^2\alpha^{p-2} + \right.$$

$$\left. \frac{(p+1)!}{(p-4)!5!}\pi^4\alpha^{p-4} - \cdots\right], \tag{34}$$

where p is greater than or equal to zero. It is advantageous to write $f_p(\alpha)$ in this form because it is finite for all $\alpha > 0$. To obtain a expression for $J_c(H)$ from Eq. (33), we express $f_p(\alpha)$ and $R(\alpha)$ as a function of H. Thus,

$$f_p(H) = \frac{(-1)^p p!}{m^{2p+2}\left[\left(\frac{H_0^*}{H}\right)^2 + \frac{\pi^2}{m^2}\right]^{p+1}}\left[a_{p0}\left(\frac{H_0^*}{H}\right)^p - \right.$$

$$\left. a_{p1}\left(\frac{H_0^*}{H}\right)^{p-2} + a_{p2}\left(\frac{H_0^*}{H}\right)^{p-4} - \cdots\right], \tag{35}$$

where $a_{p0} = \frac{(p+1)!}{p!}m^p$, $a_{p1} = \frac{(p+1)!}{(p-2)!3!}\pi^2 m^{p-2}$, $a_{p2} = \frac{(p+1)!}{(p-2)!5!}\pi^4 m^{p-4}$, and so on. It is worth commenting again that $f_p(H)$ is finite for all values of H and is defined for $p > 0$. $R(H)$ is written as

$$R(H) = 2\sum_{k=1}^{\infty}\left[\sum_{p=0}^{m-2}(-1)^{m-p}\binom{m-2}{p}\times\right.$$

$$\left. f_p(H)k^{m-p-2}\right]e^{-km\left(\frac{H_0^*}{H}\right)}. \tag{36}$$

Now Eq. (16) can be expressed as [37]:

$$J_c(H) = \frac{J_{c0}(1)^m}{(m-1)!} \left(\frac{H_0^*}{H}\right)^m \left[f_{m-2}H + R(H)\right]. \tag{37}$$

The series error estimated in $R(\alpha)$ [Eq. (34)] is [37]:

$$E_{K+1}(\alpha) \le \frac{e(m-2)!}{\pi} \left[\sum_{p=0}^{m-2} \frac{1}{(\alpha^2 + \pi^2)^{(m-p-2)/2}} \times \right.$$

$$\left. \sum_{l=0}^{p} \frac{K^{p-l}}{(p-l)!\, \alpha^{l+1}}\right] e^{-k\alpha}, \tag{38}$$

where E_{K+1} is defined as the $K+1$-order error of the series in Eq. (34).

A low applied magnetic field implies that $\alpha \gg 1$, and Eq. (37) is transformed to:

$$J_c(H) \approx J_c(0)\left(1 - \frac{\pi^2(m+1)}{6m}\frac{H}{H_0^*}\right), \tag{39}$$

where $H_0^* = \frac{\phi_0}{\langle \eta m \rangle^2} = \frac{\phi_0}{\bar{u}^2}$ is a characteristic field that determines the behavior of $J_c(H)$ in this region. In addition, $\bar{u}$ represents the mean of the width distribution function $P(u)$ involved in the transport of Cooper pairs through the sample. Eq. (39) reproduces the quasi-linear behavior that was also reported by Gonzalez et al. [21].

5. Critical current measurement

Typical J_c measurements are performed using the four-probe technique with automatic control of the sample temperature, the applied magnetic field and the bias current [14]. Details of the technique and the experimental setup are in Ref. [14] and of the synthesis and sample characterization were published elsewhere [13].

Figure (6) shows the experimental results of [38] for the critical current as a function of the applied field, together with the theoretical expression derived above for the critical field H_0^* in the figure and for $m = 2$ and 3. A very close fit to the experimental data is evident for $m = 2$ for many different applied magnetic fields.

Theoretical models of the magnetic field dependence of the transport critical current density for a polycrystalline ceramic superconductor have been studied at last years [37, 39–41]. Here we have described a tunneling critical current between grains follows a Fraunhofer diffraction pattern or a modified pattern . It is important to emphasize that we followed the same approach as in [21] and extended the analytical results to all applied magnetic fields. A characteristic field (H^*) was identified and different regimes were considered, leading to analytical expressions for $J_c(H)$: (i) analysis for low applied magnetic fields $(\alpha \gg 1)$ revealed quasi-linear behavior for $J_c(H)$ vs. H^*; (ii) for high applied magnetic fields $(\alpha \ll 1)$, $J_c(H)$ is proportional to $H^{-0.5}$, as reported in [21].

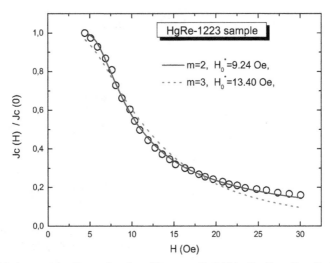

Figure 6. Critical current density as a function of the magnetic field for the $Hg_{0.80}Re_{0.20}Ba_2Ca_2Cu_3O_8$ sample ($T_c = 132$ K). The measurement was carried out at 125 K. The solid line and dot line are the theoretical fits.

Author details

C. A. C. Passos, M. S. Bolzan, M. T. D. Orlando, H. Belich Jr, J. L. Passamai Jr. and J. A. Ferreira
Physics Department, University Federal of Espirito Santo - Brazil

E. V. L. de Mello, *Physics Department, University Federal Fluminense - Brazil*

6. References

[1] Chen, M., Donzel, L., Lakner, M., Paul, W. (2004). High temperature superconductors for power applications, *J. Eur. Ceram. Soc.*, Vol. 24 (No. 6) 1815–1822.

[2] Gabovich, A. M. and Moiseev, D. P. (1986). Metal oxide superconductor BaPb1-xBixO3: unusual properties and new applications. *Sov. Phys. Usp.*, Vol. 29 (No. 12), 1135–1150.

[3] Babcock S. E. & Vargas J. L. (1995). The nature of grain-boundaries in the high-Tc superconductors, *Annual Review of Materials Science*, Vol. 25 (No. 1) 193–222.

[4] Polat, O., Sinclair, J. W., Zuev, Y. L., Thompson, J. R., Christen, D. K., Cook, S. W., Kumar, D., Yimin Chen, and Selvamanickam, V., (2011). Thickness dependence of magnetic relaxation and E-J characteristics in superconducting (Gd-Y)-Ba-Cu-O films with strong vortex pinning. *Phys. Rev. B*, Vol. 84 (No. 2), 024519-1 – 024519-13.

[5] Rosenblatt, J., Peyral, P., Raboutou, A., and Lebeau C. (1988). Coherence in 3D networks: Application to high-Tc supeconductors. *Physica B*, Vol. 152 (No. 1-2), 95–99.

[6] Raboutou, A., Rosenblatt, and J., Peyral, P., (1980). Coherence and disorder in arrays of point contacts. *Phys. Rev. Lett.*, Vol. 45 (No. 12), 1035–1039.

[7] Matsushita, T., Otabe, E.S., Fukunaga, T., Kuga, K., Yamafuji, K., Kimura, K., Hashimoto, M., (1993). Weak link property in superconducting Y-Ba-Cu-O prepared by QMG process, *IEEE Trans. Appl. Supercond.*, Vol. 3 (No. 1) 1045–1048.

[8] Polyanskii, A., Feldmann, D.M., Patnaik, S., Jiang, J., Cai, X., Larbalestier, D., DeMoranville, K., Yu, D., Parrella, R. (2001). Examination of current limiting mechanisms in monocore Bi2Sr2Ca2Cu3Ox tape with high critical current density, *IEEE Trans. Appl. Supercond.*, Vol. 11 (No. 1) 3269–3272.

[9] Ben Azzouz, F., Zouaoui, M., Mellekh, A., Annabi, M., Van Tendeloo, G., Ben Salem, M. (2007). Flux pinning by Al-based nanoparticles embedded in YBCO: A transmission electron microscopic study, *Physica C*, Vol. 455 (No. 1-2) 19–24.

[10] Ekin, J. W., Braginski, A. I., Panson, A. J., Janocko, M. A., Capone, D. W., Zaluzec, N. J., Flandermeyer, B., de Lima, O. F., Hong, M., Kwo, J., Liou, S. H. (1987). Evidence for weak link and anisotropy limitations on the transport critical current in bulk polycrystalline Y1Ba2Cu3Ox, *J. Appl. Phys*, Vol. 62 (No. 12) 4821–4828.

[11] Dupart, J. M. , Rosenblatt, J. and Baixeras J. (1977).Supercurrents and dynamic resistivities in periodic arrays of superconducting-normal contacts. *Phys. Rev. B*, Vol. 16 (No. 11), 4815–4825.

[12] Matsumoto, Y., Higuchi, K., Nishida, A., Akune, T., Sakamoto, N. (2004). Temperature dependence of critical current density and flux creep of Hg(Re)-1223 powdered specimens, *Physica C*, Vol. 412 435–439.

[13] Passos, C.A.C., Orlando, M.T.D., Fernandes, A.A.R., Oliveira, F.D.C., Simonetti, D.S.L., Fardin, J.F., Belich Jr. H., Ferreira Jr. M.M. (2005). Effects of oxygen content on the pinning energy and critical current in the granular (Hg, Re)-1223 superconductors, *Physica C*, Vol. 419 (No 1-2) 25–31.

[14] Altshuler, E., Cobas, R., Batista-Leyva, A. J., Noda,C., Flores, L. E., Martínez, C., Orlando, M. T. D. (1999). Relaxation of the transport critical current in high-Tc polycrystals, *Phys. Rev. B*, Vol. 60 (No. 5) 3673–3679.

[15] Muller, K.H. and Matthews, D.N. (1993). A model for the hysteretic critical current density in polycrystalline high-temperature superconductors, *Physica C* Vol. 206 (No. 3-4) 275–284.

[16] Peterson, R.L. and Ekin, J.W. (1988). Josephson-junction model of critical current in granular Y1Ba2Cu3O7-d superconductors, *Phys. Rev. B*, Vol. 37 (No. 16) 9848–9851.

[17] BulaevskiiL. N. , Clem J. R., Glazman L. I. (1992). Fraunhofer oscillations in a multilayer system arith Josephson coupling of layers. *Phys. Rev. B*, Vol. 46 (No. 1), 350–355.

[18] Fistul, M.V. and Giuliani, G.F. (1993). Theory of finite-size effects and vortex penetration in small Josephson junctions, *Phys. Rev. B*, Vol. 51 (No. 2) 1090–1095.

[19] Ares, O., Hart, C., Acosta, M. (1995). Transition from magnetic Fraunhofer-like to interferometric behavior in YBCO bridges with decreasing width, *Physica C*, Vol. 242 (No. 1-2) 191–196.

[20] Mezzetti, E., Gerbaldo, R., Ghigo, G., Gozzelino, L., Minetti, B., Camerlingo, C., Monaco, A., Cuttone, G., Rovelli, A. (1999). Control of the critical current density in YBa2Cu3O7-d films by means of intergrain and intragrain correlated defects, *Phys. Rev. B*, Vol. 60 (No. 10) 7623–7630.

[21] Gonzalez, J. L., Mello, E.V.L., Orlando, M.T.D., Yugue, E.S., Baggio-Saitovitch, E. (2001). Transport critical current in granular samples under high magnetic fields, *Physica C*, Vol. 364–365 347–349.

[22] Degroot, M.H. (1986). *Probability and Statistics*, Addison-Wesley Pub. Co., ISBN: 020111366X, 9780201113662. Michigan.

[23] Passos, C. A. C., Orlando, M. T. D., Oliveira, F. D. C., da Cruz, P. C. M., Passamai Jr., J. L., Orlando, C. G. P., Elói, N. A., Correa, H. P. S., Martinez, L. G. (2002). Effects of oxygen content on the properties of the Hg0.82Re0.18Ba2Ca2Cu3O8+d superconductor, *Supercond. Sci. Technol.*, Vol. 15 (No. 8) 1177–1183.

[24] Oliveira, F.D.C., Passos, C.A.C., Fardin, J.F., Simonetti, D.S.L., Passamai, J.L., Belich Jr., H., de Medeiros, E.F., Orlando, M.T.D., Ferreira, M.M. (2006). The influence of oxygen partial pressure on growth of the (Hg,Re)-1223 intergrain junction, *IEEE Trans. Appl. Supercond.*, Vol. 16 (No. 1) 15–20.

[25] Giaever, I. (1960) Eletron tunneling between 2 superconductors, *Physical Review Letters*, Vol. 5 (No.10) 464–466.

[26] Josephson, B. D. (1962) Possible new effects in superconductive tunnelling, *Physics Letters* Vol. 1 (No.7) 251–25.

[27] Mourachkine, A. (2004). *Room-Temperature Superconductivity*, Cambridge International Science, ISBN 1-904602-27-4. Cambridge.

[28] Poole, C. P. , Farach, H. A., Creswick, R. J., Prozorov R. (2007). *Superconductivity*, 2nd. Ed, Acadmemic Press Elsevier, ISBN: 978-0-12-088761-3. Amsterdam.

[29] Ambegaokar V., and Baratoff, A. (1963). Tunneling between superconductors. *Physical Review Letters* Vol. 10 486.

[30] Bruder, C., Vanotterlo A., Zimanyi G. T. (1995). Tunnel-junctions of unconventional superconductors. *Phys. Rev. B*, Vol. 51 (No. 18), 12904–12907.

[31] Rosenblatt, J. (1974). Coherent and paracoherent states in josephson-coupled granular superconductors. *Revue de Physique Appliquée*, Vol. 9 (No. 1), 217–222.

[32] Simkin,M. V. (1991) Josephson-oscillator spectrum and the reentrant phase transition in granular superconductors. *Phys. Rev. B*, Vol. 44 (No. 13), 7074–7077.

[33] Lebeau C., Raboutou, A., Peyral, P., and Rosenblatt, J. (1988). Fractal description of ferromagnetic glasses and random Josephon networks. *Physica B*, Vol. 152 (No. 1), 100–104.

[34] Matsushita, T. (2007). *Flux Pinnig in Superconductors*, Springer, ISBN-10 3-540-44514-5. New York.

[35] Spiegel, M.R. (1999). *Shcaum's Outline of Theory and Problems of Probability and Statistics*, McGraw-Hill Professional, ISBN: 0070602816, 9780070602816. New York.

[36] Abramowitz M., Stegun, I.A. (Eds.) (1972), *Handbook of Mathematical Functions with Formulas, Graphs and Mathematical Tables*, 9th printing, Dover, New York.

[37] Virgens, M. G.: MSc (Thesis in Portuguese). Physics Department, UFES (2002).

[38] Batista-Leyva, A. J., Cobas, R., Estévez-Rams, E., Orlando, M. T. D., Noda, C., Altshuler, E. (2000). Hysteresis of the critical current density in YBCO, HBCCO and BSCCO superconducting polycrystals: a comparative study, *Physica C*, Vol. 331 (No. 1) 57-66.

[39] Bogolyubov, N. A. (2010). Pattern of the induced magnetic field in high-temperature superconducting ceramics. *Physica C*, Vol. 470 (No.7-8) 361-364.

[40] Eisterer, M., Zehetmayer, M., and Weber, H.W. (2003). Current Percolation and Anisotropy in Polycrystalline MgB2. *Physics Review Letters* Vol. 90 (No. 24), 247002-1–247002-4.

[41] Preseter, M. (1996). Dynamical exponents for the current-induced percolation transition in high-Tc superconductors. *Phys. Rev. B*, Vol. 54 (No. 1), 606–618.

Chapter 8

Theory of Ferromagnetic Unconventional Superconductors with Spin-Triplet Electron Pairing

Dimo I. Uzunov

Additional information is available at the end of the chapter

1. Introduction

In the beginning of this century the unconventional superconductivity of spin-triplet type had been experimentally discovered in several itinerant ferromagnets. Since then much experimental and theoretical research on the properties of these systems has been accomplished. Here we review the phenomenological theory of ferromagnetic unconventional superconductors with spin-triplet Cooper pairing of electrons. Some theoretical aspects of the description of the phases and the phase transitions in these interesting systems, including the remarkable phenomenon of coexistence of superconductivity and ferromagnetism are discussed with an emphasis on the comparison of theoretical results with experimental data.

The spin-triplet or p-wave pairing allows parallel spin orientation of the fermion Cooper pairs in superfluid ^{3}He and unconventional superconductors [1]. For this reason the resulting unconventional superconductivity is robust with respect to effects of external magnetic field and spontaneous ferromagnetic ordering, so it may coexist with the latter. This general argument implies that there could be metallic compounds and alloys, for which the coexistence of spin-triplet superconductivity and ferromagnetism may be observed.

Particularly, both superconductivity and itinerant ferromagnetic orders can be created by the same band electrons in the metal, which means that spin-1 electron Cooper pairs participate in the formation of the itinerant ferromagnetic order. Moreover, under certain conditions the superconductivity is enhanced rather than depressed by the uniform ferromagnetic order that can generate it, even in cases when the superconductivity does not appear in a pure form as a net result of indirect electron-electron coupling.

The coexistence of superconductivity and ferromagnetism as a result of collective behavior of f-band electrons has been found experimentally for some Uranium-based intermetallic compounds as, UGe$_2$ [2–5], URhGe [6–8], UCoGe [9, 10], and UIr [11, 12]. At low temperature ($T \sim 1$ K) all these compounds exhibit thermodynamically stable phase of coexistence of spin-triplet superconductivity and itinerant (f-band) electron ferromagnetism (in short, FS phase). In UGe$_2$ and UIr the FS phase appears at high pressure ($P \sim 1$ GPa) whereas in URhGe and UCoGe, the coexistence phase persists up to ambient pressure (10^5Pa $\equiv$ 1bar).

Experiments, carried out in ZrZn$_2$ [13], also indicated the appearance of FS phase at $T <$ 1 K in a wide range of pressures ($0 < P \sim 21$ kbar). In Zr-based compounds the ferromagnetism and the p-wave superconductivity occur as a result of the collective behavior of the d-band electrons. Later experimental results [14, 15] had imposed the conclusion that bulk superconductivity is lacking in ZrZn$_2$, but the occurrence of a surface FS phase at surfaces with higher Zr content than that in ZrZn$_2$ has been reliably demonstrated. Thus the problem for the coexistence of bulk superconductivity with ferromagnetism in ZrZn$_2$ is still unresolved. This raises the question whether the FS phase in ZrZn$_2$ should be studied by surface thermodynamics methods or should it be investigated by considering that bulk and surface thermodynamic phenomena can be treated on the same footing. Taking into account the mentioned experimental results for ZrZn$_2$ and their interpretation by the experimentalists [13–15] we assume that the unified thermodynamic approach can be applied. As an argument supporting this point of view let us mention that the spin-triplet superconductivity occurs not only in bulk materials but also in quasi-two-dimensional (2D) systems – thin films and surfaces and quasi-1D wires (see, e.g., Refs. [16]). In ZrZn$_2$ and UGe$_2$ both ferromagnetic and superconducting orders vanish at the same critical pressure P_c, a fact implying that the respective order parameter fields strongly depend on each other and should be studied on the same thermodynamic basis [17].

Fig. 1 illustrates the shape of the $T - P$ phase diagrams of real intermetallic compounds. The phase transition from the normal (N) to the ferromagnetic phase (FM) (in short, N-FM transition) is shown by the line $T_F(P)$. The line $T_{FS}(P)$ of the phase transition from FM to FS (FM-FS transition) may have two or more distinct shapes. Beginning from the maximal (critical) pressure P_c, this line may extend, like in ZrZn$_2$, to all pressures $P < P_c$, including the ambient pressure P_a; see the almost straight line containing the point 3 in Fig. 1. A second possible form of this line, as known, for example, from UGe$_2$ experiments, is shown in Fig. 1 by the curve which begins at $P \sim P_c$, passes through the point 2, and terminates at some pressure $P_1 > P_a$, where the superconductivity vanishes. These are two qualitatively different physical pictures: (a) when the superconductivity survives up to ambient pressure (type I), and (b) when the superconducting states are possible only at relatively high pressure (for UGe$_2$, $P_1 \sim 1$ GPa); type II. At the tricritical points 1, 2 and 3 the order of the phase transitions changes from second order (solid lines) to first order (dashed lines). It should be emphasized that in all compounds, mentioned above, $T_{FS}(P)$ is much lower than $T_F(P)$ when the pressure P is considerably below the critical pressure P_c (for experimental data, see Sec. 8).

In Fig. 1, the circle C denotes a narrow domain around P_c at relatively low temperatures ($T \lesssim 300$ mK), where the experimental data are quite few and the predictions about the shape of the phase transition are not reliable. It could be assumed, as in the most part of the experimental papers, that ($T = 0, P = P_c$) is the zero temperature point at which both lines $T_F(P)$ and $T_{FS}(P)$ terminate. A second possibility is that these lines may join in a single (N-FS) phase transition line at some point ($T \gtrsim 0, P'_c \lesssim P_c$) above the absolute zero. In this second variant, a direct N-FS phase transition occurs, although this option exists in a very small domain of temperature and pressure variations: from point $(0, P_c)$ to point ($T \gtrsim 0, P'_c \lesssim P_c$). A third variant is related with the possible splitting of the point $(0, P_c)$, so that the N-FM line terminates at $(0, P_c)$, whereas the FM-FS line terminates at another zero temperature point

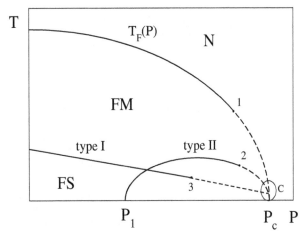

Figure 1. An illustration of $T - P$ phase diagram of p-wave ferromagnetic superconductors (details are omitted): N – normal phase, FM – ferromagnetic phase, FS – phase of coexistence of ferromagnetic order and superconductivity, $T_F(P)$ and $T_{FS}(P)$ are the respective phase transition lines: solid lines correspond to second order phase transitions, dashed lines stand for first order phase transition; 1 and 2 are tricritical points; P_c is the critical pressure, and the circle C surrounds a relatively small domain of high pressure and low temperature, where the phase diagram may have several forms depending on the particular substance. The line of the FM-FS phase transition may extend up to ambient pressure (type I ferromagnetic superconductors), or, may terminate at $T = 0$ at some high pressure $P = P_1$ (type II ferromagnetic superconductors, as indicated in the figure).

$(0, P_{0c})$; $P_{0c} \lesssim P_c$. In this case, the p-wave ferromagnetic superconductor has three points of quantum (zero temperature) phase transitions [18, 19].

These and other possible shapes of $T - P$ phase diagrams are described within the framework of the general theory of Ginzburg-Landau (GL) type [18–20] in a conformity with the experimental data; see also Ref. [21]. The same theory has been confirmed by a microscopic derivation based on a microscopic Hamiltonian including a spin-generalized BCS term and an additional Heisenberg exchange term [22].

For all compounds, cited above, the FS phase occurs only in the ferromagnetic phase domain of the $T - P$ diagram. Particularly at equilibrium, and for given P, the temperature $T_F(P)$ of the normal-to-ferromagnetic phase (or N-FM) transition is never lower than the temperature $T_{FS}(P)$ of the ferromagnetic-to-FS phase transition (FM-FS transition). This confirms the point of view that the superconductivity in these compounds is triggered by the spontaneous magnetization M, in analogy with the well-known triggering of the superfluid phase A_1 in ^{3}He at mK temperatures by the external magnetic field H. Such "helium analogy" has been used in some theoretical studies (see, e.g., Ref. [23, 24]), where Ginzburg-Landau (GL) free energy terms, describing the FS phase were derived by symmetry group arguments. The non-unitary state, with a non-zero value of the Cooper pair magnetic moment, known from the theory of unconventional superconductors and superfluidity in ^{3}He [1], has been suggested firstly in Ref. [23], and later confirmed in other studies [7, 24]; recently, the same topic was comprehensively discussed in Ref. [25].

For the spin-triplet ferromagnetic superconductors the trigger mechanism was recently examined in detail [20, 21]. The system main properties are specified by terms in the GL expansion of form $M_i \psi_j \psi_k$, which represent the interaction of the magnetization $M = \{M_j; j = 1, 2, 3\}$ with the complex superconducting vector field $\psi = \{\psi_j; j = 1, 2, 3\}$. Particularly, these terms are responsible for the appearance of superconductivity ($|\psi| > 0$) for certain T and P values. A similar trigger mechanism is familiar in the context of improper ferroelectrics [26].

A crucial feature of these systems is the nonzero magnetic moment of the spin-triplet Cooper pairs. As mentioned above, the microscopic theory of magnetism and superconductivity in non-Fermi liquids of strongly interacting heavy electrons (f and d band electrons) is either too complex or insufficiently developed to describe the complicated behavior in itinerant ferromagnetic compounds. Several authors (see [20, 21, 23–25]) have explored the phenomenological description by a self-consistent mean field theory, and here we will essentially use the thermodynamic results, in particular, results from the analysis in Refs. [20, 21]. Mean-field microscopic theory of spin-mediated pairing leading to the mentioned non-unitary superconductivity state has been developed in Ref. [17] that is in conformity with the phenomenological description that we have done.

The coexistence of s-wave (conventional) superconductivity and ferromagnetic order is a long-standing problem in condensed matter physics [27–29]. While the s-state Cooper pairs contain only opposite electron spins and can easily be destroyed by the spontaneous magnetic moment, the spin-triplet Cooper pairs possess quantum states with parallel orientation of the electron spins and therefore can survive in the presence of substantial magnetic moments. This is the basic difference in the magnetic behavior of conventional (s-state) and spin-triplet superconductivity phases. In contrast to other superconducting materials, for example, ternaty and Chevrel phase compounds, where the effect of magnetic order on s-wave superconductivity has been intensively studied in the seventies and eighties of last century (see, e.g., Refs. [27–29]), in these ferromagnetic compounds the phase transition temperature T_F to the ferromagnetic state is much higher than the phase transition temperature T_{FS} from ferromagnetic to a (mixed) state of coexistence of ferromagnetism and superconductivity. For example, in UGe$_2$ we have $T_{FS} \sim 0.8$ K versus maximal $T_F = 52$ K [2–5]. Another important difference between the ternary rare earth compounds and the intermetallic compounds (UGe$_2$, UCoGe, etc.), which are of interest in this paper, is that the experiments with the latter do not give any evidence for the existence of a standard normal-to-superconducting phase transition in zero external magnetic field. This is an indication that the (generic) critical temperature T_s of the pure superconductivity state in these intermetallic compounds is very low ($T_s \ll T_{FS}$), if not zero or even negative.

In the reminder of this paper, we present general thermodynamic treatment of systems with itinerant ferromagnetic order and superconductivity due to spin-triplet Cooper pairing of the same band electrons, which are responsible for the spontaneous magnetic moment. The usual Ginzburg-Landau (GL) theory of superconductors has been completed to include the complexity of the vector order parameter ψ, the magnetization M and new relevant energy terms [20, 21]. We outline the $T - P$ phase diagrams of ferromagnetic spin-triplet superconductors and demonstrate that in these materials two contrasting types of thermodynamic behavior are possible. The present phenomenological approach includes both mean-field and spin-fluctuation theory (SFT), as the arguments in Ref. [30]. We propose a simple, yet comprehensive, modeling of P dependence of the free energy parameters, resulting in a very good compliance of our theoretical predictions for the shape the $T - P$ phase diagrams with the experimental data (for some preliminary results, see Ref. [18, 19]).

The theoretical analysis is done by the standard methods of phase transition theory [31]. Treatment of fluctuation effects and quantum correlations [31, 32] is not included in this study. But the parameters of the generalized GL free energy may be considered either in mean-field approximation as here, or as phenomenologically renormalized parameters which are affected by additional physical phenomena, as for example, spin fluctuations.

We demonstrate with the help of present theory that we can outline different possible topologies for the $T - P$ phase diagram, depending on the values of Landau parameters, derived from the existing experimental data. We show that for spin-triplet ferromagnetic superconductors there exist two distinct types of behavior, which we denote as Zr-type (or, alternatively, type I) and U-type (or, type II); see Fig. 1. This classification of the FS, first mentioned in Ref. [18], is based on the reliable interrelationship between a quantitative criterion derived by us and the thermodynamic properties of the ferromagnetic spin-triplet superconductors. Our approach can be also applied to URhGe, UCoGe, and UIr. The results shed light on the problems connected with the order of the quantum phase transitions at ultra-low and zero temperatures. They also raise the question for further experimental investigations of the detailed structure of the phase diagrams in the high-P/low-T region.

2. Theoretical framework

Consider the GL free energy functional of the form

$$F(\psi, B) = \int_V dx \left[f_S(\psi) + f_F(M) + f_I(\psi, M) + \frac{B^2}{8\pi} - B.M \right], \tag{1}$$

where the fields ψ, M, and B are supposed to depend on the spatial vector $x \in V$ in the volume V of the superconductor. In Eq. (1), the free energy density generated by the generic superconducting subsystem (ψ) is given by

$$f_S(\psi) = f_{grad}(\psi) + a_s |\psi|^2 + \frac{b_s}{2} |\psi|^4 + \frac{u_s}{2} |\psi^2|^2 + \frac{v_s}{2} \sum_{j=1}^{3} |\psi_j|^4, \tag{2}$$

with

$$f_{grad}(\psi) = K_1 (D_i \psi_j)^* (D_i D_j) + K_2 \left[(D_i \psi_i)^* (D_j \psi_j) + (D_i \psi_j)^* (D_j \psi_i) \right] \tag{3}$$
$$+ K_3 (D_i \psi_i)^* (D_i \psi_i),$$

where a summation over the indices (i, j) is assumed, the symbol $D_j = (\hbar \partial / i \partial x_j + 2|e|A_j/c)$ of covariant differentiation is introduced, and K_j are material parameters [1]. The free energy density $f_F(M)$ of a standard ferromagnetic phase transition of second order [31] is

$$f_F(M) = c_f \sum_{j=1}^{3} |\nabla M_j|^2 + a_f M^2 + \frac{b_f}{2} M^4, \tag{4}$$

with $c_f, b_f > 0$, and $a_f = \alpha(T - T_f)$, where $\alpha_f > 0$ and T_f is the critical temperature, corresponding of the generic ferromagnetic phase transition. Finally, the energy $f_I(\boldsymbol{\psi}, \boldsymbol{M})$ produced by the possible couplings of $\boldsymbol{\psi}$ and $\boldsymbol{M}$ is given by

$$f_I(\boldsymbol{\psi}, \boldsymbol{M}) = i\gamma_0 \boldsymbol{M}.(\boldsymbol{\psi} \times \boldsymbol{\psi}^*) + \delta_0 \boldsymbol{M}^2 |\boldsymbol{\psi}|^2, \tag{5}$$

where the coupling parameter $\gamma_0 \sim J$ depends on the ferromagnetic exchange parameter $J > 0$, [23, 24] and δ_0 is the standard $\boldsymbol{M} - \boldsymbol{\psi}$ coupling parameter, known from the theory of multicritical phenomena [31] and from studies of coexistence of ferromagnetism and superconductivity in ternary compounds [27, 28].

As usual, in Eq. (2), $a_s = (T - T_s)$, where T_s is the critical temperature of the generic superconducting transition, $b_s > 0$. The parameters u_s and v_s and δ_0 may take some negative values, provided the overall stability of the system is preserved. The values of the material parameters $\mu = (T_s, T_f, \alpha_s, \alpha_f, b_s, u_s, v_s, b_f, K_j, \gamma_0$ and $\delta_0)$ depend on the choice of the substance and on intensive thermodynamic parameters, such as the temperature T and the pressure P. From a microscopic point of view, the parameters μ depend on the density of states $U_F(k_F)$ on the Fermi surface. On the other hand U_F varies with T and P. Thus the relationships $(T, P) \rightleftarrows U_F \rightleftarrows \mu$, i.e., the functional relations $\mu[U_F(T, P)]$, are of essential interest. While these relations are unknown, one may suppose some direct dependence $\mu(T, P)$. The latter should correspond to the experimental data.

The free energy (1) is quite general. It has been deduced by several reliable arguments. In order to construct Eq. (1)–(5) we have used the standard GL theory of superconductors and the phase transition theory with an account of the relevant anisotropy of the p-wave Cooper pairs and the crystal anisotropy, described by the u_s- and v_s-terms in Eq. (2), respectively. Besides, we have used the general case of cubic anisotropy, when all three components ψ_j of $\boldsymbol{\psi}$ are relevant. Note, that in certain real cases, for example, in UGe$_2$, the crystal symmetry is tetragonal, $\boldsymbol{\psi}$ effectively behaves as a two-component vector and this leads to a considerable simplification of the theory. As shown in Ref. [20], the mentioned anisotropy terms are not essential in the description of the main thermodynamic properties, including the shape of the $T - P$ phase diagram. For this reason we shall often ignore the respective terms in Eq. (2). The γ_0-term triggers the superconductivity (M-triger effect [20, 21]) while the $\delta_0 \boldsymbol{M}^2 |\boldsymbol{\psi}|^2$–term makes the model more realistic for large values of $\boldsymbol{M}$. This allows for an extension of the domain of the stable ferromagnetic order up to zero temperatures for a wide range of values of the material parameters and the pressure P. Such a picture corresponds to the real situation in ferromagnetic compounds [20].

The total free energy (1) is difficult for a theoretical investigation. The various vortex and uniform phases described by this complex model cannot be investigated within a single calculation but rather one should focus on particular problems. In Ref. [24] the vortex phase was discussed with the help of the criterion [33] for a stability of this state near the phase transition line $T_{c2}(\boldsymbol{B})$, ; see also, Ref. [34]. The phase transition line $T_{c2}(H)$ of a usual superconductor in external magnetic field $H = |\boldsymbol{H}|$ is located above the phase transition line T_s of the uniform (Meissner) phase. The reason is that T_s is defined by the equation $a_s(T) = 0$, whereas $T_{c2}(H)$ is a solution of the equation $|a_s| = \mu_B H$, where $\mu_B = |e|\hbar/2mc$ is the Bohr magneton [34]. For ferromagnetic superconductors, where $M > 0$, one should use the magnetic induction $\boldsymbol{B}$ rather than $\boldsymbol{H}$. In case of $\boldsymbol{H} = 0$ one should apply the same criterion with respect to the magnetization $\boldsymbol{M}$ for small values of $|\boldsymbol{\psi}|$ near the phase transition line $T_{c2}(M)$;

$M = |M|$. For this reason we should use the diagonal quadratic form [35] corresponding to the entire ψ^2-part of the total free energy functional (1). The lowest energy term in this diagonal quadratic part contains a coefficient a of the form $a = (a_s - \gamma_0 M - \delta M^2)$ [35]. Now the equation $a(T) = 0$ defines the critical temperature of the Meissner phase and the equation $|a_s| = \mu_B M$ stands for $T_{c2}(M)$. It is readily seen that these two equations can be written in the same form, provided the parameter γ_0 in a is substituted by $\gamma_0' = (\gamma_0 - \mu_B)$. Thus the phase transition line corresponding to the vortex phase, described by the model (1) at zero external magnetic field and generated by the magnetization M, can be obtained from the phase transition line corresponding to the uniform superconducting phase by an effective change of the value of the parameter γ_0. Both lines have the same shape and this is a particular property of the present model. The variation of the parameter γ_0 generates a family of lines.

Now we propose a possible way of theoretical treatment of the $T_{FS}(P)$ line of the FM-FS phase transition, shown in Fig. (1). This is a crucial point in our theory. The phase transition line of the uniform superconducting phase can be calculated within the thermodynamic analysis of the uniform phases, described by the free energy (1). This analysis is done in a simple variant of the free energy (1) in which the fields ψ and M do not depend on the spatial vector x. The accomplishment of such analysis will give a formula for the phase transition line $T_{FS}(P)$ which corresponds a Meissner phase coexisting with the ferromagnetic order. The theoretical result for $T_{FS}(P)$ will contain a unspecified parameter γ_0. If the theoretical line $T_{FS}(P)$ is fitted to the experimental data for the FM-FS transition line corresponding to a particular compound, the two curves will coincide for some value of γ_0, irrespectively on the structure of the FS phase. If the FS phase contains a vortex superconductivity the fitting parameter $\gamma_{0(eff)}$ should be interpreted as γ_0' but if the FS phase contains Meissner superconductivity, $\gamma_{0(eff)}$ should be identified as γ_0. These arguments justify our approach to the investigation of the experimental data for the phase diagrams of intermetallic compounds with FM and FS phases. In the remainder of this paper, we shall investigate uniform phases.

3. Model considerations

In the previous section we have justified a thermodynamic analysis of the free energy (1) in terms of uniform order parameters. Neglecting the x-dependence of ψ and M, the free energy per unit volume, $F/V = f(\psi, M)$ in zero external magnetic field ($H = 0$), can be written in the form

$$f(\psi, M) = a_s|\psi|^2 + \frac{b_s}{2}|\psi|^4 + \frac{u_s}{2}|\psi^2|^2 + \frac{v_s}{2}\sum_{j=1}^{3}|\psi_j|^4 + a_f M^2 + \frac{b_f}{2}M^4 \qquad (6)$$

$$+ i\gamma_0 M \cdot (\psi \times \psi^*) + \delta_0 M^2|\psi|^2.$$

Here we slightly modify the parameter a_f by choosing $a_f = \alpha_f[T^n - T_f^n(P)]$, where $n = 1$ gives the standard form of a_f, and $n = 2$ applies for SFT [30] and the Stoner-Wohlfarth model [36]. Previous studies [20] have shown that the anisotropy represented by the u_s and v_s terms in Eq. (6) slightly perturbs the size and shape of the stability domains of the phases, while similar effects can be achieved by varying the b_s factor in the $b_s|\psi|^4$ term. For these reasons, in the present analysis we ignore the anisotropy terms, setting $u_s = v_s = 0$, and consider $b_s \equiv b > 0$ as an effective parameter. Then, without loss of generality, we are free to choose the magnetization vector to have the form $M = (0, 0, M)$.

According to the microscopic theory of band magnetism and superconductivity the macroscopic material parameters in Eq. (6) depend in a quite complex way on the density of states at the Fermi level and related microscopic quantities [37]. That is why we can hardly use the microscopic characteristics of these complex metallic compounds in order to elucidate their thermodynamic properties, in particular, in outlining their phase diagrams in some details. However, some microscopic simple microscopic models reveal useful results, for example, the zero temperature Stoner-type model employed in Ref. [38].

We redefine for convenience the free energy (6) in a dimensionless form by $\tilde{f} = f/(b_f M_0^4)$, where $M_0 = [\alpha_f T_{f0}^n / b_f]^{1/2} > 0$ is the value of the magnetization M corresponding to the pure magnetic subsystem ($\psi \equiv 0$) at $T = P = 0$ and $T_{f0} = T_f(0)$. The order parameters assume the scaling $m = M/M_0$ and $\varphi = \psi/[(b_f/b)^{1/4}M_0]$, and as a result, the free energy becomes

$$\tilde{f} = r\phi^2 + \frac{\phi^4}{2} + tm^2 + \frac{m^4}{2} + 2\gamma m\phi_1\phi_2\sin\theta + \gamma_1 m^2\phi^2, \qquad (7)$$

where $\phi_j = |\varphi_j|$, $\phi = |\varphi|$, and $\theta = (\theta_2 - \theta_1)$ is the phase angle between the complex $\varphi_1 = \phi_1 e^{i\theta_1}$ and $\varphi_2 = \phi_2 e^{\theta_2}$. Note that the phase angle θ_3, corresponding to the third complex field component $\varphi_3 = \phi_3 e^{i\theta_3}$ does not enter explicitly in the free energy $\tilde{f}$, given by Eq. (7), which is a natural result of the continuous space degeneration. The dimensionless parameters t, r, γ and γ_1 in Eq. (7) are given by

$$t = \tilde{T}^n - \tilde{T}_f^n(P), \quad r = \kappa(\tilde{T} - \tilde{T}_s), \qquad (8)$$

where $\kappa = \alpha_s b_f^{1/2}/\alpha_f b^{1/2} T_{f0}^{n-1}$, $\gamma = \gamma_0/[\alpha_f T_{f0}^n b]^{1/2}$, and $\gamma_1 = \delta_0/(bb_f)^{1/2}$. The reduced temperatures are $\tilde{T} = T/T_{f0}$, $\tilde{T}_f(P) = T_f(P)/T_{f0}$, and $\tilde{T}_s(P) = T_s(P)/T_{f0}$.

The analysis involves making simple assumptions for the P dependence of the t, r, γ, and γ_1 parameters in Eq. (7). Specifically, we assume that only T_f has a significant P dependence, described by

$$\tilde{T}_f(P) = (1 - \tilde{P})^{1/n}, \qquad (9)$$

where $\tilde{P} = P/P_0$ and P_0 is a characteristic pressure deduced later. In ZrZn$_2$ and UGe$_2$ the P_0 values are very close to the critical pressure P_c at which both the ferromagnetic and superconducting orders vanish, but in other systems this is not necessarily the case. As we will discuss, the nonlinearity ($n = 2$) of $T_f(P)$ in ZrZn$_2$ and UGe$_2$ is relevant at relatively high P, at which the N-FM transition temperature $T_F(P)$ may not coincide with $T_f(P)$; $T_F(P)$ is the actual line of the N-FM phase transition, as shown in Fig. (1). The form (9) of the model function $\tilde{T}_f(P)$ is consistent with preceding experimental and theoretical investigations of the N-FM phase transition in ZrZn$_2$ and UGe$_2$ (see, e.g., Refs. [4, 24, 39]). Here we consider only non-negative values of the pressure P (for effects at $P < 0$, see, e.g., Ref. [44]).

The model function (9) is defined for $P \leq P_0$, in particular, for the case of $n > 1$, but we should have in mind that, in fact, the thermodynamic analysis of Eq. (7) includes the parameter t rather than $T_f(P)$. This parameter is given by

$$t(T, P) = \tilde{T}^n - 1 + \tilde{P}, \qquad (10)$$

and is well defined for any $\tilde{P}$. This allows for the consideration of pressures $P > P_0$ within the free energy (7).

The model function $\tilde{T}_f(P)$ can be naturally generalized to $\tilde{T}_f(P) = (1 - \tilde{P}^\beta)^{1/\alpha}$ but the present needs of interpretation of experimental data do not require such a complex consideration (hereafter we use Eq. (9) which corresponds to $\beta = 1$ and $\alpha = n$). Besides, other analytical forms of $\tilde{T}_f(\tilde{P})$ can also be tested in the free energy (7), in particular, expansion in powers of $\tilde{P}$, or, alternatively, in $(1 - \tilde{P})$ which satisfy the conditions $\tilde{T}_f(0) = 1$ and $\tilde{T}_f(1) = 0$. Note, that in URhGe the slope of $T_F(P) \sim T_f(P)$ is positive from $P = 0$ up to high pressures [8] and for this compound the form (9) of $\tilde{T}_f(P)$ is inconvenient. Here we apply the simplest variants of P-dependence, namely, Eqs. (9) and (10).

In more general terms, all material parameters $(r, t, \gamma, \ldots)$ may depend on the pressure. We suppose that a suitable choice of the dependence of t on P is enough for describing the main thermodynamic properties and this supposition is supported by the final results, presented in the remainder of this paper. But in some particular investigations one may need to introduce a suitable pressure dependence of other parameters.

4. Stable phases

The simplified model (7) is capable of describing the main thermodynamic properties of spin-triplet ferromagnetic superconductors. For $r > 0$, i.e., $T > T_s$, there are three stable phases [20]: (i) the normal (N-) phase, given by $\phi = m = 0$ (stability conditions: $t \geq 0$, $r \geq 0$); (ii) the pure ferromagnetic phase (FM phase), given by $m = (-t)^{1/2} > 0$, $\phi = 0$, which exists for $t < 0$ and is stable provided $r \geq 0$ and $r \geq (\gamma_1 t + \gamma|t|^{1/2})$, and (iii) the already mentioned phase of coexistence of ferromagnetic order and superconductivity (FS phase), given by $\sin\theta = \mp 1$, $\phi_3 = 0$, $\phi_1 = \phi_2 = \phi/\sqrt{2}$, where

$$\phi^2 = \kappa(\tilde{T}_s - \tilde{T}) \pm \gamma m - \gamma_1 m^2 \geq 0. \tag{11}$$

The magnetization m satisfies the equation

$$c_3 m^3 \pm c_2 m^2 + c_1 m \pm c_0 = 0 \tag{12}$$

with coefficients $c_0 = \gamma\kappa(\tilde{T} - \tilde{T}_s)$,

$$c_1 = 2\left[\tilde{T}^n + \kappa\gamma_1(\tilde{T}_s - \tilde{T}) + \tilde{P} - 1 - \frac{\gamma^2}{2}\right], \tag{13}$$

$$c_2 = 3\gamma\gamma_1, \quad c_3 = 2(1 - \gamma_1^2). \tag{14}$$

The FS phase contains two thermodynamically equivalent phase domains that can be distinguished by the upper and lower signs $(\pm)$ of some terms in Eqs. (11) and (12). The upper sign describes the domain (labelled bellow again by FS), where $m > 0$, $\sin\theta = -1$, whereas the lower sign describes the conjunct domain FS*, where $m < 0$ and $\sin\theta = 1$ (for details, see, Ref. [20]). Here we consider one of the two thermodynamically equivalent phase domains, namely, the domain FS, which is stable for $m > 0$ (FS* is stable for $m < 0$). This "one-domain approximation" correctly presents the main thermodynamic properties

N	$\tilde{T}_N$	t_N	$\tilde{P}_N(n)$
A	$\tilde{T}_s$	$\gamma^2/2$	$1 - \tilde{T}_s^n + \gamma^2/2$
B	$\tilde{T}_s + \gamma^2(2+\gamma_1)/4\kappa(1+\gamma_1)^2$	$-\gamma^2/4(1+\gamma_1)^2$	$1 - \tilde{T}_B^n - \gamma^2/4(1+\gamma_1)^2$
C	$\tilde{T}_s + \gamma^2/4\kappa(1+\gamma_1)$	0	$1 - \tilde{T}_C^n$
max	$\tilde{T}_s + \gamma^2/4\kappa\gamma_1$	$-\gamma^2/4\gamma_1^2$	$1 - \tilde{T}_m^n - \gamma^2/4\gamma_1^2$

Table 1. Theoretical results for the location $[(\tilde{T}, \tilde{P})$ - reduced coordinates] of the tricritical points A $\equiv (\tilde{T}_A, \tilde{P}_A)$ and B $\equiv (\tilde{T}_B, \tilde{P}_B)$, the critical-end point C $\equiv (\tilde{T}_C, \tilde{P}_C)$, and the point of temperature maximum, $max = (\tilde{T}_m, \tilde{P}_m)$ on the curve $\tilde{T}_{FS}(\tilde{P})$ of the FM-FS phase transitions of first and second orders (for details, see Sec. 5). The first column shows $\tilde{T}_N \equiv \tilde{T}_{(A,B,C,m)}$. The second column stands for $t_N = t_{(A,B,C,m)}$. The reduced pressure values $\tilde{P}_{(A,B,C,m)}$ of points A, B, C, and max are denoted by $\tilde{P}_N(n)$: $n = 1$ stands for the linear dependence $T_f(P)$, and $n = 2$ stands for the nonlinear $T_f(P)$ and $t(T)$, corresponding to SFT.

described by the model (6), in particular, in the case of a lack of external symmetry breaking fields. The stability conditions for the FS phase domain given by Eqs.(11) and (12) are $\gamma M \geq 0$,

$$\kappa(\tilde{T}_s - \tilde{T}) \pm \gamma m - 2\gamma_1 m^2 \geq 0, \tag{15}$$

and

$$3(1 - \gamma_1^2)m^2 + 3\gamma\gamma_1 m + \tilde{T}^n - 1 + \tilde{P} + \kappa\gamma_1(\tilde{T}_s - \tilde{T}) - \frac{\gamma^2}{2} \geq 0. \tag{16}$$

These results are valid whenever $T_f(P) > T_s(P)$, which excludes any pure superconducting phase ($\psi \neq 0, m = 0$) in accord with the available experimental data.

For $r < 0$, and $t > 0$ the models (6) and (7) exhibit a stable pure superconducting phase ($\phi_1 = \phi_2 = m = 0, \phi_3^2 = -r$) [20]. This phase may occur in the temperature domain $T_f(P) < T < T_s$. For systems, where $T_f(0) \gg T_s$, this is a domain of pressure in a very close vicinity of $P_0 \sim P_c$, where $T_F(P) \sim T_f(\tilde{P})$ decreases up to values lower than T_s. Of course, such a situation is described by the model (7) only if $T_s > 0$. This case is interesting from the experimental point of view only when $T_s > 0$ is enough above zero to enter in the scope of experimentally measurable temperatures. Up to date a pure superconducting phase has not been observed within the accuracy of experiments on the mentioned metallic compounds. For this reason, in the reminder of this paper we shall often assume that the critical temperature T_s of the generic superconducting phase transition is either non-positive $(T_s \leq 0)$, or, has a small positive value which can be neglected in the analysis of the available experimental data.

The negative values of the critical temperature T_s of the generic superconducting phase transition are generally possible and produce a variety of phase diagram topologies (Sec. 5). Note, that the value of T_s depends on the strength of the interaction mediating the formation of the spin-triplet Cooper pairs of electrons. Therefore, for the sensitiveness of such electron couplings to the crystal lattice properties, the generic critical temperature T_s depends on the pressure. This is an effect which might be included in our theoretical scheme by introducing some convenient temperature dependence of T_s. To do this we need information either from experimental data or from a comprehensive microscopic theory.

Usually, $T_s \leq 0$ is interpreted as a lack of any superconductivity but here the same non-positive values of T_s are effectively enhanced to positive values by the interaction parameter γ which triggers the superconductivity up to superconducting phase-transition temperatures $T_{FS}(P) > 0$. This is readily seen from Table 1, where we present the reduced critical temperatures on the FM-FS phase transition line $\tilde{T}_{FS}(\tilde{P})$, calculated from the present theory, namely, $\tilde{T}_m$ – the maximum of the curve $T_{FS}(P)$ (if available, see Sec. 5), the temperatures $\tilde{T}_A$ and $\tilde{T}_B$, corresponding to the tricritical points $A \equiv (\tilde{T}_A, \tilde{P}_A)$ and $B \equiv (\tilde{T}_B, \tilde{P}_B)$, and the temperature $\tilde{T}_C$, corresponding to the critical-end point $C \equiv (\tilde{T}_C, \tilde{P}_C)$. The theoretical derivation of the dependence of the multicritical temperatures $\tilde{T}_A$, $\tilde{T}_B$ and $\tilde{T}_C$ on γ, γ_1, κ, and $\tilde{T}_s$, as well as the dependence of $\tilde{T}_m$ on the same model parameters is outlined in Sec. 5. All these temperatures as well as the whole phase transition line $T_{FS}(P)$ are considerably boosted above T_s owing to positive terms of order γ^2. If $\tilde{T}_s < 0$, the superconductivity appears, provided $\tilde{T}_m > 0$, i.e., when $\gamma^2/4\kappa\gamma_1 > |\tilde{T}_s|$ (see Table 1).

5. Temperature-pressure phase diagram

Although the structure of the FS phase is quite complicated, some of the results can be obtained in analytical form. A more detailed outline of the phase domains, for example, in $T - P$ phase diagram, can be done by using suitable values of the material parameters in the free energy (7): P_0, T_{f0}, T_s, κ, γ, and γ_1. Here we present some of the analytical results for the phase transition lines and the multi-critical points. Typical shapes of phase diagrams derived directly from Eq. (7) are given in Figs. 2–7. Figure 2 shows the phase diagram calculated from Eq. (7) for parameters, corresponding to the experimental data [13] for ZrZn$_2$. Figures 3 and 4 show the low-temperature and the high-pressure parts of the same phase diagram (see Sec. 7 for details). Figures 5–7 show the phase diagram calculated for the experimental data [2, 4] of UGe$_2$ (see Sec. 8). In ZrZn$_2$, UGe$_2$, as well as in UCoGe and UIr, critical pressure P_c exists, where both superconductivity and ferromagnetic orders vanish.

As in experiments, we find out from our calculation that in the vicinity of $P_0 \sim P_c$ the FM-FS phase transition is of fist order, denoted by the solid line BC in Figs. 3, 4, 6, and 7. At lower pressure the same phase transition is of second orderq shown by the dotted lines in the same figures. The second order phase transition line $\tilde{T}_{FS}(P)$ separating the FM and FS phases is given by the solution of the equation

$$\tilde{T}_{FS}(\tilde{P}) = \tilde{T}_s + \tilde{\gamma}_1 t_{FS}(\tilde{P}) + \tilde{\gamma}[-t_{FS}(\tilde{P})]^{1/2}, \qquad (17)$$

where $t_{FS}(\tilde{P}) = t(T_{FS}, \tilde{P}) \leq 0$, $\tilde{\gamma} = \gamma/\kappa$, $\tilde{\gamma}_1 = \gamma_1/\kappa$, and $0 < \tilde{P} < \tilde{P}_B$; P_B is the pressure corresponding to the multi-critical point B, where the line $T_{FS}(P)$ terminates, as clearly shown in Figs. 4 and 7). Note, that Eq. (17) strictly coincides with the stability condition for the FM phase with respect to appearance of FS phase [20].

Additional information for the shape of this phase transition line can be obtained by the derivative $\tilde{\rho} = \partial \tilde{T}_{FS}(\tilde{P})/\partial \tilde{P}$, namely,

$$\tilde{\rho} = \frac{\tilde{\rho}_s + \tilde{\gamma}_1 - \tilde{\gamma}/2(-t_{FS})^{1/2}}{1 - n\tilde{T}_{FS}^{n-1}\left[\tilde{\gamma}_1 - \tilde{\gamma}/2[(-t_{FS})^{1/2}]\right]}, \qquad (18)$$

where $\tilde{\rho}_s = \partial \tilde{T}_s(\tilde{P})/\partial \tilde{P}$. Note, that Eq. (18) is obtained from Eqs. (10) and (17).

The shape of the line $\tilde{T}_{FS}(P)$ can vary depending on the theory parameters (see, e.g., Figs.3 and 6). For certain ratios of $\tilde{\gamma}$, $\tilde{\gamma}_1$, and values of $\tilde{\rho}_s$, the curve $\tilde{T}_{FS}(\tilde{P})$ exhibits a maximum $\tilde{T}_m = \tilde{T}_{FS}(\tilde{P}_m)$, given by $\tilde{\rho}(\tilde{\rho}_s, T_m, P_m) = 0$. This maximum is clearly seen in Figs. 6 and 7. To locate the maximum we need to know $\tilde{\rho}_s$. We have already assumed T_s does not depend on P, as explained above, which from the physical point of view means that the function $T_s(P)$ is flat enough to allow the approximation $\tilde{T}_s \approx 0$ without a substantial error in the results. From our choice of P-dependence of the free energy [Eq. (7)] parameters, it follow that $\tilde{\rho}_s = 0$.

Setting $\tilde{\rho}_s = \tilde{\rho} = 0$ in Eq. (18) we obtain

$$t(T_m, P_m) = -\frac{\tilde{\gamma}^2}{4\tilde{\gamma}_1^2},\tag{19}$$

namely, the value $t_m(T, P) = t(T_m, P_m)$ at the maximum $T_m(P_m)$ of the curve $T_{FS}(P)$. Substituting t_m back in Eq. (17) we obtain T_m, and with its help we also obtain the pressure P_m, both given in Table 1, respectively.

We want to draw the attention to a particular feature of the present theory that the coordinates T_m and P_m of the maximum (point *max*) at the curve $T_{FS}(P)$ as well as the results from various calculations with the help of Eqs. (17) and (18) are expressed in terms of the reduced interaction parameters $\tilde{\gamma}$ and $\tilde{\gamma}_1$. Thus, using certain experimental data for T_m, P_m, as well as Eqs. (17) and (18) for T_{FS}, T_s, and the derivative ρ at particular values of the pressure P, $\tilde{\gamma}$ and $\tilde{\gamma}_1$ can be calculated without any additional information, for example, for the parameter κ. This property of the model (7) is quite useful in the practical work with the experimental data.

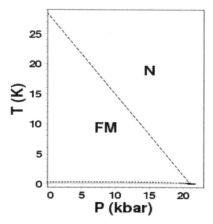

Figure 2. $T - P$ diagram of ZrZn$_2$ calculated for $T_s = 0$, $T_{f0} = 28.5$ K, $P_0 = 21$ kbar, $\kappa = 10$, $\tilde{\gamma} = 2\tilde{\gamma}_1 \approx 0.2$, and $n = 1$. The dotted line represents the FM-FS transition and the dashed line stands for the second order N-FM transition. The dotted line has a zero slope at $P = 0$. The low-temperature and high-pressure domains of the FS phase are seen more clearly in the following Figs. 3 and 4.

The conditions for existence of a maximum on the curve $T_{FS}(P)$ can be determined by requiring $\tilde{P}_m > 0$, and $\tilde{T}_m > 0$ and using the respective formulae for these quantities, shown in Table 1. This *max* always occurs in systems where $T_{FS}(0) \leq 0$ and the low-pressure part of the curve $T_{FS}(P)$ terminates at $T = 0$ for some non-negative critical pressure P_{0c} (see Sec. 6). But the *max* may occur also for some sets of material parameters, when $T_{FS}(0) > 0$ (see Fig. 3, where $P_m = 0$). All these shapes of the line $T_{FS}(P)$ are described by the model (7). Irrespectively of the particular shape, the curve $T_{FS}(P)$ given by Eq. (17) always terminates at the tricritical point (labeled B), with coordinates (P_B, T_B) (see, e.g., Figs. 4 and 7).

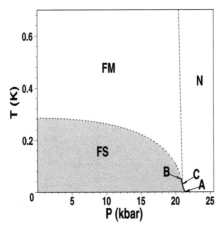

Figure 3. Details of Fig. 2 with expanded temperature scale. The points A, B, C are located in the high-pressure part ($P \sim P_c \sim 21$ kbar). The *max* point is at $P \approx 0$ kbar. The FS phase domain is shaded. The dotted line shows the second order FM-FS phase transition with $P_m \approx 0$. The solid straight line BC shows the fist-order FM-FS transition for $P > P_B$. The quite flat solid line AC shows the first order N-FS transition (the lines BC and AC are more clearly seen in Fig. 4. The dashed line stands for the second order N-FM transition.

At pressure $P > P_B$ the FM-FS phase transition is of first order up to the critical-end point C. For $P_B < P < P_C$ the FM-FS phase transition is given by the straight line BC (see, e.g., Figs. 4 and 7). The lines of all three phase transitions, N-FM, N-FS, and FM-FS, terminate at point C. For $P > P_C$ the FM-FS phase transition occurs on a rather flat smooth line of equilibrium transition of first order up to a second tricritical point A with $P_A \sim P_0$ and $T_A \sim 0$. Finally, the third transition line terminating at the point C describes the second order phase transition N-FM. The reduced temperatures $\tilde{T}_N$ and pressures $\tilde{P}_N$, $N = (A, B, C, max)$ at the three multi-critical points (A, B, and C), and the maximum $T_m(P_m)$ are given in Table 1. Note that, for any set of material parameters, $T_A < T_C < T_B < T_m$ and $P_m < P_B < P_C < P_A$.

There are other types of phase diagrams, resulting from model (7). For negative values of the generic superconducting temperature T_s, several other topologies of the $T - P$ diagram can be outlined. The results for the multicritical points, presented in Table 1, shows that, when T_s lowers below $T = 0$, T_C also decreases, first to zero, and then to negative values. When $T_C = 0$ the direct N-FS phase transition of first order disappears and point C becomes a very special zero-temperature multicritical point. As seen from Table 1, this happens for $T_s = -\gamma^2 T_f(0)/4\kappa(1+\gamma_1)$. The further decrease of T_s causes point C to fall below the zero temperature and then the zero-temperature phase transition of first order near P_c splits into two zero-temperature phase transitions: a second order N-FM transition and a first order FM-FS transition, provided T_B still remains positive.

At lower T_s also point B falls below $T = 0$ and the FM-FS phase transition becomes entirely of second order. For very extreme negative values of T_s, a very large pressure interval below P_c may occur where the FM phase is stable up to $T = 0$. Then the line $T_{FS}(P)$ will exist only for relatively small pressure values ($P \ll P_c$). This shape of the stability domain of the FS phase is also possible in real systems.

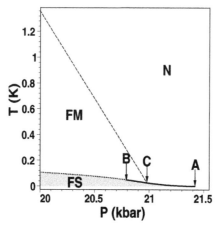

Figure 4. High-pressure part of the phase diagram of $ZrZn_2$, shown in Fig. 1. The thick solid lines AC and BC show the first-order transitions N-FS, and FM-FS, respectively. Other notations are explained in Figs. 2 and 3.

6. Quantum phase transitions

We have shown that the free energy (6) describes zero temperature phase transitions. Usually, the properties of these phase transitions essentially depend on the quantum fluctuations of the order parameters. For this reason the phase transitions at ultralow and zero temperature are called quantum phase transitions [31, 32]. The time-dependent quantum fluctuations (correlations) which describe the intrinsic quantum dynamics of spin-triplet ferromagnetic superconductors at ultralow temperatures are not included in our consideration but some basic properties of the quantum phase transitions can be outlines within the classical limit described by the free energy models (6) and (7). Let we briefly clarify this point.

The classical fluctuations are entirely included in the general GL functional (1)–(5) but the quantum fluctuations should be added in a further generalization of the theory. Generally, both classical (thermal) and quantum fluctuations are investigated by the method of the renormalization group (RG) [31], which is specially intended to treat the generalized action of system, where the order parameter fields (φ and M) fluctuate in time t and space $\vec{x}$ [31, 32]. These effects, which are beyond the scope of the paper, lead either to a precise treatment of the narrow critical region in a very close vicinity of second order phase transition lines or to a fluctuation-driven change in the phase-transition order. But the thermal fluctuations and quantum correlation effects on the thermodynamics of a given system can be unambiguously estimated only after the results from counterpart simpler theory, where these phenomena are not present, are known and, hence, the distinction in the thermodynamic properties predicted by the respective variants of the theory can be established. Here we show that the basic low-temperature and ultralow-temperature properties of the spin-triplet ferromagnetic superconductors, as given by the preceding experiments, are derived from the model (6) without any account of fluctuation phenomena and quantum correlations. The latter might be of use in a more detailed consideration of the close vicinity of quantum critical points in the phase diagrams of ferromagnetic spin-triplet superconductors. Here we show that the theory predicts quantum critical phenomena only for quite particular physical conditions whereas the low-temperature and zero-temperature phase transitions of first order are favored by both symmetry arguments and detailed thermodynamic analysis.

There is a number of experimental [9, 40] and theoretical [17, 41, 42] investigations of the problem for quantum phase transitions in unconventional ferromagnetic superconductors, including the mentioned intermetallic compounds. Some of them are based on different theoretical schemes and do not refer to the model (6). Others, for example, those in Ref. [41] reported results about the thermal and quantum fluctuations described by the model (6) before the comprehensive knowledge for the results from the basic treatment reported in the present investigation. In such cases one could not be sure about the correct interpretation of the results from the RG and the possibilities for their application to particular zero-temperature phase transitions. Here we present basic results for the zero-temperature phase transitions described by the model (6).

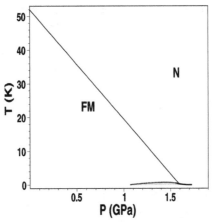

Figure 5. $T - P$ diagram of UGe$_2$ calculated taking $T_s = 0$, $T_{f0} = 52$ K, $P_0 = 1.6$ GPa, $\kappa = 4$, $\tilde{\gamma} = 0.0984$, $\tilde{\gamma}_1 = 0.1678$, and $n = 1$. The dotted line represents the FM-FS transition and the dashed line stands for the N-FM transition. The low-temperature and high-pressure domains of the FS phase are seen more clearly in the following Figs. 6 and 7.

The RG investigation [41] has demonstrated up to two loop order of the theory that the thermal fluctuations of the order parameter fields rescale the model (6) in a way which corresponds to first order phase transitions in magnetically anisotropic systems. This result is important for the metallic compounds we consider here because in all of them magnetic anisotropy is present. The uniaxial magnetic anisotropy in ZrZn$_2$ is much weaker than in UGe$_2$ but cannot be neglected when fluctuation effects are accounted for. Owing to the particular symmetry of model (6), for the case of magnetic isotropy (Heisenberg symmetry), the RG study reveals an entirely different class of (classical) critical behavior. Besides, the different spatial dimensions of the superconducting and magnetic quantum fluctuations imply a lack of stable quantum critical behavior even when the system is completely magnetically isotropic. The pointed arguments and preceding results lead to the reliable conclusion that the phase transitions, which have already been proven to be first order in the lowest-order approximation, where thermal and quantum fluctuations are neglected, will not undergo a fluctuation-driven change in the phase transition order from first to second. Such picture is described below, in Sec. 8, and it corresponds to the behavior of real compounds.

Our results definitely show that the quantum phase transition near P_c is of first order. This is valid for the whole N-FS phase transition below the critical-end point C, as well as the straight line BC. The simultaneous effect of thermal and quantum fluctuations do not change the order of the N-FS transition, and it is quite unlikely to suppose that thermal fluctuations of the superconductivity field ψ can ensure a fluctuation-driven change in the order of the FM-FS transition along the line BC. Usually, the fluctuations of ψ in low temperature superconductors are small and slightly influence the phase transition in a very narrow critical region in the vicinity of the phase-transition point. This effect is very weak and can hardly be observed in any experiment on low-temperature superconductors. Besides, the fluctuations of the magnetic induction B always tend to a fluctuation-induced first-order phase transition rather than to the opposite effect - the generation of magnetic fluctuations with infinite correlation length at the equilibrium phase-transition point and, hence, a second order phase transition [31, 43]. Thus we can quire reliably conclude that the first-order phase transitions at low-temperatures, represented by the lines BC and AC in vicinity of P_c do not change their order as a result of thermal and quantum fluctuation fluctuations.

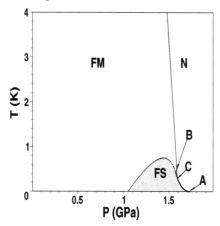

Figure 6. Low-temperature part of the $T - P$ phase diagram of UGe$_2$, shown in Fig. 5. The points A, B, C are located in the high-pressure part ($P \sim P_c \sim 1.6$ GPa). The FS phase domain is shaded. The thick solid lines AC and BC show the first-order transitions N-FS, and FM-FS, respectively. Other notations are explained in Figs. 2 and 3.

Quantum critical behavior for continuous phase transitions in spin-triplet ferromagnetic superconductors with magnetic anisotropy can therefore be observed at other zero-temperature transitions, which may occur in these systems far from the critical pressure P_c. This is possible when $T_{FS}(0) = 0$ and the $T_{FS}(P)$ curve terminates at $T = 0$ at one or two quantum (zero-temperature) critical points: $P_{0c} < P_m$ - "lower critical pressure", and $P'_{0c} > P_m$ – "upper critical pressure." In order to obtain these critical pressures one should solve Eq. (17) with respect to P, provided $T_{FS}(P) = 0$, $T_m > 0$ and $P_m > 0$, namely, when the continuous function $T_{FS}(P)$ exhibits a maximum. The critical pressure P'_{0c} is bounded in the relatively narrow interval (P_m, P_B) and can appear for some special sets of material parameters (r, t, γ, γ_1). In particular, as our calculations show, P'_{0c} do not exists for $T_s \geq 0$.

7. Criteria for type I and type II spin-triplet ferromagnetic superconductors

The analytical calculation of the critical pressures P_{0c} and P'_{0c} for the general case of $T_s \neq 0$ leads to quite complex conditions for appearance of the second critical field P'_{0c}. The correct treatment of the case $T_s \neq 0$ can be performed within the entire two-domain picture for the phase FS (see, also, Ref. [20]). The complete study of this case is beyond our aims but here we will illustrate our arguments by investigation of the conditions, under which the critical pressure P_{0c} occurs in systems with $T_s \approx 0$. Moreover, we will present the general result for $P_{0c} \geq 0$ and $P'_{0c} \geq 0$ in systems where $T_s \neq 0$.

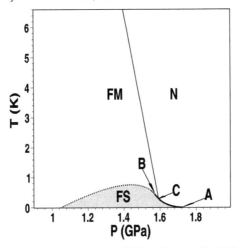

Figure 7. High-pressure part of the phase diagram of UGe$_2$, shown in Fig. 4. Notations are explained in Figs. 2, 3, 5, and 6.

Setting $T_{FS}(P_{0c}) = 0$ in Eq. (17) we obtain the following quadratic equation,

$$\tilde{\gamma}_1 m_{0c}^2 - \tilde{\gamma} m_{0c} - \tilde{T}_s = 0, \tag{20}$$

for the reduced magnetization,

$$m_{0c} = [-t(0, \tilde{P}_{0c})]^{1/2} = (1 - \tilde{P}_{0c})^{1/2} \tag{21}$$

and, hence, for $\tilde{P}_{0c}$. For $T_s \neq 0$, Eqs. (20) and (21) have two solutions with respect to $\tilde{P}_{0c}$. For some sets of material parameters these solutions satisfy the physical requirements for P_{0c} and P'_{0c} and can be identified with the critical pressures. The conditions for existence of P_{0c} and P'_{0c} can be obtained either by analytical calculations or by numerical analysis for particular values of the material parameters.

For $T_s = 0$, the trivial solution $\tilde{P}_{0c} = 1$ corresponds to $P_{0c} = P_0 > P_B$ and, hence, does not satisfy the physical requirements. The second solution,

$$\tilde{P}_{0c} = 1 - \frac{\tilde{\gamma}^2}{\tilde{\gamma}_1^2} \tag{22}$$

is positive for

$$\frac{\gamma_1}{\gamma} \geq 1 \tag{23}$$

and, as shown below, it gives the location of the quantum critical point $(T = 0, P_{0c} < P_m)$. At this quantum critical point, the equilibrium magnetization m_{0c} is given by $m_{0c} = \gamma/\gamma_1$ and is twice bigger that the magnetization $m_m = \gamma/2\gamma_1$ ([20]) at the maximum of the curve $T_{FS}(P)$.

To complete the analysis we must show that the solution (22) satisfies the condition $P_{0c} < \tilde{P}_m$. By taking $\tilde{P}_m$ from Table 1, we can show that solution (22) satisfies the condition $P_{0c} < \tilde{P}_m$ for $n = 1$, if

$$\gamma_1 < 3\kappa, \tag{24}$$

and for $n = 2$ (SFT case), when

$$\gamma < 2\sqrt{3}\kappa. \tag{25}$$

Finally, we determine the conditions under which the maximum T_m of the curve $T_{FS}(P)$ occurs at non-negative pressures. For $n = 1$, we obtain that $P_m \geq 0$ for $n = 1$, if

$$\frac{\gamma_1}{\gamma} \geq \frac{1}{2}\left(1 + \frac{\gamma_1}{\kappa}\right)^{1/2}, \tag{26}$$

whereas for $n = 2$, the condition is

$$\frac{\gamma_1}{\gamma} \geq \frac{1}{2}\left(1 + \frac{\gamma^2}{4\kappa^2}\right)^{1/2}. \tag{27}$$

Obviously, the conditions (23)-(27) are compatible with one another. The condition (26) is weaker than the condition Eq. (23), provided the inequality (24) is satisfied. The same is valid for the condition (27) if the inequality (25) is valid. In Sec. 8 we will show that these theoretical predictions are confirmed by the experimental data.

Doing in the same way the analysis of Eq. (17), some results may easily obtained for $T_s \neq 0$. In this more general case the Eq. (17) has two nontrivial solutions, which yield two possible values of the critical pressure

$$\tilde{P}_{0c(\pm)} = 1 - \frac{\gamma^2}{4\gamma_1^2}\left[1 \pm \left(1 + \frac{4\tilde{T}_s\kappa\gamma_1}{\gamma^2}\right)^{1/2}\right]^2. \tag{28}$$

The relation $\tilde{P}_{0c(-)} \geq \tilde{P}_{0c(+)}$ is always true. Therefore, to have both $\tilde{P}_{0c(\pm)} \geq 0$, it is enough to require $\tilde{P}_{0c(+)} \geq 0$. Having in mind that for the phase diagram shape, we study $\tilde{T}_m > 0$, and according to the result for $\tilde{T}_m$ in Table 1, this leads to the inequality $\tilde{T}_s > -\gamma^2/4\kappa\gamma_1$. So, we obtain that $\tilde{P}_{0c(+)} \geq 0$ will exist, if

$$\frac{\gamma_1}{\gamma} \geq 1 + \frac{\kappa\tilde{T}_s}{\gamma}, \tag{29}$$

which generalizes the condition (23).

Now we can identify the pressure $P_{0c(+)}$ with the lower critical pressure P_{0c}, and $P_{0c(-)}$ with the upper critical pressure P'_{0c}. Therefore, for wide variations in the parameters, theory (6) describes a quantum critical point P_{oc}, that exists, provided the condition (29) is satisfied. The quantum critical point $(T = 0, P_{0c})$ exists in UGe$_2$ and, perhaps, in other p-wave ferromagnetic superconductors, for example, in UIr.

Our results predict the appearance of second critical pressure – the upper critical pressure P'_{0c} that exists under more restricted conditions and, hence, can be observed in more particular systems, where $T_s < 0$. As mentioned in Sec. 5, for very extreme negative values of T_s, when $T_B < 0$, the upper critical pressure $P'_{0c} > 0$ occurs, whereas the lower critical pressure $P_{0c} > 0$ does not appear. Bue especially this situation should be investigated in a different way, namely, one should keep $T_{FS}(0)$ different from zero in Eq. (17), and consider a form of the FS phase domain in which the curve $T_{FS}(P)$ terminates at $T = 0$ for $P'_{0c} > 0$, irrespective of whether the maximum T_m exists or not. In such geometry of the FS phase domain, the maximum $T(P_m)$ may exist only in quite unusual cases, if it exists at all.

Using criteria like (23) in Sec. 8.4 we classify these superconductors in two types: (i) type I, when the condition (23) is satisfied, and (ii) type II, when the same condition does not hold. As we show in Sec. 8.2, 8.3 and 8.4, the condition (23) is satisfied by UGe$_2$ but the same condition fails for ZrZn$_2$. For this reason the phase diagrams of UGe$_2$ and ZrZn$_2$ exhibit qualitatively different shapes of the curves $T_FS(P)$. For UGe$_2$ the line $T_FS(P)$ has a maximum at some pressure $P > 0$, whereas the line $T_FS(P)$, corresponding to ZrZn$_2$, does not exhibit such maximum (see also Sec. 8).

The quantum and thermal fluctuation phenomena in the vicinities of the two critical pressures P_{0c} and P'_{0c} need a nonstandard RG treatment because they are related with the fluctuation behavior of the superconducting field ψ far below the ferromagnetic phase transitions, where the magnetization M does not undergo significant fluctuations and can be considered uniform. The presence of uniform magnetization produces couplings of M and ψ which are not present in previous RG studies and need a special analysis.

8. Application to metallic compounds

8.1. Theoretical outline of the phase diagram

In order to apply the above displayed theoretical calculations, following from free energy (7), for the outline of $T - P$ diagram of any material, we need information about the values of P_0, T_{f0}, T_s, κ γ, and γ_1. The temperature T_{f0} can be obtained directly from the experimental phase diagrams. The pressure P_0 is either identical or very close to the critical pressure P_c, for which the N-FM phase transition line terminates at $T \sim 0$. The temperature T_s of the generic superconducting transition is not available from the experiments because, as mentioned above, pure superconducting phase not coexisting with ferromagnetism has not been observed. This can be considered as an indication that T_s is very small and does not produce a measurable effect. So the generic superconducting temperature will be estimated on the basis of the following arguments. For $T_f(P) > T_s$ we must have $T_s(P) = 0$ at $P \geq P_c$, where $T_f(P) \leq 0$, and for $0 \leq P \leq P_0$, $T_s < T_C$. Therefore for materials where T_C is too small to be observed experimentally, T_s can be ignored.

As far as the shape of FM-FS transition line is well described by Eq. (17), we will make use of additional data from available experimental phase diagrams for ferroelectric superconductors. For example, in ZrZn$_2$ these are the observed values of $T_{FS}(0)$ and the slope $\rho_0 \equiv [\partial T_{FS}(P)/\partial P]_0 = (T_{f0}/P_0)\tilde{\rho}_0$ at $P = 0$; see Eq. (17). For UGe$_2$, where a maximum ($\tilde{T}_m$) is observed on the phase-transition line, we can use the experimental values of T_m, P_m, and P_{0c}. The interaction parameters $\tilde{\gamma}$ and $\tilde{\gamma}_1$ are derived using Eq. (17), and the expressions for $\tilde{T}_m$, $\tilde{P}_m$, and $\tilde{\rho}_0$, see Table 1. The parameter κ is chosen by fitting the expression for the critical-end point T_C.

8.2. ZrZn$_2$

Experiments for ZrZn$_2$ [13] gives the following values: $T_{f0} = 28.5$ K, $T_{FS}(0) = 0.29$ K, $P_0 \sim P_c = 21$ kbar. The curve $T_F(P) \sim T_f(P)$ is almost a straight line, which directly indicates that $n = 1$ is adequate in this case for the description of the P-dependence. The slope for $T_{FS}(P)$ at $P = 0$ is estimated from the condition that its magnitude should not exceed $T_{f0}/P_c \approx 0.014$ as we have assumed that is straight one, so as a result we have $-0.014 < \rho \leq 0$. This ignores the presence of a maximum. The available experimental data for ZrZn$_2$ do not give clear indication whether a maximum at (T_m, P_m) exists. If such a maximum were at $P = 0$ we would have $\rho_0 = 0$, whereas a maximum with $T_m \sim T_{FS}(0)$ and $P_m \ll P_0$ provides us with an estimated range $0 \leq \rho_0 < 0.005$. The choice $\rho_0 = 0$ gives $\tilde{\gamma} \approx 0.02$ and $\tilde{\gamma}_1 \approx 0.01$, but similar values hold for any $|\rho_0| \leq 0.003$. The multicritical points A and C cannot be distinguished experimentally. Since the experimental accuracy [13] is less than ~ 25 mK in the high-P domain ($P \sim 20 - 21$ kbar), we suppose that $T_C \sim 10$ mK, which corresponds to $\kappa \sim 10$. We employed these parameters to calculate the $T - P$ diagram using $\rho_0 = 0$ and 0.003. The differences obtained in these two cases are negligible, with both phase diagrams being in excellent agreement with experiment.

Phase diagram of ZrZn$_2$ calculated directly from the free energy (7) for $n = 1$, the above mentioned values of T_s, P_0, T_{f0}, κ, and values of $\tilde{\gamma} \approx 0.2$ and $\tilde{\gamma}_1 \approx 0.1$ which ensure $\rho_0 \approx 0$ is shown in Fig. 2. Note, that the experimental phase diagram [13] of ZrZn$_2$ looks almost exactly as the diagram in Fig. 2, which has been calculated directly from the model (7) without any approximations and simplifying assumptions. The phase diagram in Fig. 2 has the following coordinates of characteristic points: $P_A \sim P_c = 21.42$ kbar, $P_B = 20.79$ kbar, $P_C = 20.98$ kbar, $T_A = T_F(P_c) = T_{FS}(P_c) = 0$ K, $T_B = 0.0495$ K, $T_C = 0.0259$ K, and $T_{FS}(0) = 0.285$ K.

The low-T region is seen in more detail in Fig. 3, where the A, B, C points are shown and the order of the FM-FS phase transition changes from second to first order around the critical end-point C. The $T_{FS}(P)$ curve, shown by the dotted line in Fig. 3, has a maximum $T_m = 0.290$ K at $P = 0.18$ kbar, which is slightly above $T_{FS}(0) = 0.285$ K. The straight solid line BC in Fig. 3 shows the first order FM-FS phase transition which occurs for $P_B < P < P_C$. The solid AC line shows the first order N-FS phase transition and the dashed line stands for the N-FM phase transition of second order.

Although the expanded temperature scale in Fig. 3, the difference $[T_m - T_{FS}(0)] = 5$ mK is hard to see. To locate the point *max* exactly at $P = 0$ one must work with values of $\tilde{\gamma}$ and $\tilde{\gamma}_1$ of accuracy up to 10^{-4}. So, the location of the *max* for parameters corresponding to $ZrZn_2$ is very sensitive to small variations of $\tilde{\gamma}$ and $\tilde{\gamma}_1$ around the values 0.2 and 0.1, respectively. Our initial idea was to present a diagram with $T_m = T_{FS}(0) = 0.29$ K and $\rho_0 = 0$, namely, *max* exactly located at $P = 0$, but the final phase diagram slightly departs from this picture because of the mentioned sensitivity of the result on the values of the interaction parameters γ and γ_1. The theoretical phase diagram of $ZrZn_2$ can be deduced in the same way for $\rho_0 = 0.003$ and this yields $T_m = 0.301$ K at $P_m = 6.915$ kbar for initial values of $\tilde{\gamma}$ and $\tilde{\gamma}_1$ which differs from $\tilde{\gamma} = 2\tilde{\gamma}_1 = 0.2$ only by numbers of order $10^{-3} - 10^{-4}$ [18]. This result confirms the mentioned sensitivity of the location of the maximum T_m towards slight variations of the material parameters. Experimental investigations of this low-temperature/low-pressure region with higher accuracy may help in locating this maximum with better precision.

Fig. 4 shows the high-pressure part of the same phase diagram in more details. In this figure the first order phase transitions (solid lines BC and AC) are clearly seen. In fact the line AC is quite flat but not straight as the line BC. The quite interesting topology of the phase diagram of $ZrZn_2$ in the high-pressure domain ($P_B < P < P_A$) is not seen in the experimental phase diagram [13] because of the restricted accuracy of the experiment in this range of temperatures and pressures.

These results account well for the main features of the experimental behavior [13], including the claimed change in the order of the FM-FS phase transition at relatively high P. Within the present model the N-FM transition is of second order up to $P_C \sim P_c$. Moreover, if the experiments are reliable in their indication of a first order N-FM transition at much lower P values, the theory can accommodate this by a change of sign of b_f, leading to a new tricritical point located at a distinct $P_{tr} < P_C$ on the N-FM transition line. Since $T_C > 0$ a direct N-FS phase transition of first order is predicted in accord with conclusions from de Haas–van Alphen experiments [44] and some theoretical studies [40]. Such a transition may not occur in other cases where $T_C = 0$. In SFT ($n = 2$) the diagram topology remains the same but points B and C are slightly shifted to higher P (typically by about $0.01 - -0.001$ kbar).

8.3. UGe$_2$

The experimental data for UGe$_2$ indicate $T_{f0} = 52$ K, $P_c = 1.6$ GPa ($\equiv 16$ kbar), $T_m = 0.75$ K, $P_m \approx 1.15$ GPa, and $P_{0c} \approx 1.05$ GPa [2–5]. Using again the variant $n = 1$ for $T_f(P)$ and the above values for T_m and P_{0c} we obtain $\tilde{\gamma} \approx 0.0984$ and $\tilde{\gamma}_1 \approx 0.1678$. The temperature $T_C \sim 0.1$ K corresponds to $\kappa \sim 4$.

Using these initial parameters, together with $T_s = 0$, leads to the $T - P$ diagram of UGE$_2$ shown in Fig. 5. We obtain $T_A = 0$ K, $P_A = 1.723$ GPa, $T_B = 0.481$ K, $P_B = 1.563$ GPa, $T_C = 0.301$ K, and $P_C = 1.591$ GPa. Figs. 6 and 7 show the low-temperature and the high-pressure parts of this phase diagram, respectively. There is agreement with the main experimental findings, although P_m corresponding to the maximum (found at ~ 1.44 GPa in Fig. 5) is about 0.3 GPa higher than suggested experimentally [4, 5]. If the experimental plots are accurate in this respect, this difference may be attributable to the so-called (T_x) meta-magnetic phase transition in UGe$_2$, which is related to an abrupt change of the magnetization in the vicinity of P_m. Thus, one may suppose that the meta-magnetic effects, which are outside the scope of our current model, significantly affect the shape of the $T_{FS}(P)$ curve by lowering P_m (along with P_B and P_C). It is possible to achieve a lower P_m value (while leaving T_m unchanged), but this has the undesirable effect of modifying P_{c0} to a value that disagrees with experiment. In SFT ($n = 2$) the multi-critical points are located at slightly higher P (by about 0.01 GPa), as for $ZrZn_2$. Therefore, the results from the SFT theory are slightly worse than the results produced by the usual linear approximation ($n = 1$) for the parameter t.

8.4. Two types of ferromagnetic superconductors with spin-triplet electron pairing

The estimates for UGe_2 imply $\gamma_1 \kappa \approx 1.9$, so the condition for $T_{FS}(P)$ to have a maximum found from Eq. (17) is satisfied. As we discussed for $ZrZn_2$, the location of this maximum can be hard to fix accurately in experiments. However, P_{c0} can be more easily distinguished, as in the UGe_2 case. Then we have a well-established quantum (zero-temperature) phase transition of second order, i.e., a quantum critical point at some critical pressure $P_{0c} \geq 0$. As shown in Sec. 6, under special conditions the quantum critical points could be two: at the lower critical pressure $P_{0c} < P_m$ and the upper critical pressure $P'_{0c} < P_m$. This type of behavior in systems with $T_s = 0$ (as UGe_2) occurs when the criterion (23) is satisfied. Such systems (which we label as U-type) are essentially different from those such as $ZrZn_2$ where $\gamma_1 < \gamma$ and hence $T_{FS}(0) > 0$. In this latter case (Zr-type compounds) a maximum $T_m > 0$ may sometimes occur, as discussed earlier. We note that the ratio γ/γ_1 reflects a balance effect between the two ψ-M interactions. When the trigger interaction (typified by γ) prevails, the Zr-type behavior is found where superconductivity exists at $P = 0$. The same ratio can be expressed as $\gamma_0/\delta_0 M_0$, which emphasizes that the ground state value of the magnetization at $P = 0$ is also relevant. Alternatively, one may refer to these two basic types of spin-triplet ferromagnetic superconductors as "type I" (for example, for the "Zr-type compounds), and "type II" – for the U-type compounds.

As we see from this classification, the two types of spin-triplet ferromagnetic superconductors have quite different phase diagram topologies although some fragments have common features. The same classification can include systems with $T_s \neq 0$ but in this case one should use the more general criterion (29).

8.5. Other compounds

In URhGe, $T_f(0) \sim 9.5$ K and $T_{FS}(0) = 0.25$ K and, therefore, as in $ZrZn_2$, here the spin-triplet superconductivity appears at ambient pressure deeply in the ferromagnetic phase domain [6–8]. Although some similar structural and magnetic features are found in UGe_2 the results in Ref. [8] of measurements under high pressure show that, unlike the behavior of $ZrZn_2$ and UGe_2, the ferromagnetic phase transition temperature $T_F(P) \sim T_f(P)$ has a slow linear increase up to 140 kbar without any experimental indications that the N-FM transition line may change its behavior at higher pressures and show a negative slope in direction of low temperature up to a quantum critical point $T_F = 0$ at some critical pressure P_c. Such a behavior of the generic ferromagnetic phase transition temperature cannot be explained by our initial assumption for the function $T_f(P)$ which was intended to explain phase diagrams where the ferromagnetic order is depressed by the pressure and vanishes at $T = 0$ at some critical pressure P_c. The $T_{FS}(P)$ line of URhGe shows a clear monotonic negative slope to $T = 0$ at pressures above 15 kbar and the extrapolation [8] of the experimental curve $T_{FS}(P)$ tends a quantum critical point $T_{FS}(P'_{0c}) = 0$ at $P_{0c} \sim 25 - 30$ kbar. Within the framework of the phenomenological theory (6, this $T - P$ phase diagram can be explained after a modification on the $T_f(P)$-dependence is made, and by introducing a convenient nontrivial pressure dependence of the interaction parameter γ. Such modifications of the present theory are possible and follow from important physical requirements related with the behavior of the f-band electrons in URhGe. Unlike UGe$_2$, where the pressure increases the hybridization of the $5f$ electrons with band states lading to a suppression of the spontaneous magnetic moment M, in URhGe this effects is followed by a stronger effect of enhancement of the exchange coupling due to the same hybridization, and this effect leads to the slow but stable linear increase in the function $T_F(P)$[8]. These effects should be taken into account in the modeling the pressure dependence of the parameters of the theory (7) when applied to URhGe.

Another ambient pressure FS phase has been observed in experiments with UCoGe [9]. Here the experimentally derived slopes of the functions $T_F(P)$ and $T_{FS}(P)$ at relatively small pressures are opposite compared to those for URhGe and, hence, the $T - P$ phase diagram of this compound can be treated within the present theoretical scheme without substantial modifications.

Like in UGe$_2$, the FS phase in UIr [12] is embedded in the high-pressure/low-temperature part of the ferromagnetic phase domain near the critical pressure P_c which means that UIr is certainly a U-type compound. In UGe$_2$ there is one metamagnetic phase transition between two ferromagnetic phases (FM1 and FM2), in UIr there are three ferromagnetic phases and the FS phase is located in the low-T/high-P domain of the third of them - the phase FM3. There are two metamagnetic-like phase transitions: FM1-FM2 transition which is followed by a drastic decrease of the spontaneous magnetization when the the lower-pressure phase FM1 transforms to FM2, and a peak of the ac susceptibility but lack of observable jump of the magnetization at the second (higher pressure) "metamagnetic" phase transition from FM2 to FM3. Unlike the picture for UGe$_2$, in UIr both transitions, FM1-FM2 and FM2-FM3 are far from the maximum $T_m(P_m)$ so in this case one can hardly speculate that the *max* is produced by the nearby jump of magnetization. UIr seems to be a U-type spin-triplet ferromagnetic superconductor.

9. Final remarks

Finally, even in its simplified form, this theory has been shown to be capable of accounting for a wide variety of experimental behavior. A natural extension to the theory is to add a M^6 term which provides a formalism to investigate possible metamagnetic phase transitions [45] and extend some first order phase transition lines. Another modification of this theory, with regard to applications to other compounds, is to include a P dependence for some of the other GL parameters. The fluctuation and quantum correlation effects can be considered by the respective field-theoretical action of the system, where the order parameters ψ and M are not uniform but rather space and time dependent. The vortex (spatially non-uniform) phase due to the spontaneous magnetization M is another phenomenon which can be investigated by a generalization of the theory by considering nonuniform order parameter fields ψ and M (see, e.g., Ref. [28]). Note that such theoretical treatments are quite complex and require a number of approximations. As already noted in this paper the magnetic fluctuations stimulate first order phase transitions for both finite and zero phase-transition temperatures.

Author details

Dimo I. Uzunov
Collective Phenomena Laboratory, G. Nadjakov Institute of Solid State Physics, Bulgarian Academy of Sciences, BG-1784 Sofia, Bulgaria.
Pacs: 74.20.De, 74.25.Dw, 64.70.Tg

10. References

[1] D. Vollhardt and P. Wölfle, it The Superfluid Phases of Helium 3 (Taylor & Francis, London, 1990); D. I. Uzunov, in: *Advances in Theoretical Physics*, edited by E. Caianiello (World Scientific, Singapore, 1990), p. 96; M. Sigrist and K. Ueda, Rev. Mod. Phys. 63, 239 (1991).

[2] S. S. Saxena, P. Agarwal, K. Ahilan, F. M. Grosche, R. K. W. Haselwimmer, M.J. Steiner, E. Pugh, I. R. Walker, S.R. Julian, P. Monthoux, G. G. Lonzarich, A. Huxley. I. Sheikin, D. Braithwaite, and J. Flouquet, Nature 406, 587 (2000).

[3] A. Huxley, I. Sheikin, E. Ressouche, N. Kernavanois, D. Braithwaite, R. Calemczuk, and J. Flouquet, Phys. Rev. B63, 144519 (2001).

[4] N. Tateiwa, T. C. Kobayashi, K. Hanazono, A. Amaya, Y. Haga. R. Settai, and Y. Onuki, J. Phys. Condensed Matter 13, L17 (2001).

[5] A. Harada, S. Kawasaki, H. Mukuda, Y. Kitaoka, Y. Haga, E. Yamamoto, Y. Onuki, K. M. Itoh, E. E. Haller, and H. harima, Phys. Rev. B 75, 140502 (2007).

[6] D. Aoki, A. Huxley, E. Ressouche, D. Braithwaite, J. Flouquet, J-P.. Brison, E. Lhotel, and C. Paulsen, Nature 413, 613 (2001).

[7] F. Hardy, A. Huxley, Phys. Rev. Lett. 94, 247006 (2005).

[8] F. Hardy, A. Huxley, J. Flouquet, B. Salce, G. Knebel, D. Braithwate, D. Aoki, M. Uhlarz, and C. Pfleiderer, Physica B 359-361 1111 (2005).

[9] N. T. Huy, A. Gasparini, D. E. de Nijs, Y. Huang, J. C. P. Klaasse, T. Gortenmulder, A. de Visser, A. Hamann, T. Görlach, and H. v. Löhneysen, Phys. Rev. Lett. 99, 067006 (2007).

[10] N. T. Huy, D. E. de Nijs, Y. K. Huang, and A. de Visser, Phys. Rev. Lett. 100, 077001 (2008).

[11] T. Akazawa, H. Hidaka, H. Kotegawa, T. C. Kobayashi, T. Fujiwara, E. Yamamoto, Y. Haga, R. Settai, and Y. Onuki, Physica B 359-361, 1138 (2005).

[12] T. C. Kobayashi,S. Fukushima, H. Hidaka, H. Kotegawa, T. Akazawa, E. Yamamoto, Y. Haga, R. Settai, and Y. Onuki, Physica B 378-361, 378 (2006).

[13] C. Pfleiderer, M. Uhlatz, S. M. Hayden, R. Vollmer, H. v. Löhneysen, N. R. Berhoeft, and G. G. Lonzarich, Nature 412, 58 (2001).

[14] E. A. Yelland, S. J. C. Yates, O. Taylor, A. Griffiths, S. M. Hayden, and A. Carrington, Phys. Rev. B 72, 184436 (2005).

[15] E. A. Yelland, S. M. Hayden, S. J. C. Yates, C. Pfleiderer, M. Uhlarz, R. Vollmer, H. v Löhneysen, N. R. Bernhoeft, R. P. Smith, S. S. Saxena, and N. Kimura, Phys. Rev. B72, 214523 (2005).

[16] C. J. Bolesh and T. Giamarchi, Phys. Rev. Lett. 71, 024517 (2005); R. D. Duncan, C. Vaccarella, and C. A. S. de Melo, Phys. Rev. B 64, 172503 (2001).

[17] A. H. Nevidomskyy, Phys. Rev. Lett. 94, 097003 (2005).

[18] M. G. Cottam, D. V. Shopova and D. I. Uzunov, Phys. Lett. A 373, 152 (2008).

[19] D. V. Shopova and D. I. Uzunov, Phys. Rev. B 79, 064501 (2009).

[20] D. V. Shopova and D. I. Uzunov, Phys. Rev. 72, 024531 (2005); Phys. Lett. A 313, 139 (2003).

[21] D. V. Shopova and D. I. Uzunov, in: *Progress in Ferromagnetism Research*, ed. by V. N. Murray (Nova Science Publishers, New York, 2006), p. 223; D. V. Shopova and D. I. Uzunov, J. Phys. Studies , 4, 426 (2003) 426; D. V. Shopova and D. I. Uzunov, Compt. Rend Acad. Bulg. Sci. 56, 35 (2003) 35; D. V. Shopova, T. E. Tsvetkov, and D. I. Uzunov, Cond. Matter Phys. 8, 181 (2005) 181; D. V. Shopova, and D. I. Uzunov, Bulg. J. of Phys. 32, 81 (2005).

[22] E. K. Dahl and A. Sudbø, Phys. Rev. B 75, 1444504 (2007).

[23] K. Machida and T. Ohmi, Phys. Rev. Lett. 86, 850 (2001).

[24] M. B. Walker and K. V. Samokhin, Phys. Rev. Lett. 88, 207001 (2002); K. V. Samokhin and M. B. Walker, Phys. Rev. B 66, 024512 (2002); Phys. Rev. B 66, 174501 (2002).

[25] J. Linder, A. Sudbø, Phys. Rev. B 76, 054511 (2007); J. Linder, I. B. Sperstad, A. H. Nevidomskyy, M. Cuoco, and A. Sodbø, Phys. Rev. 77, 184511 (2008); J. Linder, T. Yokoyama, and A. Sudbø, Phys. Rev. B 78, 064520 (2008); J. Linder, A. H. Nevidomskyy, A. Sudbø, Phys. Rev. B 78, 172502 (2008).

[26] R. A. Cowley, Adv. Phys. 29, 1 (1980); J-C. Tolédano and P. Tolédano, *The Landau Theory of Phase Transitions* (World Scientific, Singapore, 1987).

[27] S. V. Vonsovsky, Yu. A. Izyumov, and E. Z. Kurmaev, *Superconductivity of Transition Metals* (Springer Verlag, Berlin, 1982).

[28] L. N. Bulaevskii, A. I Buzdin, M. L. Kulić, and S. V. Panyukov, Adv. Phys. 34, 175 (1985); Sov. Phys. Uspekhi, 27, 927 (1984). A. I. Buzdin and L. N. Bulaevskii, Sov. Phys. Uspekhi 29, 412 (1986).

[29] E. I. Blount and C. M. Varma, Phys. Rev. Lett. 42, 1079 (1979).

[30] K. K. Murata and S. Doniach, Phys. Rev. Lett. 29, 285 (1972); G. G. Lonzarich and L. Taillefer, J. Phys. C: Solid State Phys. 18, 4339 (1985); T. Moriya, J. Phys. Soc. Japan 55, 357 (1986); H. Yamada, Phys. Rev. B 47, 11211 (1993).

[31] D. I. Uzunov, *Theory of Critical Phenomena*, Second Edition (World Scientific, Singapore, 2010).

[32] D. V. Shopova and D. I. Uzunov, Phys. Rep. C 379, 1 (2003).

[33] A. A. Abrikosov, Zh. Eksp. Teor. Fiz. 32, 1442 (1957) [Sov. Phys. JETP 5q 1174 (1957)].

[34] E. M. Lifshitz and L. P. Pitaevskii, *Statistical Physics, II Part* (Pergamon Press, London, 1980) [*Landau-Lifshitz Course in Theoretical Physics, Vol. IX*].

[35] H. Belich, O. D. Rodriguez Salmon, D. V. Shopova and D. I. Uzunov, Phys. Lett. A 374, 4161 (2010); H. Belich and D. I. Uzunov, Bulg. J. Phys. 39, 27 (2012).

[36] E. P. Wohlfarth, J. Appl. Phys. 39, 1061 (1968); Physica B&C 91B, 305 (1977).

[37] P. Misra, *Heavy-Fermion Systems*, (Elsevier, Amsterdam, 2008).

[38] K. G. Sandeman, G. G. Lonzarich, and A. J. Schofield, Phys. Rev. Lett. 90, 167005 (2003).

[39] T. F. Smith, J. A. Mydosh, and E. P. Wohlfarth, Phys. rev. Lett. 27, 1732 (1971); G. Oomi, T. Kagayama, K. Nishimura, S. W. Yun, and Y. Onuki, Physica B 206, 515 (1995).

[40] M. Uhlarz, C. Pfleiderer, and S. M. Hayden, Phys. Rev. Lett. 93, 256404 (2004).

[41] D. I. Uzunov, Phys. Rev. B74, 134514 (2006); Europhys. Lett. 77, 20008 (2007).

[42] D. Belitz, T. R. Kirkpatrick, J. Rollbühler, Phys. Rev. Lett. 94, 247205 (2005); G. A. Gehring, Europhys. Lett. 82, 60004 (2008).

[43] B. I. Halperin, T. C. Lubensky, and S. K. Ma, Phys. Rev. Lett. 32, 292 (1974); J-H. Chen, T. C. Lubensky, and D. R. Nelson, *Phys. Rev.* B17, 4274 (1978).

[44] N. Kimura *et al.*, Phys. Rev. Lett. 92 , 197002 (2004).

[45] A. Huxley, I. Sheikin, and D. Braithwaite, Physica B 284-288, 1277 (2000).

Chapter 9

Path-Integral Description of Cooper Pairing

Jacques Tempere and Jeroen P.A. Devreese

Additional information is available at the end of the chapter

1. Introduction

Before we start applying path integration to treat Cooper pairing and superfluidity, it is a good idea to quickly review the concepts behind path integration. There are many textbooks providing plentiful details, such as Feynman's seminal text [1] and Kleinert's comprehensive compendium [2], and other works listed in the bibliography [3–5]. We will assume that the reader is already familiar with the basics of path-integral theory, so if the following paragraphs are not merely reminders to you, it is probably better to first consult these textbooks.

Quantum mechanics, according to the path-integral formalism, rests on two axioms. The first axiom, the superposition axiom, states that the amplitude of any process is a weighed sum of the amplitudes of all possible possibilities for the process to occur. These "possible possibilities" should be interpreted as the alternatives that cannot be distinguished by the experimental setup under consideration. For example, the amplitude for a particle to go from a starting point "A" to a final point "B" is a weighed sum of the amplitudes of all the paths that this particle can take to get to "B" from "A". The second axiom assigns to the weight the complex value $\exp\{iS/\hbar\}$ where S is the action functional. In our example, each path $x(t)$ that the particle can take to go from A to B gets a weight $\exp\{iS[x(t)]/\hbar\}$ since the action is the time integral of the Lagrangian. There is a natural link with quantum-statistical mechanics: in the path-integral formalism, quantum statistical averages are expressed as the same weighed averages but now the weight is a real value $\exp\{-S[x(\tau)]/\hbar\}$ and the path is taken in imaginary time $\tau = it$.

In the example of the above paragraph, we considered a particle which could take many different paths from A to B. However, the same axioms can be applied to fields. As an example we take a complex scalar field $\phi_{x,t}$, where x and t denote position and time respectively. Let us mentally discretize space-time, and to make things easy, we assume there are only five moments in time and five places to sit. In this simple universe, the field $\phi_{x,t}$ is represented by a set of 25 complex numbers, i.e. an element of $\mathbb{C}^{25}$ if $\mathbb{C}$ is the set of complex numbers. Summing over all possible realizations of the fields corresponds to integrating over $\mathbb{C}^{25}$, a 25-fold integral over complex variables, or a 50-fold integral over real variables. Writing

$\phi_{\mathbf{x},t} = u_{\mathbf{x},t} + iv_{\mathbf{x},t}$ with u and v real, the summation over all possible possibilities for $\phi_{\mathbf{x},t}$ is written as

$$\int \mathcal{D}\phi_{\mathbf{x},t} := \prod_{x=1}^{5}\prod_{t=1}^{5} \int du_{\mathbf{x},t} \int dv_{\mathbf{x},t}. \tag{1}$$

The notation with calligraphic $\mathcal{D}$ indicates the path-integral sum, and we keep this notation for actual continuous spacetime, that we may see as a limit of a finer and finer grid of spacetime points (our 5x5 grid is obviously very crude). Although this is, strictly speaking, no longer a sum over paths, it is still called a path integral because the description is based on the same axiomatic view as outlined in the previous paragraph.

Each particular realization of $\phi_{\mathbf{x},t}$ again gets assigned a weight, where now we need a functional of $\phi_{\mathbf{x},t}$ (or, in our example, a function of 25 complex variables). Again, we use the action functional

$$S[\phi_{\mathbf{x},t}] = \int \mathcal{L}(\dot{\phi}_{\mathbf{x},t}, \phi_{\mathbf{x},t})dt \tag{2}$$

to construct the weight, where $\mathcal{L}$ is the Lagrangian of the field theory suitable for $\phi_{\mathbf{x},t}$. A central quantity to calculate is the statistical partition sum

$$\mathcal{Z} = \int \mathcal{D}\phi_{\mathbf{x},\tau} \, \exp\{-S[\phi_{\mathbf{x},\tau}]/\hbar\}, \tag{3}$$

where $\tau = it$ indicates imaginary times required for the quantum statistical expression, running from $\tau = 0$ to $\tau = \hbar\beta$ with $\beta = 1/(k_B T)$ the inverse temperature. Bose gases in condensed matter are described by complex scalar fields like $\phi_{\mathbf{x},t}$, and the path integral can basically only be solved analytically when the action functional is quadratic in form, i.e. when

$$S[\phi_{\mathbf{x},\tau}]/\hbar = \int d\mathbf{x} \int dt \int d\mathbf{x}' \int dt' \, \phi_{\mathbf{x},\tau}\mathbb{A}(\mathbf{x},t;\mathbf{x}',t')\phi_{\mathbf{x}',\tau'}. \tag{4}$$

In our simple universe, $\mathbb{A}$ would be a 25×25 matrix, and the path integral would reduce to a product of 25 complex Gaussian integrals, leading to

$$\mathcal{Z} \propto \frac{1}{\det(\mathbb{A})}. \tag{5}$$

The proportionality is written here because every integration also gives a (physically unimportant) factor π that can be absorbed in the integration measure, if needed.

Fermionic systems, such as the electrons in a metal or ultracold fermionic atoms in a magnetic trap, cannot be described by complex scalar fields: fermionic fields should anticommute [6]. If we axiomatically impose anticommutation onto scalar variables, we obtain Grassmann variables [7]. Since there is also a spin degree of freedom, the fermionic fields require a spin index σ as well as spacetime indices $\mathbf{x}, \tau$: $\psi_{\mathbf{x},\tau,\sigma}$. As Grassmann variables anticommute, we have $\psi_{\mathbf{x},\tau,\sigma}^2 = 0$. Integrals over Grassmann variables ("Berezin-Grassmann integrals" [8]) are defined by $\int d\psi_{\mathbf{x},\tau,\sigma} = 0$ and $\int \psi_{\mathbf{x},\tau,\sigma}d\psi_{\mathbf{x},\tau,\sigma} = 1$. As was the case for bosonic fields, also for fermionic fields there is only one generic path integration that can be done analytically, namely that with a quadratic action. A quadratic action functional in Grassmann fields is written in general form as

$$\mathcal{S}[\bar{\psi}_{\mathbf{x},\tau,\sigma}, \psi_{\mathbf{x},\tau,\sigma}]/\hbar = \sum_{\sigma} \int d\mathbf{x} \int dt \sum_{\sigma'} \int d\mathbf{x}' \int dt' \, \bar{\psi}_{\mathbf{x},\tau,\sigma} \mathbb{A}(\mathbf{x}, t; \mathbf{x}', t') \psi_{\mathbf{x}',\tau',\sigma'}, \tag{6}$$

where $\bar{\psi}_{\mathbf{x},\tau,\sigma}$ and $\psi_{\mathbf{x},\tau,\sigma}$ are different Grassmann variables (so in our example, we would need 50 pairs of Grassmann elements). The result is

$$\mathcal{Z} = \int \mathcal{D}\bar{\psi}_{\mathbf{x},\tau,\sigma} \int \mathcal{D}\psi_{\mathbf{x},\tau,\sigma} \exp\left\{-\mathcal{S}[\bar{\psi}_{\mathbf{x},\tau,\sigma}, \psi_{\mathbf{x},\tau,\sigma}]\right\} = \det(\mathbb{A}). \tag{7}$$

Despite having only analytic results for quadratic action functionals, the path-integral technique is nevertheless a very versatile tool and has become in fact the main tool to study field theory [9]. The trick usually consists in finding suitable transformations and approximations to bring the path integral into the same form as that with quadratic action functionals.

2. The action functional for the atomic Fermi gas

In the study of superconductivity, ultracold quantum gases offer a singular advantage over condensed matter systems in that their system parameters can be tuned experimentally with a high degree of precision and over a wide range. For example, the interaction strength between fermionic atoms is tunable by an external magnetic field. This field can be used to vary the scattering length over a Feshbach resonance, from a large negative to a large positive value. In the limit of large negative scatting lengths, a cloud of ultracold fermionic atoms will undergo Cooper pairing, and exhibit superfluidity when cooled below the critical temperature. On the other side of the resonance, at large positive scattering lengths, a molecular bound state gets admixed to the scattering state, and a Bose-Einstein condensate (BEC) of fermionic dimers can form. With the magnetically tunable s-wave scattering length, the entire crossover region between the Bardeen-Cooper-Schrieffer (BCS) superfluid and the molecular BEC can be investigated in a way that is thus far not possible in solids.

Also the amount of atoms in each hyperfine state can be tuned experimentally with high precision. Typically, fermionic atoms (such as ^{40}K or ^{6}Li) are trapped in two different hyperfine states. These two hyperfine spin states provide the "spin-up" and "spin-down" partners that form the Cooper pairs. Unlike in metals, in quantum gases the individual amounts of "spin-up" and "spin-down" components of the Fermi gas can be set independently (using evaporative cooling and Rabi oscillations). This allows to investigate how Cooper pairing (and the ensuing superfluidity) is frustrated when there is no equal amount of spin ups and spin downs, i.e. in the so-called "(spin-)imbalanced Fermi gas". In a superconducting metal, the magnetic field that would be required to provide a substantial imbalance between spin-up and spin-down electrons is simply expelled by the Meissner effect, so that the imbalanced situation cannot be studied. The particular question of the effect of spin-imbalance is of great current interest [10], and we will keep our treatment general enough to include this effect.

Finally, the geometry of the gas is adaptable. Counterpropagating laser beams can be used to make periodic potentials for the atoms, called "optical lattices". Imposing such a lattice in just one direction transforms the atomic cloud into a stack of pancake-shaped clouds, with tunable tunneling amplitude between pancakes. The confinement in the out-of-pancake direction can be made tight enough to allow to study the physics of the two-dimensional system effectively. Imposing an optical lattice in more than one direction, all manner of crystals can be formed, and it becomes possible to engineer an experimental realization of the Bose and Fermi Hubbard models, for example. The joint tunability of the number of atoms, temperature, dimensionality, and interaction strength has in the past couple of decades turned quantum gases into powerful quantum simulators of condensed matter models [11]. In this respect, ultracold quantum gases are also being used to enlarge our knowledge of superconductivity, through the study of pairing and superfluidity.

A key aspect of ultracold quantum gases is that the interatomic interaction can be characterized by a single number, the s-wave scattering length mentioned above. In fact, the requirement to use the label "ultracold" is that the typical wave length associated with the atomic motion is much larger than the range of the interatomic potential, so that higher partial waves in the scattering process are frozen out. This aspect allows to simplify the treatment of the interatomic interactions tremendously. Rather than using a complicated interatomic potential, we can use a contact pseudopotential $V(\mathbf{r} - \mathbf{r}') = g\delta(\mathbf{r} - \mathbf{r}')$, and adapt its strength g such that the model- or pseudopotential has the same s-wave scattering length as the true potential. Doing so requires some care [12, 13], and using the Lippmann-Schwinger equation up to second order results in the following expression for the renormalized strength g of the contact interaction (in the three dimensional case):

$$\frac{1}{g} = \frac{m}{4\pi\hbar^2 a_s} - \int \frac{d\mathbf{k}}{(2\pi)^3} \frac{1}{\hbar^2 k^2/m}. \tag{8}$$

Moreover, for a Fermi gas there is an additional simplification: due to the obligation of antisymmetrizing the wave function the s-wave scattering amplitude between fermions with the same (hyperfine) spin is zero. This means that in an ultracold Fermi gas, only atoms with different spin states interact. This allows us to write the action functional for the Fermi gas of atoms with mass m as

$$\mathcal{S}[\{\bar{\psi}_{\mathbf{x},\tau,\sigma}, \psi_{\mathbf{x},\tau,\sigma}\}] = \int_0^{\hbar\beta} d\tau \int d\mathbf{x} \sum_\sigma \left[\bar{\psi}_{\mathbf{x},\tau,\sigma} \left(\hbar \frac{\partial}{\partial \tau} - \frac{\hbar^2}{2m} \nabla_\mathbf{x}^2 - \mu_\sigma \right) \psi_{\mathbf{x},\tau,\sigma} \right]$$
$$+ \int_0^{\hbar\beta} d\tau \int d\mathbf{x} \int d\mathbf{y} \, \bar{\psi}_{\mathbf{x},\tau,\uparrow} \bar{\psi}_{\mathbf{y},\tau,\downarrow} \, g\delta(\mathbf{x} - \mathbf{y}) \, \psi_{\mathbf{y},\tau,\downarrow} \psi_{\mathbf{x},\tau,\uparrow}. \tag{9}$$

Here, $\beta = 1/(k_B T)$ is again the inverse temperature. The quadratic term corresponds to the free particle Lagrangian. The amounts of spin up $\sigma =\uparrow$ and spin down $\sigma =\downarrow$ are set by the chemical potentials $\mu_\uparrow$ and $\mu_\downarrow$, respectively. It is the total particle density $n_\uparrow + n_\downarrow$ that is used to define a Fermi wave vector: $k_F = (3\pi^2(n_\uparrow + n_\downarrow))^{1/3}$ in three dimensions, and $k_F = (2\pi(n_\uparrow + n_\downarrow))^{1/2}$ in two dimensions. The quartic term corresponds to the interaction part, and we have used the contact pseudopotential.

To keep our notations simple and make integrals and variables dimensionless, we will introduce natural units of k_F, the Fermi wave number, and E_F, the Fermi energy. Also we use $\omega_F = E_F/\hbar$ for frequency unit and $T_F = E_F/k_B$ for temperature unit [so this all comes down to setting $\hbar, 2m, k_F, k_B = 1$]. Note that since the fields have units of volume$^{-1/2}$, the action has units of $\hbar$. Moreover, g has units of energy times volume, since $g\delta(\mathbf{x} - \mathbf{y})$ is a potential energy. Introducing the dimensionless space/time variables $\mathbf{x}' = \mathbf{x}k_F$ and $\tau' = E_F\tau/\hbar$, we get

$$S\left[\{\bar{\psi}_{\mathbf{x},\tau,\sigma}, \psi_{\mathbf{x},\tau,\sigma}\}\right] = \int_0^{T_F/T} \frac{\hbar d\tau'}{E_F} \int d\mathbf{x}' \sum_\sigma \left[\bar{\psi}_{\mathbf{x}',\tau',\sigma}\left(E_F\frac{\partial}{\partial\tau'} - \frac{\hbar^2 k_F^2}{2m}\nabla^2_{\mathbf{x}'} - \mu_\sigma\right)\psi_{\mathbf{x}',\tau',\sigma}\right]$$

$$+ \int_0^{T_F/T} \frac{\hbar d\tau'}{E_F} \int d\mathbf{x}' \int d\mathbf{y}' \; \bar{\psi}_{\mathbf{x}',\tau',\uparrow}\bar{\psi}_{\mathbf{y}',\tau',\downarrow}\, g\left[k_F^3\delta(\mathbf{x}' - \mathbf{y}')\right]\psi_{\mathbf{y}',\tau',\downarrow}\psi_{\mathbf{x}',\tau',\uparrow}. \quad (10)$$

Since $\bar{\psi}_{\mathbf{x},\tau,\sigma}\psi_{\mathbf{x},\tau,\sigma}d\mathbf{x}$ was dimensionless to start with, it must be equal to the corresponding expression with the primed variables. Introducing $\mu_\sigma' = \mu_\sigma/E_F$, $\beta' = \beta E_F$ and $g' = gk_F^3/E_F$, and using that $E_F = (\hbar k_F)^2/(2m)$ we get

$$S\left[\{\bar{\psi}_{\mathbf{x},\tau,\sigma}, \psi_{\mathbf{x},\tau,\sigma}\}\right] = \hbar\int_0^{\beta'} d\tau' \int d\mathbf{x}' \sum_\sigma \left[\bar{\psi}_{\mathbf{x}',\tau',\sigma}\left(\frac{\partial}{\partial\tau'} - \nabla^2_{\mathbf{x}'} - \mu_\sigma'\right)\psi_{\mathbf{x}',\tau',\sigma}\right]$$

$$+ \hbar g' \int_0^{\beta'} d\tau' \int d\mathbf{x}' \; \bar{\psi}_{\mathbf{x}',\tau',\uparrow}\bar{\psi}_{\mathbf{x}',\tau',\downarrow}\psi_{\mathbf{x}',\tau',\downarrow}\psi_{\mathbf{x}',\tau',\uparrow}. \quad (11)$$

Finally, note that, in our units (with $\mathbf{k}' = \mathbf{k}/k_F$) the renormalized strength of the contact potential is

$$\frac{1}{g'} = \frac{1}{8\pi k_F a_s} - \int \frac{d\mathbf{k}'}{(2\pi)^3}\frac{1}{2(k')^2}. \quad (12)$$

Dropping the primes, we get the starting point of our treatment:

$$\mathcal{Z} = \int \mathcal{D}\bar{\psi}_{\mathbf{x},\tau,\sigma}\mathcal{D}\psi_{\mathbf{x},\tau,\sigma}\exp\left\{-\int_0^\beta d\tau \int d\mathbf{x} \sum_\sigma \bar{\psi}_{\mathbf{x},\tau,\sigma}\left(\frac{\partial}{\partial\tau} - \nabla^2_{\mathbf{x}} - \mu_\sigma\right)\psi_{\mathbf{x},\tau,\sigma}\right.$$

$$\left. -g\int_0^\beta d\tau \int d\mathbf{x} \; \bar{\psi}_{\mathbf{x},\tau,\uparrow}\bar{\psi}_{\mathbf{x},\tau,\downarrow}\psi_{\mathbf{x},\tau,\downarrow}\psi_{\mathbf{x},\tau,\uparrow}\right\}. \quad (13)$$

This is, if you will, the statement of the problem that we want to investigate. Working with the operator version of quantum mechanics, you would state your starting Hamiltonian – in the path-integral formalism, you have to state your starting action functional. Ours describes a gas of fermionic particles, of two spin species (with the possibility of spin imbalance), with contact interactions between particles of different spins. The goal of our calculation is to obtain the free energy of the system (so that we have access to its thermodynamics), and to identify the superfluid phase (and its order parameter). This is done in five main steps, and in the following subsections we go through them in detail.

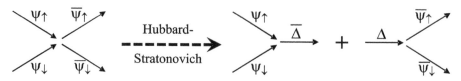

Figure 1. Illustration of the Hubbard-Stratonovich transformation: the interaction term originally comprised of a product of four fermion fields is decomposed in a term that represents two fermions pairing up, a term for the propagation of the pairs (not shown) and a term for the pair breaking up into two fermions.

3. Step 1: Hubbard-Stratonovich fields and the Nambu spinors

The Hubbard-Stratonovich transformation is based on the Gaussian integral formula for completing the squares:

$$
\exp\left\{ -g \int_0^\beta d\tau \int d\mathbf{x}\, \bar{\psi}_{\mathbf{x},\tau,\uparrow}\bar{\psi}_{\mathbf{x},\tau,\downarrow}\psi_{\mathbf{x},\tau,\downarrow}\psi_{\mathbf{x},\tau,\uparrow} \right\} = \int \mathcal{D}\bar{\Delta}_{\mathbf{x},\tau}\mathcal{D}\Delta_{\mathbf{x},\tau}
$$

$$
\exp\left\{ \int_0^\beta d\tau \int d\mathbf{x}\, \left[\frac{\bar{\Delta}_{\mathbf{x},\tau}\Delta_{\mathbf{x},\tau}}{g} + \bar{\Delta}_{\mathbf{x},\tau}\psi_{\mathbf{x},\tau,\downarrow}\psi_{\mathbf{x},\tau,\uparrow} + \Delta_{\mathbf{x},\tau}\bar{\psi}_{\mathbf{x},\tau,\uparrow}\bar{\psi}_{\mathbf{x},\tau,\downarrow} \right] \right\}. \tag{14}
$$

In this formula, the auxiliary fields $\bar{\Delta}_{\mathbf{x},\tau}, \Delta_{\mathbf{x},\tau}$ do not have a spin index and are complex *bosonic* fields and not Grassmann variables. We interpret this bosonic field as the field of the fermion pairs, as illustrated in figure (1).

Using this in our starting point, expression (13), we get

$$
\mathcal{Z} = \int \mathcal{D}\bar{\psi}_{\mathbf{x},\tau,\sigma}\mathcal{D}\psi_{\mathbf{x},\tau,\sigma} \int \mathcal{D}\bar{\Delta}_{\mathbf{x},\tau}\mathcal{D}\Delta_{\mathbf{x},\tau} \exp\left\{ -\int_0^\beta d\tau \int d\mathbf{x}\, \left[-\frac{\bar{\Delta}_{\mathbf{x},\tau}\Delta_{\mathbf{x},\tau}}{g} \right.\right.
$$
$$
\left.\left. + \sum_\sigma \bar{\psi}_{\mathbf{x},\tau,\sigma}\left(\frac{\partial}{\partial\tau} - \nabla_{\mathbf{x}}^2 - \mu_\sigma \right)\psi_{\mathbf{x},\tau,\sigma} - \bar{\Delta}_{\mathbf{x},\tau}\psi_{\mathbf{x},\tau,\downarrow}\psi_{\mathbf{x},\tau,\uparrow} - \Delta_{\mathbf{x},\tau}\bar{\psi}_{\mathbf{x},\tau,\uparrow}\bar{\psi}_{\mathbf{x},\tau,\downarrow} \right] \right\}. \tag{15}
$$

The resulting action is quadratic in the fermion fields, and can be integrated out easily when we introduce Nambu notation. This combines a spin-up and a spin-down Fermi field into a new spinor, the Nambu spinor, given by

$$
\eta_{\mathbf{x},\tau} = \begin{pmatrix} \psi_{\mathbf{x},\tau,\uparrow} \\ \bar{\psi}_{\mathbf{x},\tau,\downarrow} \end{pmatrix} \quad \text{and} \quad \bar{\eta}_{\mathbf{x},\tau} = \begin{pmatrix} \bar{\psi}_{\mathbf{x},\tau,\uparrow} & \psi_{\mathbf{x},\tau,\downarrow} \end{pmatrix}, \tag{16}
$$

or, in component form

$$
\begin{cases} \eta_{\mathbf{x},\tau,1} = \psi_{\mathbf{x},\tau,\uparrow} \\ \eta_{\mathbf{x},\tau,2} = \bar{\psi}_{\mathbf{x},\tau,\downarrow} \end{cases} \text{and} \begin{cases} \bar{\eta}_{\mathbf{x},\tau,1} = \bar{\psi}_{\mathbf{x},\tau,\uparrow} \\ \bar{\eta}_{\mathbf{x},\tau,2} = \psi_{\mathbf{x},\tau,\downarrow} \end{cases}.
$$

Note that we have to take care about the measure of integration. The Grassmann path integral means by definition

$$\int \mathcal{D}\bar{\psi}_{\mathbf{x},\tau,\sigma} \mathcal{D}\psi_{\mathbf{x},\tau,\sigma} := \prod_{\mathbf{x},\tau,\sigma} \left(\int d\bar{\psi}_{\mathbf{x},\tau,\sigma} \int d\psi_{\mathbf{x},\tau,\sigma} \right)$$

$$= \prod_{\mathbf{x},\tau} \left(\int d\bar{\psi}_{\mathbf{x},\tau,\uparrow} \int d\psi_{\mathbf{x},\tau,\uparrow} \int d\bar{\psi}_{\mathbf{x},\tau,\downarrow} \int d\psi_{\mathbf{x},\tau,\downarrow} \right).$$

The factors (between brackets) in this product can be swapped as long as we keep the two fields together: so if we keep $d\bar{\psi}_{\mathbf{x},\tau,\sigma}d\psi_{\mathbf{x},\tau,\sigma}$ pairs together, the order of the $\{\mathbf{x},\tau,\sigma\}$ does not matter. The Grassmann path integral over the η spinors means by definition

$$\int \mathcal{D}\bar{\eta}_{\mathbf{x},\tau} \mathcal{D}\eta_{\mathbf{x},\tau} := \prod_{\mathbf{x},\tau} \left(\int d\bar{\eta}_{\mathbf{x},\tau,1} \int d\eta_{\mathbf{x},\tau,1} \int d\bar{\eta}_{\mathbf{x},\tau,2} \int d\eta_{\mathbf{x},\tau,2} \right).$$

Now replace component by component, to get

$$\int \mathcal{D}\bar{\eta}_{\mathbf{x},\tau} \mathcal{D}\eta_{\mathbf{x},\tau} = \prod_{\mathbf{x},\tau} \left(\int d\bar{\psi}_{\mathbf{x},\tau,\uparrow} \int d\psi_{\mathbf{x},\tau,\uparrow} \int d\psi_{\mathbf{x},\tau,\downarrow} \int d\bar{\psi}_{\mathbf{x},\tau,\downarrow} \right)$$

$$= \prod_{\mathbf{x},\tau} \left(-\int d\bar{\psi}_{\mathbf{x},\tau,\uparrow} \int d\psi_{\mathbf{x},\tau,\uparrow} \int d\bar{\psi}_{\mathbf{x},\tau,\downarrow} \int d\psi_{\mathbf{x},\tau,\downarrow} \right).$$

There is, for every $\mathbf{x}, \tau$, a minus sign when compared to the measure of $\mathcal{D}\bar{\psi}_{\mathbf{x},\tau,\sigma}\mathcal{D}\psi_{\mathbf{x},\tau,\sigma}$:

$$\int \mathcal{D}\bar{\psi}_{\mathbf{x},\tau,\sigma} \mathcal{D}\psi_{\mathbf{x},\tau,\sigma} \rightarrow \int \mathcal{D}\bar{\eta}_{\mathbf{x},\tau} \mathcal{D}\eta_{\mathbf{x},\tau} \left[\prod_{\mathbf{x},\tau}(-1) \right].$$

This is important when taking the integrals. Indeed, for the Gaussian integral with coefficients $\mathbb{A}_{\mathbf{x},\tau}$ (these are 2×2 matrices since our Nambu spinors have 2 components):

$$\int \mathcal{D}\bar{\psi}_{\mathbf{x},\tau,\sigma} \mathcal{D}\psi_{\mathbf{x},\tau,\sigma} \exp\left\{ -\sum_{\mathbf{x},\tau} \bar{\eta}_{\mathbf{x},\tau} \cdot \mathbb{A}_{\mathbf{x},\tau} \cdot \eta_{\mathbf{x},\tau} \right\}$$

$$= \prod_{\mathbf{x},\tau} \int d\bar{\psi}_{\mathbf{x},\tau,\uparrow} \int d\psi_{\mathbf{x},\tau,\uparrow} \int d\bar{\psi}_{\mathbf{x},\tau,\downarrow} \int d\psi_{\mathbf{x},\tau,\downarrow} \exp\{ -\bar{\eta}_{\mathbf{x},\tau} \cdot \mathbb{A}_{\mathbf{x},\tau} \cdot \eta_{\mathbf{x},\tau} \}$$

$$= \prod_{\mathbf{x},\tau} (-1) \int d\bar{\eta}_{\mathbf{x},\tau,1} \int d\eta_{\mathbf{x},\tau,1} \int d\bar{\eta}_{\mathbf{x},\tau,2} \int d\eta_{\mathbf{x},\tau,2} \exp\{ -\bar{\eta}_{\mathbf{x},\tau} \cdot \mathbb{A}_{\mathbf{x},\tau} \cdot \eta_{\mathbf{x},\tau} \}$$

$$= \prod_{\mathbf{x},\tau} (-1) \det_{\sigma}(\mathbb{A}_{\mathbf{x},\tau}).$$

Note that the determinant here is only the determinant over the 2×2 matrix between the Nambu spinors, it is the "spinor determinant", indicated by a σ subscript. By exponentiating the logarithm we can write this as

$$\prod_{\mathbf{x},\tau} (-1) \det_{\sigma}(\mathbb{A}_{\mathbf{x},\tau}) = \prod_{\mathbf{x},\tau} \exp\{ \ln[-\det_{\sigma}(\mathbb{A}_{\mathbf{x},\tau})] \} = \exp\left\{ \sum_{\mathbf{x},\tau} \ln[-\det_{\sigma}(\mathbb{A}_{\mathbf{x},\tau})] \right\},$$

where with the sum we mean the trace:

$$\sum_{\mathbf{x},\tau} \ln[-\det_{\sigma}(\mathbb{A}_{\mathbf{x},\tau})] = \text{Tr}\{ \ln[-\det_{\sigma}(\mathbb{A}_{\mathbf{x},\tau})] \}, \tag{17}$$

which really is nothing else but

$$\int \mathcal{D}\bar{\psi}_{\mathbf{x},\tau,\sigma} \mathcal{D}\psi_{\mathbf{x},\tau,\sigma} \exp\left\{ -\sum_{\mathbf{x},\tau} \bar{\eta}_{\mathbf{x},\tau} \cdot \mathbb{A}_{\mathbf{x},\tau} \cdot \eta_{\mathbf{x},\tau} \right\} = \exp\left\{ \sum_{\mathbf{x},\tau} \ln\left[-\det_\sigma (\mathbb{A}_{\mathbf{x},\tau}) \right] \right\}. \tag{18}$$

So, the swap in order gives the minus sign in front of the determinant. Indeed, sometimes the integration measure does matter.

4. Step 2: Performing the Grassmann integrations

Now we still have to figure out what the matrix between the Nambu spinors is before we can perform the integrations over the Grassmann fields in (15). For reasons that become clear in the light of Green's functions, we will not call this matrix $\mathbb{A}$, but instead we will call it $-\mathbf{G}^{-1}$. and prove that

$$-\mathbf{G}^{-1} = \begin{pmatrix} \dfrac{\partial}{\partial \tau} - \nabla_{\mathbf{x}}^2 - \mu_\uparrow & -\Delta_{\mathbf{x},\tau} \\ -\bar{\Delta}_{\mathbf{x},\tau} & \dfrac{\partial}{\partial \tau} + \nabla_{\mathbf{x}}^2 + \mu_\downarrow \end{pmatrix}. \tag{19}$$

We do this by expanding as follows

$$\bar{\eta}_{\mathbf{x},\tau} \cdot \left(-\mathbf{G}^{-1} \right) \cdot \eta_{\mathbf{x},\tau} = \left(\bar{\psi}_{\mathbf{x},\tau,\uparrow} \; \psi_{\mathbf{x},\tau,\downarrow} \right) \cdot \begin{pmatrix} \dfrac{\partial}{\partial \tau} - \nabla_{\mathbf{x}}^2 - \mu_\uparrow & -\Delta_{\mathbf{x},\tau} \\ -\bar{\Delta}_{\mathbf{x},\tau} & \dfrac{\partial}{\partial \tau} + \nabla_{\mathbf{x}}^2 + \mu_\downarrow \end{pmatrix} \cdot \begin{pmatrix} \psi_{\mathbf{x},\tau,\uparrow} \\ \bar{\psi}_{\mathbf{x},\tau,\downarrow} \end{pmatrix}. \tag{20}$$

This gives

$$\bar{\eta}_{\mathbf{x},\tau} \cdot \left(-\mathbf{G}^{-1} \right) \cdot \eta_{\mathbf{x},\tau} = \bar{\psi}_{\mathbf{x},\tau,\uparrow} \left(\dfrac{\partial}{\partial \tau} - \nabla_{\mathbf{x}}^2 - \mu_\uparrow \right) \psi_{\mathbf{x},\tau,\uparrow} - \Delta_{\mathbf{x},\tau} \bar{\psi}_{\mathbf{x},\tau,\uparrow} \bar{\psi}_{\mathbf{x},\tau,\downarrow}$$

$$- \bar{\Delta}_{\mathbf{x},\tau} \psi_{\mathbf{x},\tau,\downarrow} \psi_{\mathbf{x},\tau,\uparrow} + \psi_{\mathbf{x},\tau,\downarrow} \left(\dfrac{\partial}{\partial \tau} + \nabla_{\mathbf{x}}^2 + \mu_\downarrow \right) \bar{\psi}_{\mathbf{x},\tau,\downarrow}. \tag{21}$$

The first three terms in (21) are exactly as in the action in (15). The last term is more difficult since it involves a derivative that is positioned between two Grassmann variables. We know that when two Grassmann variables are swapped, they get a minus sign. To see what we should do with derivatives, we have to jump a bit ahead of ourselves and do the Fourier transforms. Then we get rid of the operator character of the derivatives, and we can clearly see what the rules are. Starting from (see next section):

$$\psi_{\mathbf{x},\tau,\sigma} = \frac{1}{\sqrt{\beta V}} \sum_n \sum_{\mathbf{k}} e^{-i\omega_n \tau + i\mathbf{k}\cdot\mathbf{x}} \psi_{\mathbf{k},n,\sigma}, \tag{22}$$

$$\bar{\psi}_{\mathbf{x},\tau,\sigma} = \frac{1}{\sqrt{\beta V}} \sum_n \sum_{\mathbf{k}} e^{i\omega_n \tau - i\mathbf{k}\cdot\mathbf{x}} \bar{\psi}_{\mathbf{k},n,\sigma}, \tag{23}$$

we know what the derivative will mean

$$\psi_{\mathbf{x},\tau,\sigma} \left(\frac{\partial}{\partial \tau} \right) \bar{\psi}_{\mathbf{x},\tau,\sigma} \rightarrow \psi_{\mathbf{k},n,\sigma} \left(i\omega_n \bar{\psi}_{\mathbf{k},n,\sigma} \right).$$

Now there is no trouble in swapping the Grassmann variables, this just results in a minus sign:

$$\psi_{\mathbf{k},n,\sigma} \left(i\omega_n \bar{\psi}_{\mathbf{k},n,\sigma} \right) = \bar{\psi}_{\mathbf{k},n,\sigma} \left(-i\omega_n \right) \psi_{\mathbf{k},n,\sigma}.$$

But now we find that this is just equal to

$$\bar{\psi}_{x,\tau,\sigma}\left(\frac{\partial}{\partial\tau}\right)\psi_{x,\tau,\sigma} \to \bar{\psi}_{k,n,\sigma}\left(-i\omega_n\psi_{k,n,\sigma}\right).$$

So the rule is: if there is an odd-degree derivative sandwiched between two conjugate Grassmann variables, and these are swapped, there is no sign change. For second derivatives, there is again a sign change, for third derivatives again no change,.... If we now apply this newly gained knowledge to the fourth term of (21) we arrive at

$$\psi_{x,\tau,\downarrow}\frac{\partial}{\partial\tau}\bar{\psi}_{x,\tau,\downarrow} + \psi_{x,\tau,\downarrow}\left(\nabla_x^2 + \mu_\downarrow\right)\bar{\psi}_{x,\tau,\downarrow} = -\bar{\psi}_{x,\tau,\downarrow}\left(-\frac{\partial}{\partial\tau}\right)\psi_{x,\tau,\downarrow} - \bar{\psi}_{x,\tau,\downarrow}\left(\nabla_x^2 + \mu_\downarrow\right)\psi_{x,\tau,\downarrow}$$

$$= \bar{\psi}_{x,\tau,\downarrow}\left(\frac{\partial}{\partial\tau} - \nabla_x^2 - \mu_\downarrow\right)\psi_{x,\tau,\downarrow}.$$

Now we see that the fourth term in (21) is indeed also equal to the corresponding term in (15).

Note that in general, the inverse Green's matrix $-G_{x,\tau}^{-1}$ is not diagonal in x, τ (for interactions other than the delta function). We should treat $-G^{-1}$ as an operator in e.g. position representation, and write

$$\left\langle x',\tau'\left|-G^{-1}\right|x,\tau\right\rangle = \left\langle x',\tau'|x,\tau\right\rangle\begin{pmatrix}\frac{\partial}{\partial\tau} - \nabla_x^2 - \mu_\uparrow & -\Delta_{x,\tau} \\ -\bar{\Delta}_{x,\tau} & \frac{\partial}{\partial\tau} + \nabla_x^2 + \mu_\downarrow\end{pmatrix}. \tag{24}$$

The matrix gives the operator in position (and time) representation. If it were not diagonal there would be terms like $\bar{\eta}_{x',\tau'}\left\langle x',\tau'\left|-G^{-1}\right|x,\tau\right\rangle\eta_{x,\tau}$ in the Gaussian integral and we would need to first diagonalize the whole spacetime matrix. Luckily, we are using a contact potential. Note that when the system is not interacting, there will be no pairs, and we get in position representation

$$-G_0^{-1} \to \begin{pmatrix}\frac{\partial}{\partial\tau} - \nabla_x^2 - \mu_\uparrow & 0 \\ 0 & \frac{\partial}{\partial\tau} + \nabla_x^2 + \mu_\downarrow\end{pmatrix}. \tag{25}$$

So, in

$$\mathcal{Z} = \int D\bar{\Delta}_{x,\tau}D\Delta_{x,\tau}\int D\bar{\psi}_{x,\tau,\sigma}D\psi_{x,\tau,\sigma}\exp\left\{-\int_0^\beta d\tau\int dx\left[-\frac{\bar{\Delta}_{x,\tau}\Delta_{x,\tau}}{g} + \bar{\eta}_{x,\tau}\cdot\left(-G_{x,\tau}^{-1}\right)\cdot\eta_{x,\tau}\right]\right\} \tag{26}$$

we can use our earlier result, expression (18) to obtain

$$\mathcal{Z} = \int D\bar{\Delta}_{x,\tau}D\Delta_{x,\tau}\exp\left\{-\int_0^\beta d\tau\int dx\left[-\frac{\bar{\Delta}_{x,\tau}\Delta_{x,\tau}}{g} - \ln\left[-\det_\sigma\left(-G_{x,\tau}^{-1}\right)\right]\right]\right\}. \tag{27}$$

This result does not contain any approximation (apart from the choice of starting Lagrangian). But, $-G^{-1}$ depends on $\Delta_{x,\tau}$ and contains a bunch of derivatives too, so we have no way to calculate the logarithm of that (remember that the determinant is here the spinor determinant over the 2×2 matrix but, as noted earlier, there is no problem with that). We will need to go to reciprocal space. Rather than using the spacetime coordinates $\mathbf{x}, \tau$ we work in the space of wave numbers $\mathbf{k}$ and Matsubara frequencies $\omega_n = (2n+1)\pi/\beta$ for fermions and $\omega_n = 2n\pi/\beta$ for bosons (both with $n \in \mathbb{Z}$).

5. Intermezzo 1: The long road to reciprocal space

The goal of this section is to rewrite (27) in reciprocal space, so that we can trace over the wave numbers and (Matsubara) frequencies rather than positions and times. The starting point for fermions is

$$\langle \mathbf{x}, \tau | \mathbf{k}, n \rangle = \frac{\exp\{i\mathbf{k} \cdot \mathbf{x}\}}{\sqrt{V}} \frac{e^{-i\omega_n \tau}}{\sqrt{\beta}}. \tag{28}$$

For bosons we replace $\omega_n = (2n+1)\pi/\beta$ by $\omega_n = 2n\pi/\beta$. The available wave numbers are the same for bosons as for fermions, they are given by $\{k_x, k_y, k_z\} = (2\pi/L)\{n_x, n_y, n_z\}$ with $n_x, n_y, n_z \in \mathbb{Z}$. There are various valid choices for normalizing the plane waves, and that tends to lead to confusion in the results found in the literature. That is why we will go through quite some detail to follow the effects of the choice of normalization that we have made here. We basically want the reciprocal space kets to obey the completeness relation

$$\mathbb{I} = \sum_{\mathbf{k},n} |\mathbf{k}, n\rangle \langle \mathbf{k}, n|. \tag{29}$$

The spacetime kets obey

$$\mathbb{I} = \int_0^\beta d\tau \int d\mathbf{x} \, |\mathbf{x}, \tau\rangle \langle \mathbf{x}, \tau|. \tag{30}$$

This leads to the orthonormality relations

$$\langle \mathbf{k}, n | \mathbf{k}', n' \rangle = \delta(\mathbf{k} - \mathbf{k}')\delta_{nn'}. \tag{31}$$

The consistency of these relations can be proven by inserting expression (30) for $\mathbb{I}$ between the ket and the bra in (31), and using $\int_0^\beta d\tau = \beta$ and $\int d\mathbf{x} = V$. This gives us an integral representation of the delta function,

$$\int_0^\beta d\tau \int d\mathbf{x} \, \exp\{-i(\omega_{n'} - \omega_n)\tau + i(\mathbf{k}' - \mathbf{k}) \cdot \mathbf{x}\} = V\beta \, \delta(\mathbf{k} - \mathbf{k}')\delta_{nn'}. \tag{32}$$

Similarly, we have the orthogonality relation

$$\langle \mathbf{x}, \tau | \mathbf{x}', \tau' \rangle = \delta(\mathbf{x} - \mathbf{x}')\delta(\tau - \tau'). \tag{33}$$

Enforcing the consistency of inserting expression (29) in between the bra and the ket in (33), leads to the following integral for the delta function:

$$\frac{1}{V\beta} \sum_{\mathbf{k},n} \exp\{-i\omega_n(\tau - \tau') + i\mathbf{k} \cdot (\mathbf{x} - \mathbf{x}')\} = \delta(\mathbf{x} - \mathbf{x}')\delta(\tau - \tau'). \tag{34}$$

Note the factor $1/(\beta V)$ in front of the summation. It is easy to check, by the way, that the definitions for the Fourier transform of the Grassmann variables in the previous paragraph are consistent with the definitions of the Fourier transforms given here. How do the fields transform under this? If we introduce $\Delta_{\mathbf{q},m} = \langle \mathbf{q}, m | \Delta \rangle$ and assume that $\Delta_{\mathbf{x},\tau} = \langle \mathbf{x}, \tau | \Delta \rangle$ then

$$\Delta_{\mathbf{x},\tau} = \sum_{\mathbf{q},m} \frac{e^{-i\omega_m\tau+i\mathbf{q}\cdot\mathbf{x}}}{\sqrt{\beta V}} \Delta_{\mathbf{q},m} \tag{35}$$

$$\Leftrightarrow \Delta_{\mathbf{q},m} = \int_0^\beta d\tau \int d\mathbf{x} \frac{e^{i\omega_m\tau-i\mathbf{q}\cdot\mathbf{x}}}{\sqrt{\beta V}} \Delta_{\mathbf{x},\tau}. \tag{36}$$

For the mind in need of consistency checks, note that substituting (35) in (36) gives back (34). And substituting (36) into (35) gives back (32). The same relations hold between $\psi_{\mathbf{x},\tau,\sigma}$ and $\psi_{\mathbf{k},n,\sigma}$ – we used those already in (22). We also introduce $\bar{\Delta}_{\mathbf{q},m} = \langle \Delta | \mathbf{q}, m \rangle$ and assume that $\bar{\Delta}_{\mathbf{x},\tau} = \langle \Delta | \mathbf{x}, \tau \rangle$, then we have

$$\bar{\Delta}_{\mathbf{x},\tau} = \sum_{\mathbf{q},m} \frac{e^{i\omega_m\tau-i\mathbf{q}\cdot\mathbf{x}}}{\sqrt{\beta V}} \bar{\Delta}_{\mathbf{q},m} \tag{37}$$

$$\Leftrightarrow \bar{\Delta}_{\mathbf{q},m} = \int_0^\beta d\tau \int d\mathbf{x} \frac{e^{-i\omega_m\tau+i\mathbf{q}\cdot\mathbf{x}}}{\sqrt{\beta V}} \bar{\Delta}_{\mathbf{x},\tau}. \tag{38}$$

Also here we have the same relations linking $\bar{\psi}_{\mathbf{x},\tau,\sigma}$ and $\bar{\psi}_{\mathbf{k},n,\sigma}$, cf. expression (23). Note that here we have distributed the $\sqrt{\beta V}$ factors evenly over the Fourier and the inverse Fourier. As a result, we will have to tag a factor $\sqrt{\beta V}$ along later. If we keep the βV completely in the Fourier transform (38) or (36) then this does not appear.

Everything is now is set for the Fourier transformation of the partition function given in (26). The first term in the action of (26) is re-expressed with (37) and (35) as

$$\int_0^\beta d\tau \int d\mathbf{x}\, \bar{\Delta}_{\mathbf{x},\tau}\Delta_{\mathbf{x},\tau} = \int_0^\beta d\tau \int d\mathbf{x} \left(\sum_{\mathbf{q},m} \frac{e^{i\omega_m\tau-i\mathbf{q}\cdot\mathbf{x}}}{\sqrt{\beta V}} \bar{\Delta}_{\mathbf{q},m} \right) \left(\sum_{\mathbf{q}',m'} \frac{e^{-i\omega_{m'}\tau+i\mathbf{q}'\cdot\mathbf{x}}}{\sqrt{\beta V}} \Delta_{\mathbf{q}',m'} \right)$$

$$= \sum_{\mathbf{q},m}\sum_{\mathbf{q}',m'} \left[\frac{1}{\beta V} \int_0^\beta d\tau \int d\mathbf{x}\, e^{i(\omega_m-\omega_{m'})\tau-i(\mathbf{q}-\mathbf{q}')\cdot\mathbf{x}} \right] \bar{\Delta}_{\mathbf{q},m}\Delta_{\mathbf{q}',m'}$$

$$= \sum_{\mathbf{q},m} \bar{\Delta}_{\mathbf{q},m}\Delta_{\mathbf{q},m}, \tag{39}$$

where we have used (32) in the last step. This is merely Parseval's rule.

Now we want to calculate the reciprocal space representation of the second part of the action in (26) given by

$$\int_0^\beta d\tau \int d\mathbf{x} \left[\bar{\eta}_{\mathbf{x},\tau} \cdot \left(-\mathbf{G}_{\mathbf{x},\tau}^{-1} \right) \cdot \eta_{\mathbf{x},\tau} \right]$$

$$= \int_0^\beta d\tau \int d\mathbf{x} \left(\bar{\psi}_{\mathbf{x},\tau,\uparrow} \left(\frac{\partial}{\partial\tau} - \nabla_\mathbf{x}^2 - \mu_\uparrow \right) \psi_{\mathbf{x},\tau,\uparrow} + \psi_{\mathbf{x},\tau,\downarrow} \left(\frac{\partial}{\partial\tau} + \nabla_\mathbf{x}^2 + \mu_\downarrow \right) \bar{\psi}_{\mathbf{x},\tau,\downarrow} \right.$$

$$\left. -\Delta_{\mathbf{x},\tau} \bar{\psi}_{\mathbf{x},\tau,\uparrow} \bar{\psi}_{\mathbf{x},\tau,\downarrow} - \bar{\Delta}_{\mathbf{x},\tau} \psi_{\mathbf{x},\tau,\downarrow} \psi_{\mathbf{x},\tau,\uparrow} \right). \tag{40}$$

Every term in (40) can be transformed to reciprocal space using the rules and conventions that we stated above. The first term in (40) becomes

$$
\int_0^\beta d\tau \int d\mathbf{x}\, \bar{\psi}_{\mathbf{x},\tau,\uparrow} \left(\frac{\partial}{\partial \tau} - \nabla_\mathbf{x}^2 - \mu_\uparrow \right) \psi_{\mathbf{x},\tau,\uparrow}
$$

$$
= \sum_{\mathbf{k},n} \sum_{\mathbf{k}',n'} \frac{1}{\beta V} \int_0^\beta d\tau \int d\mathbf{x} \left(e^{i\omega_n \tau - i\mathbf{k}.\mathbf{x}} \bar{\psi}_{\mathbf{k},n,\uparrow} \right) \left(-i\omega_{n'} + (k')^2 - \mu_\uparrow \right) \left(e^{-i\omega_{n'}\tau + i\mathbf{k}'.\mathbf{x}} \psi_{\mathbf{k}',n',\uparrow} \right)
$$

$$
= \sum_{\mathbf{k},n} \left(-i\omega_n + k^2 - \mu_\uparrow \right) \bar{\psi}_{\mathbf{k},n,\uparrow} \psi_{\mathbf{k},n,\uparrow}. \tag{41}
$$

The second term transforms completely analogously:

$$
\int_0^\beta d\tau \int d\mathbf{x}\, \psi_{\mathbf{x},\tau,\downarrow} \left(\frac{\partial}{\partial \tau} + \nabla_\mathbf{x}^2 + \mu_\downarrow \right) \bar{\psi}_{\mathbf{x},\tau,\downarrow}
$$

$$
= \sum_{\mathbf{k},n} \left(i\omega_n - k^2 + \mu_\downarrow \right) \psi_{\mathbf{k},n,\downarrow} \bar{\psi}_{\mathbf{k},n,\downarrow}. \tag{42}
$$

In the interaction terms, all three fields have to be transformed

$$
- \int_0^\beta d\tau \int d\mathbf{x}\, \Delta_{\mathbf{x},\tau} \bar{\psi}_{\mathbf{x},\tau,\uparrow} \bar{\psi}_{\mathbf{x},\tau,\downarrow}
$$

$$
= -\frac{1}{\sqrt{\beta V}} \sum_{\mathbf{k},n} \sum_{\mathbf{k}',n'} \left[\frac{1}{\sqrt{\beta V}} \int_0^\beta d\tau \int d\mathbf{x} \left(e^{i(\omega_n + \omega_{n'})\tau - i(\mathbf{k}+\mathbf{k}').\mathbf{x}} \Delta_{\mathbf{x},\tau} \right) \right] \bar{\psi}_{\mathbf{k},n,\uparrow} \bar{\psi}_{\mathbf{k}',n',\downarrow}
$$

$$
= -\frac{1}{\sqrt{\beta V}} \sum_{\mathbf{k},n} \sum_{\mathbf{k}',n'} \Delta_{\mathbf{k}+\mathbf{k}',n+n'}\, \bar{\psi}_{\mathbf{k},n,\uparrow} \bar{\psi}_{\mathbf{k}',n',\downarrow}, \tag{43}
$$

and analogously for the fourth term we arrive at

$$
- \int_0^\beta d\tau \int d\mathbf{x}\, \bar{\Delta}_{\mathbf{x},\tau} \psi_{\mathbf{x},\tau,\downarrow} \psi_{\mathbf{x},\tau,\uparrow} = -\frac{1}{\sqrt{\beta V}} \sum_{\mathbf{k},n} \sum_{\mathbf{k}',n'} \bar{\Delta}_{\mathbf{k}+\mathbf{k}',n+n'}\, \psi_{\mathbf{k},n,\downarrow} \psi_{\mathbf{k}',n',\uparrow}. \tag{44}
$$

In (43) and (44) we used (36) and (38) respectively, together with the fact that the sum of two fermionic Matsubara frequencies results in a bosonic Matsubara frequency. Putting all results together, the partition sum in reciprocal space equals

$$
\mathcal{Z} = \int \mathcal{D}\bar{\psi}_{\mathbf{k},n,\sigma} \mathcal{D}\psi_{\mathbf{k},n,\sigma} \int \mathcal{D}\bar{\Delta}_{\mathbf{q},m} \mathcal{D}\Delta_{\mathbf{q},m} \exp \left(\sum_{\mathbf{q},m} \frac{\bar{\Delta}_{\mathbf{q},m} \Delta_{\mathbf{q},m}}{g} \right.
$$

$$
\left. - \sum_{\mathbf{k}',n'} \sum_{\mathbf{k}'',n''} \bar{\eta}_{\mathbf{k}',n'} \langle \mathbf{k}',n' | -G^{-1} | \mathbf{k}'',n'' \rangle \eta_{\mathbf{k}'',n''} \right), \tag{45}
$$

where the reciprocal space representation of the inverse Green's function is given by:

$$
\langle \mathbf{k},n | -G^{-1} | \mathbf{k}',n' \rangle = \langle \mathbf{k},n | \mathbf{k}',n' \rangle \begin{pmatrix} -i\omega_n + k^2 - \mu_\uparrow & 0 \\ 0 & i\omega_n - k^2 + \mu_\downarrow \end{pmatrix}
$$

$$
+ \frac{1}{\sqrt{\beta V}} \begin{pmatrix} 0 & -\Delta_{\mathbf{k}+\mathbf{k}',n+n'} \\ -\bar{\Delta}_{\mathbf{k}+\mathbf{k}',n+n'} & 0 \end{pmatrix}. \tag{46}
$$

where the following Nambu spinors were used

$$\eta_{\mathbf{k},n} = \begin{pmatrix} \psi_{\mathbf{k},n,\uparrow} \\ \bar{\psi}_{\mathbf{k},n,\downarrow} \end{pmatrix} \text{ en } \bar{\eta}_{\mathbf{k},n} = \begin{pmatrix} \bar{\psi}_{\mathbf{k},n,\uparrow} & \psi_{\mathbf{k},n,\downarrow} \end{pmatrix}. \tag{47}$$

Only the first part, the non-interacting part, is diagonal. The part related to the fermion pair field is not diagonal in reciprocal space, in the sense that components of the inverse Green's function with $\mathbf{k}',n'$ different from $\mathbf{k},n$ are nonzero, so we cannot use the result (18). In principle this is still a quadratic integral and we could do it, but it would involve taking a determinant (or logarithm) of $\langle \mathbf{k}',n' | -\mathbf{G}^{-1} | \mathbf{k}'',n'' \rangle$. This is a $\infty \times \infty$ matrix with 2×2 matrices as its elements. So let us postpone this horror, and see if we can get somewhere with approximations.

6. Step 3: the saddle-point approximation

We have performed step 2, the Grassmann integrations, and even have obtained two equivalent expressions of the result, in position space and in reciprocal space. The result still contains a path integral over the bosonic pair fields $\bar{\Delta}_{\mathbf{q},m}, \Delta_{\mathbf{q},m}$, and this integral cannot be done analytically. Then why go through all the trouble of introducing these fields, and doing the fermionic integrals, if in the end we are left with another path integral that cannot be done exactly? The advantage of having rewritten the action into the form with the bosonic fields, is that we can use additional information about this bosonic pair field. Indeed, if we want to investigate the superfluid state, where the pairs are Bose condensed, we know that the field will be dominated by one contribution, that of the $\mathbf{q} = 0$ term. This is similar to the assumption of Bogoliubov in his famous treatment of the helium superfluid. In that case, Bogoliubov proposed to shift the bosonic operators over a $\mathbf{q} = 0$ contribution, so that the shifted operators could be seen as small fluctuations. This scheme is what we will apply now to treat the Bose fluid of pairs described by $\bar{\Delta}_{\mathbf{q},m}, \Delta_{\mathbf{q},m}$.

As mentioned, the simplest approximation is to assume all pairs are condensed in the $\mathbf{q} = 0$, $m = 0$ state and set for the two pair fields

$$\Delta_{\mathbf{q},m} = \sqrt{\beta V} \delta(\mathbf{q}) \delta_{m,0} \times \Delta \tag{48}$$

$$\bar{\Delta}_{\mathbf{q},m} = \sqrt{\beta V} \delta(\mathbf{q}) \delta_{m,0} \times \Delta^*. \tag{49}$$

We introduce the factor $\sqrt{\beta V}$ for the ease of calculation and to give Δ units of energy. By applying the saddle point (48) and (49) the partition function (45) becomes

$$\mathcal{Z}_{sp} = \int \mathcal{D}\bar{\psi}_{\mathbf{k},n,\sigma} \mathcal{D}\psi_{\mathbf{k},n,\sigma} \exp\left(\frac{\beta V}{g} |\Delta|^2 - \sum_{\mathbf{k},n} \left(-i\omega_n + k^2 - \mu_\uparrow \right) \bar{\psi}_{\mathbf{k},n,\uparrow} \psi_{\mathbf{k},n,\uparrow} \right.$$

$$\left. - \sum_{\mathbf{k},n} \left(-i\omega_n - k^2 + \mu_\downarrow \right) \psi_{-\mathbf{k},-n,\downarrow} \bar{\psi}_{-\mathbf{k},-n,\downarrow} + \sum_{\mathbf{k},n} \left(\Delta\, \bar{\psi}_{\mathbf{k},n,\uparrow} \bar{\psi}_{-\mathbf{k},-n,\downarrow} + \Delta^*\, \psi_{-\mathbf{k},-n,\downarrow} \psi_{\mathbf{k},n,\uparrow} \right) \right). \tag{50}$$

We explicitly write out the action without Nambu spinors here to make an important and subtle point, which otherwise is quickly overlooked. The alert reader will have noticed that we have re-indexed the indices in the spin-down two-particle term: $\mathbf{k} \to -\mathbf{k}$ and $n \to -n$.

This re-indexation is necessary in order to write the action in Nambu spinor notation. Here the following Nambu spinors will be used

$$\eta_{\mathbf{k},n} = \begin{pmatrix} \psi_{\mathbf{k},n,\uparrow} \\ \bar{\psi}_{-\mathbf{k},-n,\downarrow} \end{pmatrix} \quad \text{en} \quad \bar{\eta}_{\mathbf{k},n} = \begin{pmatrix} \bar{\psi}_{\mathbf{k},n,\uparrow} & \psi_{-\mathbf{k},-n,\downarrow} \end{pmatrix}. \tag{51}$$

This then leads to the inverse Green's function in the saddle-point approximation

$$\left\langle \mathbf{k}',n' \left| -G_{sp}^{-1} \right| \mathbf{k},n \right\rangle = \left\langle \mathbf{k},n | \mathbf{k}',n' \right\rangle \begin{pmatrix} -i\omega_n + k^2 - \mu_\uparrow & -\Delta \\ -\Delta^* & -i\omega_n - k^2 + \mu_\downarrow \end{pmatrix}$$

$$= \left\langle \mathbf{k},n | \mathbf{k}',n' \right\rangle \left(-G_{sp}^{-1} \right)'_{\mathbf{k},n}. \tag{52}$$

Our choice has made the inverse Green's function diagonal, since none of the terms $\Delta_{\mathbf{k}'+\mathbf{k},n'+n}$ with $\mathbf{k}'+\mathbf{k} \neq 0$ or $n'+n \neq 0$ survive due to the delta functions in (48), and similarly for $\bar{\Delta}$.

In the saddle-point approximation to the partition sum we no longer have to perform the integrations over the bosonic degrees (it is this integration which was approximated):

$$\mathcal{Z}_{sp} = \int \mathcal{D}\bar{\psi}_{\mathbf{k},n,\sigma} \mathcal{D}\psi_{\mathbf{k},n,\sigma} \exp\left\{ \frac{\beta V}{g} |\Delta|^2 - \sum_{\mathbf{k},n} \bar{\eta}_{\mathbf{k},n} \cdot \left(-G_{sp}^{-1} \right) \cdot \eta_{\mathbf{k},n} \right\}. \tag{53}$$

We can now perform the Grassmann integration using expression (7):

$$\int \mathcal{D}\bar{\psi}_{\mathbf{k},n,\sigma} \mathcal{D}\psi_{\mathbf{k},n,\sigma} \exp\left\{ -\sum_{\mathbf{k},n} \bar{\eta}_{\mathbf{k},n} \cdot \left(-G_{sp}^{-1} \right)_{\mathbf{k},n} \cdot \eta_{\mathbf{k},n} \right\} = \exp\left\{ \sum_{\mathbf{k},n} \ln\left[-\det_\sigma \left(-G_{sp}^{-1} \right)_{\mathbf{k},n} \right] \right\}. \tag{54}$$

This spinor-determinant is given by

$$-\det_\sigma \left(-G_{sp}^{-1} \right)_{\mathbf{k},n} = -\det \begin{pmatrix} -i\omega_n + k^2 - \mu_\uparrow & -\Delta \\ -\Delta^* & -i\omega_n - k^2 + \mu_\downarrow \end{pmatrix}$$

$$= -\left(-i\omega_n + k^2 - \mu_\uparrow \right)\left(-i\omega_n - k^2 + \mu_\downarrow \right) + |\Delta|^2. \tag{55}$$

Introducing

$$\mu = (\mu_\uparrow + \mu_\downarrow)/2, \tag{56}$$

$$\zeta = (\mu_\uparrow - \mu_\downarrow)/2, \tag{57}$$

$$E_{\mathbf{k}} = \sqrt{\left(k^2 - \mu\right)^2 + |\Delta|^2}, \tag{58}$$

we can rewrite this as

$$-\det_\sigma \left(-G_{sp}^{-1} \right)_{\mathbf{k},n} = (i\omega_n + \zeta - E_{\mathbf{k}})(-i\omega_n - \zeta - E_{\mathbf{k}}). \tag{59}$$

This can easily be checked by substituting and seeing that both expressions are indeed equal (and equal to $\omega_n^2 + \left(k^2 - \mu\right)^2 + |\Delta|^2 - \zeta^2 - 2i\omega_n\zeta$). So, the saddle-point partition sum becomes

$$\mathcal{Z}_{sp} = \exp\left\{ \frac{\beta V}{g} |\Delta|^2 + \sum_{\mathbf{k},n} \ln\left[(i\omega_n + \zeta - E_{\mathbf{k}})(-i\omega_n - \zeta - E_{\mathbf{k}}) \right] \right\}. \tag{60}$$

The partition function is also linked to the thermodynamic potential through the well-known formula

$$\mathcal{Z}_{sp} = e^{-\beta F_{sp}(T,V,\mu_\uparrow,\mu_\downarrow)}.$$ (61)

Note that this is the free energy as a function of the chemical potentials. It is related to the usual free energy through

$$F(T,V,\mu_\uparrow,\mu_\downarrow) = F(T,V,N_\uparrow,N_\downarrow) - \mu_\uparrow N_\uparrow - \mu_\downarrow N_\downarrow$$ (62)

$$\text{with} \quad \begin{cases} N_\uparrow = -\dfrac{\partial F(T,V,\mu_\uparrow,\mu_\downarrow)}{\partial\mu_\uparrow}\bigg|_{T,V,\mu_\downarrow} \\[2ex] N_\downarrow = -\dfrac{\partial F(T,V,\mu_\uparrow,\mu_\downarrow)}{\partial\mu_\downarrow}\bigg|_{T,V,\mu_\uparrow} \end{cases} ,$$ (63)

or, in μ and ζ:

$$F(T,V,\mu,\zeta) = F(T,V,N,\delta N) - \mu N - \zeta \delta N,$$ (64)

where $N = N_\uparrow + N_\downarrow$ and $\delta N = N_\uparrow - N_\downarrow$. F is commonly referred to as the "thermodynamic potential". In the following notations, we use $\Omega(T,V,\mu,\zeta) = F(T,V,\mu,\zeta)/V$ for the thermodynamic potential *per unit volume*,

$$\Omega(T,V,\mu,\zeta) = \frac{1}{V}F(T,V,N,\delta N) - \mu n - \zeta \delta n,$$ (65)

where $n = N/V$ and $\delta n = \delta N/V$ are the total density and the density difference

$$n = -\frac{\partial\Omega(T,V,\mu,\zeta)}{\partial\mu}\bigg|_{T,V,\zeta}$$ (66)

$$\delta n = -\frac{\partial\Omega(T,V,\mu,\zeta)}{\partial\zeta}\bigg|_{T,V,\mu}.$$ (67)

We give these expressions explicitly here to avoid the subtle difficulties related to identifying the dependent variables. For the saddle-point action we get:

$$\mathcal{Z}_{sp} = e^{-\beta V \Omega_{sp}(T,\mu,\zeta)}.$$ (68)

We dropped the explicit dependence on V because it also drops from the expression for Ω_{sp}. From (60) we get the saddle-point thermodynamic potential per unit volume

$$\Omega_{sp}(T,\mu,\zeta;\Delta) = -\frac{1}{g}|\Delta|^2 - \frac{1}{V}\sum_{\mathbf{k}}\frac{1}{\beta}\sum_n \ln\left[(i\omega_n + \zeta - E_{\mathbf{k}})(-i\omega_n - \zeta - E_{\mathbf{k}})\right].$$ (69)

Note that since the density of $\mathbf{k}$-states in reciprocal space is $V/(2\pi)^3$ we can replace

$$\frac{1}{V}\sum_{\mathbf{k}} \to \int\frac{d\mathbf{k}}{(2\pi)^3}.$$ (70)

We do not have to take the Matsubara sums now. For example we might want to take derivatives first.

In the sum over Matsubara frequencies we sum up terms with all $n \in \mathbb{Z}$. We can re-order the terms in this sum. In particular, we can set $n' = -1 - n$ and sum over all n'. For this substitution $i\omega_{n'} = i\omega_{-n-1} = -i\omega_n$. This means that in a general Matsubara summation

$$\frac{1}{\beta} \sum_{n=-\infty}^{\infty} f(i\omega_n) = \frac{1}{\beta} \sum_{n=-\infty}^{\infty} f(-i\omega_n) \tag{71}$$

must hold. Introducing

$$i\nu_n = i\omega_n + \zeta \tag{72}$$

as shifted Matsubara frequencies, and $\xi_{\mathbf{k}} = k^2 - \mu$, we get

$$\Omega_{sp}(T, \mu, \zeta; \Delta) = -\frac{1}{g} |\Delta|^2 - \int \frac{d\mathbf{k}}{(2\pi)^3} \left\{ \frac{1}{\beta} \sum_{n=-\infty}^{\infty} \ln\left[-(i\nu_n + E_{\mathbf{k}})(i\nu_n - E_{\mathbf{k}}) \right] - \xi_{\mathbf{k}} \right\}. \tag{73}$$

We use this notation to make an important point. The thermodynamic potential per unit volume Ω_{sp} will depend on our choice of saddle-point value, but this is not one of the thermodynamic variables like T, μ and ζ. If we want to treat Δ as a separate input for Ω_{sp}, we will write this as $\Omega_{sp}(T, \mu, \zeta; \Delta)$ to emphasize the distinction between Δ and the true thermodynamic variables. We extract the dependence of Δ on the thermodynamic variables from the gap equation

$$\left. \frac{\partial \Omega_{sp}(T, \mu, \zeta; \Delta)}{\partial \Delta} \right|_{T,\mu,\zeta} = 0 \quad \longrightarrow \quad \Delta(T, \mu, \zeta)$$

and we have to insert this result back into $\Omega_{sp}(T, \mu, \zeta; \Delta(T, \mu, \zeta)) = \Omega_{sp}(T, \mu, \zeta)$ when applying thermodynamic relations. For example, the number equations are given by the thermodynamic relations

$$n_{sp} = -\left. \frac{\partial \Omega_{sp}(T, \mu, \zeta)}{\partial \mu} \right|_{T,\zeta} \tag{74}$$

$$\delta n_{sp} = -\left. \frac{\partial \Omega_{sp}(T, \mu, \zeta)}{\partial \zeta} \right|_{T,\mu}. \tag{75}$$

If we want to treat Δ as a separate variable then we need to use the chain rule

$$n_{sp} = -\left. \frac{\partial \Omega_{sp}[T, \mu, \zeta; \Delta]}{\partial \mu} \right|_{T,\zeta,\Delta} - \left. \frac{\partial \Omega_{sp}[T, \mu, \zeta; \Delta]}{\partial \Delta} \right|_{T,\zeta,\mu} \times \left. \frac{\partial \Delta(T, \mu, \zeta)}{\partial \mu} \right|_{T,\zeta}. \tag{76}$$

Now you might wonder what all the fuss is about, since we just imposed

$$\left. \frac{\partial \Omega_{sp}[T, \mu, \zeta; \Delta]}{\partial \Delta} \right|_{T,\zeta,\mu} = 0, \tag{77}$$

so that here

$$n_{sp} = -\left. \frac{\partial \Omega_{sp}[T, \mu, \zeta; \Delta]}{\partial \mu} \right|_{T,\zeta,\Delta}. \tag{78}$$

We have been cautious with the partial derivatives and thermodynamic relations, and this might seem superfluous for the saddle-point approximation, but it will become very important when we are adding fluctuation corrections to the thermodynamic potential.

7. Results at the saddle-point level

To obtain the results at the saddle-point level, we need to perform the Matsubara summations in the thermodynamic potential:

$$\frac{1}{\beta}\sum_{n=-\infty}^{\infty}\ln\left[-(i\nu_n - E_{\mathbf{k}})(i\nu_n + E_{\mathbf{k}})\right] = \frac{1}{\beta}\ln\left[2\cosh(\beta E_{\mathbf{k}}) + 2\cosh(\beta\zeta)\right] + \sum_{n=-\infty}^{\infty}\ln\left(\nu_n^2\right). \quad (79)$$

Due to the presence of the logarithm this Matsubara sum is divergent. Luckily however, the divergent part, which does not depend on the system parameters, can be neatly isolated, as shown in (79). This means that the free energy can be regularized by subtracting this unphysical term. This then results in:

$$\Omega_{sp}(T,\mu,\zeta;\Delta) = -\frac{1}{g}|\Delta|^2 - \frac{1}{V}\sum_{\mathbf{k}}\left\{\frac{1}{\beta}\ln\left[2\cosh(\beta E_{\mathbf{k}}) + 2\cosh(\beta\zeta)\right] - \zeta_{\mathbf{k}}\right\}. \quad (80)$$

Using the renormalized contact potential strength, this becomes

$$\Omega_{sp}(T,\mu,\zeta;\Delta) = -\frac{1}{8\pi k_F a_s}|\Delta|^2 - \frac{1}{V}\sum_{\mathbf{k}}\left\{\frac{1}{\beta}\ln\left[2\cosh(\beta E_{\mathbf{k}}) + 2\cosh(\beta\zeta)\right] - \zeta_{\mathbf{k}} - \frac{|\Delta|^2}{2k^2}\right\}. \quad (81)$$

We kept $(k_F a_s)$ as the measure of interaction strength, explicitly writing k_F although we have used it as a length unit, as if $(k_F a_s)$ were a single symbol for the interaction strength. In three dimensions, this is reduced to

$$\Omega_{sp}(T,\mu,\zeta;\Delta) = -\frac{1}{8\pi k_F a_s}|\Delta|^2 - \int\frac{d\mathbf{k}}{(2\pi)^3}\left\{\frac{1}{\beta}\ln\left[2\cosh(\beta E_{\mathbf{k}}) + 2\cosh(\beta\zeta)\right] - \zeta_{\mathbf{k}} - \frac{|\Delta|^2}{2k^2}\right\}. \quad (82)$$

Since we have isotropy, this is

$$\Omega_{sp}(T,\mu,\zeta;\Delta) = -\frac{1}{8\pi k_F a_s}|\Delta|^2 - \frac{1}{2\pi^2}\int_0^\infty dk\, k^2\left\{\frac{1}{\beta}\ln\left[2\cosh(\beta E_{\mathbf{k}}) + 2\cosh(\beta\zeta)\right] - \zeta_{\mathbf{k}} - \frac{|\Delta|^2}{2k^2}\right\}. \quad (83)$$

This result is shown in figure (2).

In the limit of low temperature we have a simplification:

$$\lim_{\beta\to\infty}\frac{1}{\beta}\ln\left[2\cosh(\beta E_{\mathbf{k}}) + 2\cosh(\beta\zeta)\right] = \max\left(E_{\mathbf{k}},|\zeta|\right). \quad (84)$$

Also note that in the limit of large k, the logarithm behaves as

$$\lim_{k\to\infty}\ln\left[2\cosh(\beta E_{\mathbf{k}}) + 2\cosh(\beta\zeta)\right] = \zeta_{\mathbf{k}} + \frac{|\Delta|^2}{2k^2} + \mathcal{O}(k^{-4}), \quad (85)$$

so that it is precisely the extra terms $\zeta_{\mathbf{k}} + |\Delta|^2/(2k^2)$ that keep the integrand from diverging and keep the saddle-point free energy that we calculate here finite. Remember that $\zeta_{\mathbf{k}} = k^2 - \mu$ and $E_{\mathbf{k}} = \sqrt{\zeta_{\mathbf{k}}^2 + \Delta^2}$.

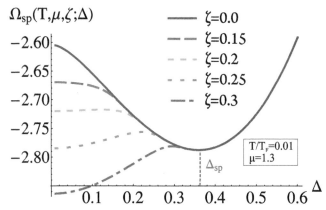

Figure 2. The saddle-point free energy per unit volume is shown as a function of the gap parameter Δ, for different values of the imbalance chemical potential ζ, at temperature $T/T_F = 0.01$ and average chemical potential $\mu = 1.3$. As the imbalance chemical potential ζ is increased, the normal state minimum at $\Delta = 0$ develops and becomes the global minimum above a critical imbalance level. The superfluid state minimum at $\Delta \neq 0$ is not influenced by ζ, indicating that the superfluid state is balanced, i.e. has an equal amount of spin-up and spin-down components. The excess component in this state must be expelled, leading to phase separation.

From this result, and from

$$\frac{\partial E_{\mathbf{k}}}{\partial \Delta} = \frac{\Delta}{E_{\mathbf{k}}} \tag{86}$$

we can obtain the gap equation:

$$\frac{\partial \Omega_{sp}}{\partial \Delta} = 0$$

$$\Longleftrightarrow -\frac{1}{k_F a_s} = \frac{2}{\pi} \int_0^\infty dk \left[\frac{\sinh(\beta E_{\mathbf{k}})}{\cosh(\beta E_{\mathbf{k}}) + \cosh(\beta \zeta)} \frac{k^2}{E_{\mathbf{k}}} - 1 \right]. \tag{87}$$

For every temperature and every μ, ζ, this equation can be solved to obtain $\Delta_{sp}(T, \mu, \zeta)$, the value of Δ that indeed minimizes Ω_{sp} (as can be checked from the sign of the second derivative, or by visual inspection of the plot of Ω_{sp} as a function of Δ at fixed T, μ, ζ, see figure (2)). The resulting saddle-point gap is illustrated in figure (3) and figure (4).

The presence of imbalance ($\zeta \neq 0$) only affects the energies $E_{\mathbf{k}} < \zeta$ in the temperature zero limit, since in that limit

$$\lim_{\beta \to \infty} \frac{\sinh(\beta E_{\mathbf{k}})}{\cosh(\beta \zeta) + \cosh(\beta E_{\mathbf{k}})} = \begin{cases} 1 \text{ for } |\zeta| < E_{\mathbf{k}} \\ 0 \text{ for } |\zeta| > E_{\mathbf{k}} \end{cases}. \tag{88}$$

The presence of a gap results in $E_{\mathbf{k}} > \Delta$. Thus, as long as $\zeta < \Delta$, the imbalance (at T=0) will not affect the gap equation, and we retrieve the Clogston limit for superconductivity. This argument is only valid in the BCS limit, because for the BEC/BCS crossover we still need to solve (independently) the number equation that is coupled to the gap equation. Results for this are illustrated in figure (4). Only in the BCS limit we can set $\mu = E_F$ (see figure 5) and not worry about the overall chemical potential further.

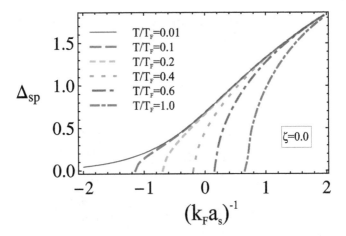

Figure 3. The value of the gap parameter that minimizes the saddle-point free energy is shown as a function of the interaction strength $1/(k_F a_s)$ for $\zeta = 0$. In the limit of $(k_F a_s)^{-1} \to -\infty$, the Bardeen-Cooper-Schrieffer (BCS) result is retrieved, whereas in the limit of $(k_F a_s)^{-1} \to +\infty$, a Bose-Einstein condensate (BEC) of tightly bound Cooper pairs is formed. The different curves show the effect of increasing the temperature: the BCS state is more strongly affected than the BEC state. However, in the BEC state, phase fluctuations (discussed in the next sections) will become the dominant mechanism to destroy superfluidity, rather than the breakup of the Cooper pairs as in BCS.

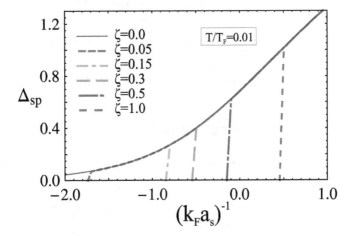

Figure 4. The value of the gap parameter that minimizes the saddle-point free energy is shown as a function of the interaction strength $1/(k_F a_s)$, as in figure 3. In the current figure, the different curves correspond to different values of the imbalance chemical potential ζ. At the saddle-point level, it becomes clear that introducing imbalance between 'spin-up' and 'spin-down' components leads to the appearance of a critical interaction parameter below which pairing is suppressed.

When working with a fixed number of particles, the chemical potentials need to be related to these numbers of particles through the number equations. These are again found from the thermodynamic potential, and from

$$-\frac{\partial E_{\mathbf{k}}}{\partial \mu} = \frac{\zeta_{\mathbf{k}}}{E_{\mathbf{k}}} \tag{89}$$

we get

$$n_{sp} = -\left.\frac{\partial \Omega_{sp}[T, \mu, \zeta; \Delta]}{\partial \mu}\right|_{T, \zeta, \Delta} = \frac{1}{2\pi^2}\int\limits_0^\infty dk\, k^2 \left\{ 1 - \frac{\sinh(\beta E_{\mathbf{k}})}{\cosh(\beta E_{\mathbf{k}}) + \cosh(\beta\zeta)}\frac{\zeta_{\mathbf{k}}}{E_{\mathbf{k}}} \right\} \tag{90}$$

and

$$\delta n_{sp} = -\left.\frac{\partial \Omega_{sp}[T, \mu, \zeta; \Delta]}{\partial \zeta}\right|_{T, \mu, \Delta} = \frac{1}{2\pi^2}\int\limits_0^\infty dk\, k^2 \frac{\sinh(\beta\zeta)}{\cosh(\beta E_{\mathbf{k}}) + \cosh(\beta\zeta)}. \tag{91}$$

Now note that we did indeed work at a fixed number of particles from the start, in introducing our units $k_F = (3\pi^2 n)^{1/3}$. Hence, these two number equations become

$$\frac{1}{3\pi^2} = \frac{1}{2\pi^2}\int\limits_0^\infty dk\, k^2 \left\{ 1 - \frac{\sinh(\beta E_{\mathbf{k}})}{\cosh(\beta E_{\mathbf{k}}) + \cosh(\beta\zeta)}\frac{\zeta_{\mathbf{k}}}{E_{\mathbf{k}}} \right\} \tag{92}$$

$$\frac{1}{3\pi^2}\frac{\delta n_{sp}}{n_{sp}} = \frac{1}{2\pi^2}\int\limits_0^\infty dk\, k^2 \frac{\sinh(\beta\zeta)}{\cosh(\beta E_{\mathbf{k}}) + \cosh(\beta\zeta)} \tag{93}$$

and we have to solve the gap equation in conjunction with these two number equations: all three have to be satisfied. The third one can be solved and merely fixes ζ as a function of δn_{sp}, but the first number equation together with the gap equation are coupled in the two remaining unknowns μ, Δ. Solutions for Δ are shown in figures (3) and (4). Figure (5) shows results for the chemical potential.

8. Intermezzo 2: Partition sum phase factor

Before we look at fluctuations beyond mean field, there is an important remark to be made: we can no longer delay looking at non-diagonal Gaussian integrations. Remember, when we do not restrict ourselves to the saddle-point value, we need to evaluate (see Eq. (45))

$$\int \mathcal{D}\bar{\psi}_{\mathbf{k},n,\sigma}\mathcal{D}\psi_{\mathbf{k},n,\sigma}\exp\left\{ -\sum_{\mathbf{k},n}\sum_{\mathbf{k}',n'}\bar{\eta}_{\mathbf{k}',n'}\cdot\left\langle\mathbf{k}',n'\left|-\mathbf{G}^{-1}\right|\mathbf{k},n\right\rangle\cdot\eta_{\mathbf{k},n} \right\}$$

$$= \left[\prod_{\mathbf{k},n}(-1)\right]\int\mathcal{D}\bar{\eta}_{\mathbf{k},n}\mathcal{D}\eta_{\mathbf{k},n}\exp\left\{ -\sum_{\mathbf{k},n}\sum_{\mathbf{k}',n'}\bar{\eta}_{\mathbf{k}',n'}\cdot\left\langle\mathbf{k}',n'\left|-\mathbf{G}^{-1}\right|\mathbf{k},n\right\rangle\cdot\eta_{\mathbf{k},n} \right\}. \tag{94}$$

Remember that the factors (-1) are coming from the change in integration measure. To keep track of it, we will give it its own symbol

$$\mathcal{X} = \prod_{\mathbf{k},n}(-1). \tag{95}$$

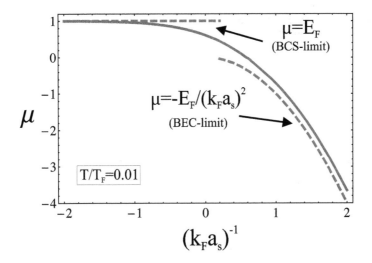

Figure 5. The chemical potential μ is shown as a function of the interaction parameter $1/(k_F a_s)$, for a low temperature and a balanced gas ($\zeta = 0$). In the regime of negative a_s the chemical potential tends to the Fermi energy, as in the BCS theory. For positive a_s, the chemical potential tends to the binding energy of the strongly bound 'Cooper molecules' that Bose-Einstein condense to form a superfluid.

The matrix $\langle \mathbf{k}', n' | -\mathbf{G}^{-1} | \mathbf{k}, n \rangle$ has rows and columns indexed by $\mathbf{k}, n$ and $\mathbf{k}', n'$. Each element in the matrix is itself a 2×2 matrix, and we could think of $\langle \mathbf{k}', n' | -\mathbf{G}^{-1} | \mathbf{k}, n \rangle = \mathbb{A}_{\mathbf{k}', n'; \mathbf{k}, n}$ as an $N \times N$ 'metamatrix' of matrices.

On the other hand, we could consider $-\mathbf{G}^{-1}$ as a $2N \times 2N$ matrix of scalars, indexed by $\mathbf{k}, n, j$ where $j = 1, 2$ indexes the spinor components. That way we need to consider the Nambu integral

$$\int \mathcal{D}\bar{\eta}_{\mathbf{k},n} \mathcal{D}\eta_{\mathbf{k},n} \exp\left\{ -\sum_{\mathbf{k},n,j} \sum_{\mathbf{k}',n',j'} \bar{\eta}_{\mathbf{k}',n',j'} \left(-\mathbf{G}^{-1}\right)_{\mathbf{k}',n',j';\mathbf{k},n,j} \eta_{\mathbf{k},n,j} \right\}, \tag{96}$$

but this is just a standard Gaussian Grassmann integration. We could solve it by diagonalizing the $2N \times 2N$ matrix $-\mathbf{G}^{-1}$ with some $2N \times 2N$ unitary transformation matrix to prove

$$\int \mathcal{D}\bar{\eta}_{\mathbf{k},n} \mathcal{D}\eta_{\mathbf{k},n} \exp\left\{ -\sum_{\mathbf{k},n,j} \sum_{\mathbf{k}',n',j'} \bar{\eta}_{\mathbf{k}',n',j'} \left(-\mathbf{G}^{-1}\right)_{\mathbf{k}',n',j';\mathbf{k},n,j} \eta_{\mathbf{k},n,j} \right\} = \det\left(-\mathbf{G}^{-1}\right), \tag{97}$$

where the determinant is now of the $2N \times 2N$ matrix. More matrix trickery is coming. We start innocently, by claiming

$$\det\left(-\mathbf{G}^{-1}\right) = \exp\left\{ \ln\left[\det\left(-\mathbf{G}^{-1}\right)\right]\right\} \tag{98}$$

Now, remember that

- The determinant of any matrix is equal to the product of its eigenvalues.

- The logarithm of a $2N \times 2N$ matrix $\mathbb{B}$ is another $2N \times 2N$ matrix $\mathbb{C}$ such that $\mathbb{B} = \exp(\mathbb{C}) = \mathbb{I} + \sum_{n=1}^{\infty} (\mathbb{C})^n / n!$.
- The logarithm of a diagonal matrix is a diagonal matrix with the logarithm of the diagonal elements.

From these statements it can be shown that the logarithm of the determinant of a matrix equals the trace of the logarithm of the matrix. In other words

$$\ln\left[\det\left(-\mathbf{G}^{-1}\right)\right] = \mathrm{Tr}\left[\ln\left(-\mathbf{G}^{-1}\right)\right], \tag{99}$$

where the trace is taken over all $\mathbf{k}, n, j$ values. Combining (97) with (98) and (99), we get

$$\int \mathcal{D}\bar{\psi}_{\mathbf{k},n,\sigma} \mathcal{D}\psi_{\mathbf{k},n,\sigma} \exp\left\{-\sum_{\mathbf{k},n\mathbf{k}',n'} \bar{\eta}_{\mathbf{k}',n'} \cdot \left\langle \mathbf{k}',n' \left| -\mathbf{G}^{-1} \right| \mathbf{k},n \right\rangle \cdot \eta_{\mathbf{k},n}\right\} = \mathcal{X}\exp\left\{\mathrm{Tr}\left[\ln\left(-\mathbf{G}^{-1}\right)\right]\right\}. \tag{100}$$

Expression (100) allows to write compactly the result of doing the Grassmann integrations in (45) and obtain the following exact result

$$\mathcal{Z} = \mathcal{X}\int \mathcal{D}\bar{\Delta}_{\mathbf{q},m}\mathcal{D}\Delta_{\mathbf{q},m} \exp\left\{\sum_{\mathbf{q},m} \frac{\bar{\Delta}_{\mathbf{q},m}\Delta_{\mathbf{q},m}}{g} + \mathrm{Tr}\left[\ln\left(-\mathbf{G}^{-1}\right)\right]\right\}. \tag{101}$$

Replacing $-\mathbf{G}^{-1}$ by $-\mathbf{G}_{sp}^{-1}$ gives back the saddle-point result,

$$\mathcal{Z}_{sp} = \mathcal{X}\exp\left\{\frac{\beta V |\Delta|^2}{g} + \mathrm{Tr}\left[\ln\left(-\mathbf{G}_{sp}^{-1}\right)\right]\right\}. \tag{102}$$

9. Step 4: Adding fluctuations beyond mean field

If we want to improve on the saddle-point solution, we set

$$\Delta_{\mathbf{q},m} = \sqrt{\beta V}\delta(\mathbf{q})\delta_{m,0}\Delta + \phi_{\mathbf{q},m} \tag{103}$$

$$\bar{\Delta}_{\mathbf{q},m} = \sqrt{\beta V}\delta(\mathbf{q})\delta_{m,0}\Delta^* + \bar{\phi}_{\mathbf{q},m}. \tag{104}$$

This is like the Bogoliubov shift in the second quantized theory of helium or Bose gases. We have a condensate contribution $\propto \Delta$ and add to it small fluctuations $\phi_{\mathbf{q},m}$. Then we will expand up to second order in the fluctuations $\phi_{\mathbf{q},m}, \bar{\phi}_{\mathbf{q},m}$ and get a quadratic (bosonic) path integral that we can perform exactly. This line summarized the long tough program ahead in a single sentence.

We can choose our fluctuations around the saddle point differently. Rather than using two conjugate fields, we can choose to vary amplitude and phase, and employ fields $|\phi|_{\mathbf{q},m} e^{i\theta_{\mathbf{q},m}}$ so that

$$\Delta_{\mathbf{q},m} = \sqrt{\beta V}\delta(\mathbf{q})\delta_{m,0}\Delta + |\phi|_{\mathbf{q},m} e^{i\theta_{\mathbf{q},m}} \tag{105}$$

$$\bar{\Delta}_{\mathbf{q},m} = \sqrt{\beta V}\delta(\mathbf{q})\delta_{m,0}\Delta^* + |\phi|_{\mathbf{q},m} e^{-i\theta_{\mathbf{q},m}}. \tag{106}$$

This leads to a hydrodynamic description as a function of the density of fluctuations and the phase field. Basically this is a change in the integration variables representing the Gaussian fluctuation contribution to the free energy, and we expect the same results. We will work here with (103) and (104). Plugging this into the expression (46) for the inverse Green's function we can write

$$-\mathbf{G}^{-1} = -\mathbf{G}_{sp}^{-1} + \mathbf{F}, \tag{107}$$

with like before, a diagonal piece for the saddle-point contribution,

$$\left\langle \mathbf{k}',n' \left| -\mathbf{G}_{sp}^{-1} \right| \mathbf{k},n \right\rangle = \left\langle \mathbf{k},n | \mathbf{k}',n' \right\rangle \begin{pmatrix} -i\omega_n + k^2 - \mu_\uparrow & -\Delta \\ -\Delta^* & -i\omega_n - k^2 + \mu_\downarrow \end{pmatrix} \tag{108}$$

and now the additional piece

$$\left\langle \mathbf{k}',n' \left| \mathbf{F} \right| \mathbf{k},n \right\rangle = \frac{1}{\sqrt{\beta V}} \begin{pmatrix} 0 & -\phi_{\mathbf{k}'-\mathbf{k},n'-n} \\ -\bar{\phi}_{\mathbf{k}-\mathbf{k}',n-n'} & 0 \end{pmatrix}. \tag{109}$$

Here again the change in sign in the index of the ϕ terms $\bar{\phi}_{\mathbf{k}+\mathbf{k}',n+n'} \to \bar{\phi}_{\mathbf{k}-\mathbf{k}',n-n'}$ is due to a re-indexation of the $\mathbf{k}$ and n indices, similar to expression (50). Now we are ready to expand $\ln\left(-\mathbf{G}^{-1}\right)$ in the exact expression (101) for the partition function, using matrix algebra

$$\ln\left(-\mathbf{G}^{-1}\right) = \ln\left(-\mathbf{G}_{sp}^{-1} + \mathbf{F}\right) = \ln\left[-\mathbf{G}_{sp}^{-1} + \mathbf{G}_{sp}^{-1}\mathbf{G}_{sp}\mathbf{F}\right]$$

$$= \ln\left[-\mathbf{G}_{sp}^{-1}\left(\mathbb{I} - \mathbf{G}_{sp}\mathbf{F}\right)\right] = \ln\left(-\mathbf{G}_{sp}^{-1}\right) + \ln\left(\mathbb{I} - \mathbf{G}_{sp}\mathbf{F}\right). \tag{110}$$

So, the result for the (quadratic approximation to the) partition sum using the "Bogoliubov shifted fields" becomes

$$\mathcal{Z}_q = \mathcal{X} \int \mathcal{D}\bar{\phi}_{\mathbf{q},m} \mathcal{D}\phi_{\mathbf{q},m} \exp\left\{ \sum_{\mathbf{q},m} \frac{1}{g} \left(\sqrt{\beta V}\Delta^*\delta(\mathbf{q})\delta_{m,0} + \bar{\phi}_{\mathbf{q},m} \right) \left(\sqrt{\beta V}\Delta\delta(\mathbf{q})\delta_{m,0} + \phi_{\mathbf{q},m} \right) \right.$$

$$\left. + \mathrm{Tr}\left[\ln\left(-\mathbf{G}_{sp}^{-1}\right)\right] + \mathrm{Tr}\left[\ln\left(\mathbb{I} - \mathbf{G}_{sp}\mathbf{F}\right)\right] \right\}. \tag{111}$$

We can bring all the stuff that does not depend on the fluctuations in front of the integrals

$$\mathcal{Z}_q = \mathcal{X} \exp\left\{ \frac{\beta V |\Delta|^2}{g} + \mathrm{Tr}\left[\ln\left(-\mathbf{G}_{sp}^{-1}\right)\right] \right\}$$

$$\times \int \mathcal{D}\bar{\phi}_{\mathbf{q},m} \mathcal{D}\phi_{\mathbf{q},m} \exp\left\{ \frac{\sqrt{\beta V}}{g} \left(\Delta^*\phi_{0,0} + \Delta\bar{\phi}_{0,0}\right) + \sum_{\mathbf{q},m} \frac{\bar{\phi}_{\mathbf{q},m}\phi_{\mathbf{q},m}}{g} + \mathrm{Tr}\left[\ln\left(\mathbb{I} - \mathbf{G}_{sp}\mathbf{F}\right)\right] \right\}. \tag{112}$$

The factor on the first line is nothing else but $\mathcal{Z}_{sp}$, expression (102). This absorbs the nasty factor $\mathcal{X}$ and results in $\exp\left\{-\beta V \Omega_{sp}\right\}$.

Only at this point we make an approximation and claim that $\mathbf{F}$ is small enough to state

$$\ln\left(\mathbb{I} - \mathbf{G}_{sp}\mathbf{F}\right) \approx -\mathbf{G}_{sp}\mathbf{F} - \frac{1}{2}\mathbf{G}_{sp}\mathbf{F}\mathbf{G}_{sp}\mathbf{F} \tag{113}$$

We neglect the terms from $\left(G_{sp}\mathbb{F}\right)^3$ and higher in orders of $\mathbb{F}$. We get

$$Z_q = Z_{sp} \int \mathcal{D}\bar{\phi}_{\mathbf{q},m}\mathcal{D}\phi_{\mathbf{q},m} \exp\left\{ \frac{\sqrt{\beta V}}{g}\left(\Delta^*\phi_{0,0} + \Delta\bar{\phi}_{0,0}\right) - \mathrm{Tr}\left[G_{sp}\mathbb{F}\right] \right.$$

$$\left. + \sum_{\mathbf{q},m} \frac{\bar{\phi}_{\mathbf{q},m}\phi_{\mathbf{q},m}}{g} - \frac{1}{2}\mathrm{Tr}\left[G_{sp}\mathbb{F}G_{sp}\mathbb{F}\right] \right\}. \tag{114}$$

What about the terms linear in the fluctuation fields ? They have to vanish. If we correctly determined the saddle point, then that means that the derivative (i.e. small derivations) vanishes. This then results in:

$$Z_q = Z_{sp} \int \mathcal{D}\bar{\phi}_{\mathbf{q},m}\mathcal{D}\phi_{\mathbf{q},m} \exp\left\{ \sum_{\mathbf{q},m} \frac{\bar{\phi}_{\mathbf{q},m}\phi_{\mathbf{q},m}}{g} - \frac{1}{2}\mathrm{Tr}\left[G_{sp}\mathbb{F}G_{sp}\mathbb{F}\right] \right\} \tag{115}$$

and so the quantity that we need to calculate is $\mathrm{Tr}\left[G_{sp}\mathbb{F}G_{sp}\mathbb{F}\right]$. The remaining path integral will be denoted by Z_{fl}, the partition sum of fluctuations, so that $Z_q = Z_{sp}Z_{fl}$.

10. Intermezzo 3: More matrix misery

The trace is taken over all $\mathbf{k}, n$ values and over the Nambu components. We get

$$\mathrm{Tr}\left[G_{sp}\mathbb{F}G_{sp}\mathbb{F}\right] = \sum_{\mathbf{k},n} \mathrm{tr}_\sigma \left\langle \mathbf{k},n \left| G_{sp}\mathbb{F}G_{sp}\mathbb{F}\right| \mathbf{k},n \right\rangle, \tag{116}$$

with tr_σ the trace of the 2×2 spinor matrix. The matrix $\left\langle \mathbf{k}',n' \left| G_{sp}^{-1}\right| \mathbf{k},n \right\rangle$ is not so hard to invert, since it is diagonal in $\mathbf{k}, n$ indices. We start from

$$\left\langle \mathbf{k}',n' \left| G_{sp}^{-1}\right| \mathbf{k},n \right\rangle = \left\langle \mathbf{k}',n'|\mathbf{k},n\right\rangle \begin{pmatrix} i\omega_n - k^2 + \mu_\uparrow & \Delta \\ \Delta^* & i\omega_n + k^2 - \mu_\downarrow \end{pmatrix}. \tag{117}$$

The diagonal elements in $\mathbf{k}, n$ need to be inverted. It is not hard to invert a two by two matrix:

$$\left\langle \mathbf{k}',n' \left| G_{sp}\right| \mathbf{k},n \right\rangle = \left\langle \mathbf{k}',n'|\mathbf{k},n\right\rangle \begin{pmatrix} i\omega_n - k^2 + \mu_\uparrow & \Delta \\ \Delta^* & i\omega_n + k^2 - \mu_\downarrow \end{pmatrix}^{-1}$$

$$= \frac{\left\langle \mathbf{k}',n'|\mathbf{k},n\right\rangle}{(i\omega_n + E_{\mathbf{k}} + \zeta)(i\omega_n - E_{\mathbf{k}} + \zeta)} \begin{pmatrix} i\omega_n + k^2 - \mu_\downarrow & -\Delta \\ -\Delta^* & i\omega_n - k^2 + \mu_\uparrow \end{pmatrix}. \tag{118}$$

We already had

$$\left\langle \mathbf{k}',n' \left| \mathbb{F}\right| \mathbf{k},n \right\rangle = \frac{1}{\sqrt{\beta V}} \begin{pmatrix} 0 & -\phi_{\mathbf{k}'-\mathbf{k},n'-n} \\ -\bar{\phi}_{\mathbf{k}-\mathbf{k}',n-n'} & 0 \end{pmatrix}. \tag{119}$$

Then,

$$\left\langle \mathbf{k},n \left| G_{sp}\mathbb{F}\right| \mathbf{k}',n' \right\rangle = \frac{1}{\sqrt{\beta V}} \frac{-1}{(i\omega_n + E_{\mathbf{k}} + \zeta)(i\omega_n - E_{\mathbf{k}} + \zeta)}$$

$$\times \begin{pmatrix} i\omega_n + k^2 - \mu_\downarrow & -\Delta \\ -\Delta^* & i\omega_n - k^2 + \mu_\uparrow \end{pmatrix} \cdot \begin{pmatrix} 0 & \phi_{\mathbf{k}-\mathbf{k}',n-n'} \\ \bar{\phi}_{\mathbf{k}'-\mathbf{k},n'-n} & 0 \end{pmatrix}. \tag{120}$$

This equals

$$
\langle \mathbf{k}, n \left| \mathbb{G}_{sp}\mathbb{F} \right| \mathbf{k}', n' \rangle = \frac{1}{\sqrt{\beta V}} \frac{-1}{(i\omega_n + E_\mathbf{k} + \zeta)(i\omega_n - E_\mathbf{k} + \zeta)}
$$
$$
\times \begin{pmatrix} -\Delta \bar{\phi}_{\mathbf{k}'-\mathbf{k}, n'-n} & (i\omega_n + k^2 - \mu_\downarrow)\, \phi_{\mathbf{k}-\mathbf{k}', n-n'} \\ (i\omega_n - k^2 + \mu_\uparrow)\, \bar{\phi}_{\mathbf{k}'-\mathbf{k}, n'-n} & -\Delta^* \phi_{\mathbf{k}-\mathbf{k}', n-n'} \end{pmatrix}. \tag{121}
$$

Now we can use this to calculate

$$
\langle \mathbf{k}, n \left| \mathbb{G}_{sp}\mathbb{F}\mathbb{G}_{sp}\mathbb{F} \right| \mathbf{k}, n \rangle = \sum_{\mathbf{k}', n'} \langle \mathbf{k}, n \left| \mathbb{G}_{sp}\mathbb{F} \right| \mathbf{k}', n' \rangle \langle \mathbf{k}', n' \left| \mathbb{G}_{sp}\mathbb{F} \right| \mathbf{k}, n \rangle. \tag{122}
$$

This is

$$
\langle \mathbf{k}, n \left| \mathbb{G}_{sp}\mathbb{F}\mathbb{G}_{sp}\mathbb{F} \right| \mathbf{k}, n \rangle =
$$
$$
= \frac{1}{\beta V} \sum_{\mathbf{k}', n'} \frac{1}{(i\omega_n + \zeta + E_\mathbf{k})(i\omega_n + \zeta - E_\mathbf{k})} \frac{1}{(i\omega_{n'} + \zeta + E_{\mathbf{k}'})(i\omega_{n'} + \zeta - E_{\mathbf{k}'})}
$$
$$
\times \begin{pmatrix} -\Delta \bar{\phi}_{\mathbf{k}'-\mathbf{k}, n'-n} & (i\omega_n + k^2 - \mu_\downarrow)\, \phi_{\mathbf{k}-\mathbf{k}', n-n'} \\ (i\omega_n - k^2 + \mu_\uparrow)\, \bar{\phi}_{\mathbf{k}'-\mathbf{k}, n'-n} & -\Delta^* \phi_{\mathbf{k}-\mathbf{k}', n-n'} \end{pmatrix}
$$
$$
\cdot \begin{pmatrix} -\Delta \bar{\phi}_{\mathbf{k}-\mathbf{k}', n-n'} & (i\omega_{n'} + k'^2 - \mu_\downarrow)\, \phi_{\mathbf{k}'-\mathbf{k}, n'-n} \\ (i\omega_{n'} - k'^2 + \mu_\uparrow)\, \bar{\phi}_{\mathbf{k}-\mathbf{k}', n-n'} & -\Delta^* \phi_{\mathbf{k}'-\mathbf{k}, n'-n} \end{pmatrix}. \tag{123}
$$

Since we need to take the trace over the resulting 2×2 product, we only need to calculate the upper left element of the matrix multiplication, namely

$$
\Delta^2 \bar{\phi}_{\mathbf{k}'-\mathbf{k}, n'-n} \bar{\phi}_{\mathbf{k}-\mathbf{k}', n-n'} + \left(i\omega_n + k^2 - \mu_\downarrow\right) \phi_{\mathbf{k}-\mathbf{k}', n-n'} \left(i\omega_{n'} - k'^2 + \mu_\uparrow\right) \bar{\phi}_{\mathbf{k}-\mathbf{k}', n-n'} \tag{124}
$$

and the lower right element

$$
\left(i\omega_n - k^2 + \mu_\uparrow\right) \bar{\phi}_{\mathbf{k}'-\mathbf{k}, n'-n} \left(i\omega_{n'} + k'^2 - \mu_\downarrow\right) \phi_{\mathbf{k}'-\mathbf{k}, n'-n} + (\Delta^*)^2 \phi_{\mathbf{k}-\mathbf{k}', n-n'} \phi_{\mathbf{k}'-\mathbf{k}, n'-n}. \tag{125}
$$

We use $(i\omega_n + k^2 - \mu_\downarrow) = (i\omega_n + \xi_\mathbf{k} + \zeta)$ with $\xi_\mathbf{k} = k^2 - \mu$, and the similar relations, to get

$$
\sum_{\mathbf{k}, n} \mathrm{tr}_\sigma \langle \mathbf{k}, n \left| \mathbb{G}_{sp}\mathbb{F}\mathbb{G}_{sp}\mathbb{F} \right| \mathbf{k}, n \rangle
$$
$$
= \frac{1}{\beta V} \sum_{\mathbf{k}, n} \sum_{\mathbf{k}', n'} \frac{1}{(i\omega_n + \zeta + E_\mathbf{k})(i\omega_n + \zeta - E_\mathbf{k})} \frac{1}{(i\omega_{n'} + \zeta + E_{\mathbf{k}'})(i\omega_{n'} + \zeta - E_{\mathbf{k}'})}
$$
$$
\times \Big[\Delta^2 \bar{\phi}_{\mathbf{k}'-\mathbf{k}, n'-n} \bar{\phi}_{\mathbf{k}-\mathbf{k}', n-n'} + (i\omega_n + \zeta + \xi_\mathbf{k})(i\omega_{n'} + \zeta - \xi_{\mathbf{k}'}) \phi_{\mathbf{k}-\mathbf{k}', n-n'} \bar{\phi}_{\mathbf{k}-\mathbf{k}', n-n'}
$$
$$
+ (i\omega_n + \zeta - \xi_\mathbf{k})(i\omega_{n'} + \zeta + \xi_{\mathbf{k}'}) \bar{\phi}_{\mathbf{k}'-\mathbf{k}, n'-n} \phi_{\mathbf{k}'-\mathbf{k}, n'-n} + (\Delta^*)^2 \phi_{\mathbf{k}-\mathbf{k}', n-n'} \phi_{\mathbf{k}'-\mathbf{k}, n'-n} \Big]. \tag{126}
$$

This looks confusing. We rename some summation indices

$$
\mathbf{q} = \mathbf{k} - \mathbf{k}' \to \mathbf{k} = \mathbf{q} + \mathbf{k}'
$$
$$
m = n - n' \to n = m + n' \tag{127}
$$

and introduce

$$iv_n = i\omega_n + \zeta \tag{128}$$

to clear up the mess

$$\sum_{\mathbf{k},n} \mathrm{tr}_\sigma \left\langle \mathbf{k}, n \left| G_{sp} \mathbb{F} G_{sp} \mathbb{F} \right| \mathbf{k}, n \right\rangle$$

$$= \frac{1}{\beta V} \sum_{\mathbf{q},m} \sum_{\mathbf{k}',n'} \frac{1}{\left(iv_{m+n'} + E_{\mathbf{q}+\mathbf{k}'} \right) \left(iv_{m+n'} - E_{\mathbf{q}+\mathbf{k}'} \right)} \frac{1}{\left(iv_{n'} + E_{\mathbf{k}'} \right) \left(iv_{n'} - E_{\mathbf{k}'} \right)}$$

$$\times \left[\Delta^2 \bar{\phi}_{-\mathbf{q},-m} \bar{\phi}_{\mathbf{q},m} + \left(iv_{m+n'} + \xi_{\mathbf{q}+\mathbf{k}'} \right) \left(iv_{n'} - \xi_{\mathbf{k}'} \right) \phi_{\mathbf{q},m} \bar{\phi}_{\mathbf{q},m} \right.$$

$$\left. + \left(iv_{m+n'} - \xi_{\mathbf{q}+\mathbf{k}'} \right) \left(iv_{n'} + \xi_{\mathbf{k}'} \right) \bar{\phi}_{-\mathbf{q},-m} \phi_{-\mathbf{q},-m} + (\Delta^*)^2 \phi_{\mathbf{q},m} \phi_{-\mathbf{q},-m} \right]. \tag{129}$$

Now we drop the primes and see a bilinear form in between the square brackets (the order does not matter for bosonic fields):

$$\sum_{\mathbf{k},n} \mathrm{tr}_\sigma \left\langle \mathbf{k}, n \left| G_{sp} \mathbb{F} G_{sp} \mathbb{F} \right| \mathbf{k}, n \right\rangle \tag{130}$$

$$= \frac{1}{\beta V} \sum_{\mathbf{q},m} \sum_{\mathbf{k},n} \frac{1}{\left(iv_{m+n} + E_{\mathbf{q}+\mathbf{k}} \right) \left(iv_{m+n} - E_{\mathbf{q}+\mathbf{k}} \right)} \frac{1}{\left(iv_n + E_{\mathbf{k}} \right) \left(iv_n - E_{\mathbf{k}} \right)}$$

$$\times \left(\bar{\phi}_{\mathbf{q},m} \ \phi_{-\mathbf{q},-m} \right) \cdot \begin{pmatrix} \left(iv_{m+n} + \xi_{\mathbf{q}+\mathbf{k}} \right) \left(iv_n - \xi_{\mathbf{k}} \right) & \Delta^2 \\ (\Delta^*)^2 & \left(iv_{m+n} - \xi_{\mathbf{q}+\mathbf{k}} \right) \left(iv_n + \xi_{\mathbf{k}} \right) \end{pmatrix} \cdot \begin{pmatrix} \phi_{\mathbf{q},m} \\ \bar{\phi}_{-\mathbf{q},-m} \end{pmatrix}.$$

Now we can calculate

$$\mathcal{Z}_{fl} = \int \mathcal{D}\bar{\phi}_{\mathbf{q},m} \mathcal{D}\phi_{\mathbf{q},m} \exp \left\{ \sum_{\mathbf{q},m} \frac{\bar{\phi}_{\mathbf{q},m} \phi_{\mathbf{q},m}}{g} - \frac{1}{2} \mathrm{Tr} \left[G_{sp} \mathbb{F} G_{sp} \mathbb{F} \right] \right\}. \tag{131}$$

In the first term the sum runs over positive and negative $\mathbf{q}$ but this can be symmetrized through

$$\sum_{\mathbf{q},m} \frac{\bar{\phi}_{\mathbf{q},m} \phi_{\mathbf{q},m}}{g} = \frac{1}{2} \sum_{\mathbf{q},m} \frac{1}{g} \left(\bar{\phi}_{-\mathbf{q},-m} \phi_{-\mathbf{q},-m} + \bar{\phi}_{\mathbf{q},m} \phi_{\mathbf{q},m} \right)$$

$$= \frac{1}{2} \sum_{\mathbf{q},m} \left(\bar{\phi}_{\mathbf{q},m} \ \phi_{-\mathbf{q},-m} \right) \cdot \frac{1}{g} \begin{pmatrix} 1 & 0 \\ 0 & 1 \end{pmatrix} \cdot \begin{pmatrix} \phi_{\mathbf{q},m} \\ \bar{\phi}_{-\mathbf{q},-m} \end{pmatrix}. \tag{132}$$

Using this result and (130) we can write the partition sum in the following "Gaussian integral" form

$$\mathcal{Z}_{fl} = \int \mathcal{D}\bar{\phi}_{\mathbf{q},m} \mathcal{D}\phi_{\mathbf{q},m} \exp \left\{ -\frac{1}{2} \sum_{\mathbf{q},m} \left(\bar{\phi}_{\mathbf{q},m} \ \phi_{-\mathbf{q},-m} \right) \cdot \begin{pmatrix} M_{11}(\mathbf{q}, i\omega_m) & M_{12}(\mathbf{q}, i\omega_m) \\ M_{21}(\mathbf{q}, i\omega_m) & M_{22}(\mathbf{q}, i\omega_m) \end{pmatrix} \cdot \begin{pmatrix} \phi_{\mathbf{q},m} \\ \bar{\phi}_{-\mathbf{q},-m} \end{pmatrix} \right\}, \tag{133}$$

where

$$M_{11}(\mathbf{q}, i\omega_m) = \frac{1}{\beta V} \sum_{\mathbf{k},n} \frac{\left(iv_{m+n} + \xi_{\mathbf{q}+\mathbf{k}} \right)}{\left(iv_{m+n} + E_{\mathbf{q}+\mathbf{k}} \right) \left(iv_{m+n} - E_{\mathbf{q}+\mathbf{k}} \right)} \frac{\left(iv_n - \xi_{\mathbf{k}} \right)}{\left(iv_n + E_{\mathbf{k}} \right) \left(iv_n - E_{\mathbf{k}} \right)} - \frac{1}{g} \tag{134}$$

and
$$M_{22}(\mathbf{q}, i\omega_m) = \frac{1}{\beta V} \sum_{\mathbf{k},n} \frac{\left(iv_{m+n} - \xi_{\mathbf{q}+\mathbf{k}}\right)}{\left(iv_{m+n} + E_{\mathbf{q}+\mathbf{k}}\right)\left(iv_{m+n} - E_{\mathbf{q}+\mathbf{k}}\right)} \frac{(iv_n + \xi_{\mathbf{k}})}{(iv_n + E_{\mathbf{k}})(iv_n - E_{\mathbf{k}})} - \frac{1}{g} \quad (135)$$

and
$$M_{12}(\mathbf{q}, i\omega_m) = \frac{1}{\beta V} \sum_{\mathbf{k},n} \frac{1}{\left(iv_{m+n} + E_{\mathbf{q}+\mathbf{k}}\right)\left(iv_{m+n} - E_{\mathbf{q}+\mathbf{k}}\right)} \frac{1}{(iv_n + E_{\mathbf{k}})(iv_n - E_{\mathbf{k}})} \Delta^2 \quad (136)$$

$$M_{21}(\mathbf{q}, i\omega_m) = \frac{1}{\beta V} \sum_{\mathbf{k},n} \frac{1}{\left(iv_{m+n} + E_{\mathbf{q}+\mathbf{k}}\right)\left(iv_{m+n} - E_{\mathbf{q}+\mathbf{k}}\right)} \frac{1}{(iv_n + E_{\mathbf{k}})(iv_n - E_{\mathbf{k}})} (\Delta^*)^2. \quad (137)$$

You will have noticed that we let the matrix elements depend on the bosonic Matsubara frequencies $i\omega_m = i(2\pi m/\beta)$. If we remember our short hand notations

$$iv_{m+n} = i\omega_m + i\omega_n + \zeta \quad (138)$$
$$iv_n = i\omega_n + \zeta \quad (139)$$

then it is clear where these appear. The matrix $\mathbb{M}$ is at the heart of our treatment of the fluctuations. It acts like a bosonic Green's function for the pair fields. It has some symmetry properties which can be derived from (134)–(137) by shifting the summation variables. We find

$$M_{11}(\mathbf{q}, i\omega_m) = M_{22}(-\mathbf{q}, -i\omega_m). \quad (140)$$

Moreover, since our saddle-point result shows us that $|\Delta|$ is fixed by the gap equation but we can choose the phase independently, we put Δ real and get

$$M_{12}(\mathbf{q}, i\omega_m) = M_{21}(\mathbf{q}, i\omega_m) = M_{21}(-\mathbf{q}, -i\omega_m) = M_{12}(-\mathbf{q}, -i\omega_m). \quad (141)$$

We still need to take sums over fermionic Matsubara frequencies, and integrals over $\mathbf{k}$ in the matrix elements, this will be a bundle of joy for those who love complex analysis and residue calculus.

More generally, we will have to study the poles of the components $M_{ij}(\mathbf{q}, z)$ in the complex plane to find excitations: poles in the Green's functions give us the quasiparticle spectrum.

11. Step 5: Integrating out the fluctuations

The bosonic path integral is easier than the fermionic one, since we do not have to worry about the signs. We do have to worry not to double-count the fields. Since we sum over all $\mathbf{q}$, and seeing the symmetry properties of $\mathbb{M}$, each term appears twice. We can restrict ourselves to half the $\mathbf{q}, m$ space and still get all the terms:

$$\frac{1}{2}\sum_{\mathbf{q},m} \left(\bar{\phi}_{\mathbf{q},m} \; \phi_{-\mathbf{q},-m}\right) \cdot \mathbb{M}(\mathbf{q}, i\omega_m) \cdot \begin{pmatrix} \phi_{\mathbf{q},m} \\ \bar{\phi}_{-\mathbf{q},-m} \end{pmatrix} = \sum_{\substack{\mathbf{q},m \\ q_z \geq 0}} \left(\bar{\phi}_{\mathbf{q},m} \; \phi_{-\mathbf{q},-m}\right) \cdot \mathbb{M}(\mathbf{q}, i\omega_m) \cdot \begin{pmatrix} \phi_{\mathbf{q},m} \\ \bar{\phi}_{-\mathbf{q},-m} \end{pmatrix}.$$
$$(142)$$

Similarly,
$$\int \mathcal{D}\bar{\phi}_{\mathbf{q},m}\mathcal{D}\phi_{\mathbf{q},m} = \prod_{\mathbf{q},m} \int d\bar{\phi}_{\mathbf{q},m}d\phi_{\mathbf{q},m} = \prod_{\substack{\mathbf{q},m \\ q_z \geq 0}} \int d\bar{\phi}_{\mathbf{q},m}d\phi_{\mathbf{q},m} \int d\bar{\phi}_{-\mathbf{q},-m}d\phi_{-\mathbf{q},-m}. \quad (143)$$

Therefore, when we integrate

$$\mathcal{Z}_{fl} = \int \mathcal{D}\bar{\phi}_{\mathbf{q},m} \mathcal{D}\phi_{\mathbf{q},m} \exp\left\{ -\frac{1}{2}\sum_{\mathbf{q},m} \left(\bar{\phi}_{\mathbf{q},m} \ \phi_{-\mathbf{q},-m} \right) \cdot \mathbb{M}(\mathbf{q}, i\omega_m) \cdot \begin{pmatrix} \phi_{\mathbf{q},m} \\ \bar{\phi}_{-\mathbf{q},-m} \end{pmatrix} \right\} \tag{144}$$

we get

$$\mathcal{Z}_{fl} = \prod_{\substack{\mathbf{q},m \\ q_z \geqslant 0}} \frac{\pi^4}{\|\mathbb{M}(\mathbf{q}, i\omega_m)\|}, \tag{145}$$

with

$$\|\mathbb{M}(\mathbf{q}, i\omega_m)\| = M_{11}(\mathbf{q}, i\omega_m) M_{22}(\mathbf{q}, i\omega_m) - M_{12}(\mathbf{q}, i\omega_m) M_{21}(\mathbf{q}, i\omega_m) = \Gamma(\mathbf{q}, i\omega_m). \tag{146}$$

This can be rewritten as

$$\mathcal{Z}_{fl} = \exp\left\{ \sum_{\mathbf{q},m,(q_z \geqslant 0)} \ln\left[\frac{\pi^4}{\Gamma(\mathbf{q}, i\omega_m)} \right] \right\} = \exp\left\{ \frac{1}{2}\sum_{\mathbf{q},m} \ln\left[\frac{\pi^4}{\Gamma(\mathbf{q}, i\omega_m)} \right] \right\}. \tag{147}$$

The π^4, unlike the minus sign we had before, gives an irrelevant factor for the bosons, shifting the overall free energy by a constant amount. We obtain

$$\mathcal{Z}_{fl} \propto \exp\left\{ -\frac{1}{2}\sum_{\mathbf{q},m} \ln\left[\Gamma(\mathbf{q}, i\omega_m)\right] \right\}. \tag{148}$$

The total thermodynamic potential per unit volume Ω can now be divided in a saddle-point contribution Ω_{sp} and a fluctuation contribution Ω_{fl}: the latter satisfies

$$\mathcal{Z}_{fl} = \exp\left\{ -\beta V \Omega_{fl} \right\} \tag{149}$$

and since $\mathcal{Z} = \mathcal{Z}_{sp}\mathcal{Z}_{fl}$ leads to $\Omega = \Omega_{sp} + \Omega_{fl}$ we find

$$\Omega_{fl} = \frac{1}{2\beta V} \sum_{\mathbf{q},m} \ln\left[\Gamma(\mathbf{q}, i\omega_m)\right]$$
$$\text{with } \Gamma(\mathbf{q}, i\omega_m) = M_{11}(\mathbf{q}, i\omega_m) M_{11}(-\mathbf{q}, -i\omega_m) - M_{12}^2(\mathbf{q}, i\omega_m) \tag{150}$$

where we used the symmetry properties of the fluctuation matrix (140) and (141).

The equations determining our system become the following. For the gap equation we still have the saddle-point condition:

$$\left. \frac{\partial \Omega_{sp}(T, \mu, \zeta; \Delta)}{\partial \Delta} \right|_{T, \mu, \zeta} = 0 \quad \longrightarrow \quad \Delta(T, \mu, \zeta), \tag{151}$$

but we have to insert this in the full thermodynamic potential (including fluctuations) $\Omega(T, \mu, \zeta; \Delta(T, \mu, \zeta)) = \Omega(T, \mu, \zeta)$ to apply thermodynamic relations. For example, the number equations are given by the thermodynamic relations

$$n = -\left. \frac{\partial \Omega(T, \mu, \zeta)}{\partial \mu} \right|_{T, \zeta} \tag{152}$$

$$\delta n = -\left. \frac{\partial \Omega(T, \mu, \zeta)}{\partial \zeta} \right|_{T, \mu}. \tag{153}$$

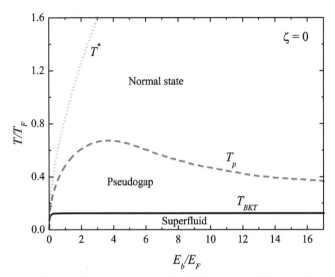

Figure 6. This figure illustrates the importance of including fluctuations. Nearly nowhere is this more evident than in the case of a two-dimensional Fermi gas (see [15]). Without taking into account fluctuations, the saddle point result predicts pairing in the 2D Fermi gas at T^*. Including fluctuations, this temperature is reduced to T_p. But the mere presence of pairing is still insufficient to guarantee superfluidity. Indeed, superfluidity is further suppressed by phase fluctuations and occurs only below T_{BKT}, the Berezinskii-Kosterlitz-Thouless temperature.

For the density $n = n_{sp} + n_{fl}$ we derive the free energy with respect to μ keeping only ζ and T constant, and for the density difference $\delta n = \delta n_{sp} + \delta n_{fl}$ we derive to ζ keeping only μ and T constant. Now remember our previous discussion: when we keep Δ as a variable we must use

$$n = - \left.\frac{\partial \Omega [T,\mu,\zeta;\Delta]}{\partial \mu}\right|_{T,\zeta,\Delta} - \left.\frac{\partial \Omega [T,\mu,\zeta;\Delta]}{\partial \Delta}\right|_{T,\zeta,\mu} \times \left.\frac{\partial \Delta(T,\mu,\zeta)}{\partial \mu}\right|_{T,\zeta}. \tag{154}$$

Only $\partial \Omega_{sp} [T,\mu,\zeta;\Delta] /\partial\Delta$ will vanish: the partial derivative of Ω_{fl} will remain. So we get

$$n = - \left.\frac{\partial \Omega [T,\mu,\zeta;\Delta]}{\partial \mu}\right|_{T,\zeta,\Delta} - \left.\frac{\partial \Omega_{fl} [T,\mu,\zeta;\Delta]}{\partial \Delta}\right|_{T,\zeta,\mu} \times \left.\frac{\partial \Delta(T,\mu,\zeta)}{\partial \mu}\right|_{T,\zeta}. \tag{155}$$

We can write this as a sum of saddle-point density and fluctuation density

$$n = n_{sp} + n_{fl}$$

$$\text{with} \begin{cases} n_{sp} = - \left.\dfrac{\partial \Omega_{sp} [T,\mu,\zeta;\Delta]}{\partial \mu}\right|_{T,\zeta,\Delta} \\[2mm] n_{fl} = - \left.\dfrac{\partial \Omega_{fl} [T,\mu,\zeta;\Delta]}{\partial \mu}\right|_{T,\zeta,\Delta} - \left.\dfrac{\partial \Omega_{fl} [T,\mu,\zeta;\Delta]}{\partial \Delta}\right|_{T,\zeta,\mu} \times \left.\dfrac{\partial \Delta(T,\mu,\zeta)}{\partial \mu}\right|_{T,\zeta} \end{cases}. \tag{156}$$

In these relations the derivatives are taken with respect to a thermodynamic variable while keeping the others constant. Note that Δ is an order parameter but not a thermodynamic variable for these relations.

12. Conclusions

At this point, we can summarize the full set of equations for the problem, including fluctuations. As a function of temperature $T\ (= 1/(k_B \beta))$, chemical potentials μ, ζ and gap Δ, the thermodynamic potential per unit volume is given by

$$\Omega(T, \mu, \zeta; \Delta) = \Omega_{sp}(T, \mu, \zeta; \Delta) + \Omega_{fl}(T, \mu, \zeta; \Delta) \tag{157}$$

where the saddle-point and quadratic fluctuation contributions are

$$\Omega_{sp}(T, \mu, \zeta; \Delta) = -\frac{1}{g}|\Delta|^2 - \frac{1}{\beta V} \sum_{\mathbf{k},n} \ln\left[-(i\nu_n + E_{\mathbf{k}})(i\nu_n - E_{\mathbf{k}})\right], \tag{158}$$

$$\Omega_{fl}(T, \mu, \zeta; \Delta) = \frac{1}{2\beta V} \sum_{\mathbf{q},m} \ln\left[M_{11}(\mathbf{q}, i\omega_m) M_{11}(-\mathbf{q}, -i\omega_m) - M_{12}^2(\mathbf{q}, i\omega_m)\right]. \tag{159}$$

Here the fluctuation matrix elements are

$$M_{11}(\mathbf{q}, i\omega_m) = \frac{1}{\beta V} \sum_{\mathbf{k},n} \frac{\left(i\nu_{m+n} + \xi_{\mathbf{q}+\mathbf{k}}\right)}{\left(i\nu_{m+n} + E_{\mathbf{q}+\mathbf{k}}\right)\left(i\nu_{m+n} - E_{\mathbf{q}+\mathbf{k}}\right)} \frac{(i\nu_n - \xi_{\mathbf{k}})}{(i\nu_n + E_{\mathbf{k}})(i\nu_n - E_{\mathbf{k}})} - \frac{1}{g}, \tag{160}$$

$$M_{12}(\mathbf{q}, i\omega_m) = \frac{1}{\beta V} \sum_{\mathbf{k},n} \frac{1}{\left(i\nu_{m+n} + E_{\mathbf{q}+\mathbf{k}}\right)\left(i\nu_{m+n} - E_{\mathbf{q}+\mathbf{k}}\right)} \frac{1}{(i\nu_n + E_{\mathbf{k}})(i\nu_n - E_{\mathbf{k}})} \Delta^2. \tag{161}$$

In these expressions

$$i\nu_n = i\omega_n + \zeta = i(2n+1)\pi/\beta + \zeta \tag{162}$$

are the shifted fermionic Matsubara frequencies and $\omega_m = 2m\pi/\beta$ are the bosonic Matsubara frequencies. Furthermore, $E_{\mathbf{k}} = \sqrt{\xi_{\mathbf{k}}^2 + |\Delta|^2}$ is the Bogoliubov dispersion with $\xi_{\mathbf{k}} = k^2 - \mu$. Note that the Matsubara summations in the matrix elements can be performed, simplifying the results so that only a $\mathbf{k}$-integral is left. The fermionic Matsubara summation in M_{11} results in

$$M_{11}(\mathbf{q}, i\omega_m) = -\frac{1}{g} + \frac{1}{V} \sum_{\mathbf{k}} \frac{\sinh(\beta E_{\mathbf{k}})}{\cosh(\beta E_{\mathbf{k}}) + \cosh(\beta \zeta)} \frac{1}{2E_{\mathbf{k}}}$$

$$\times \left(\frac{\left(i\omega_m + \xi_{\mathbf{q}+\mathbf{k}} - E_{\mathbf{k}}\right)(E_{\mathbf{k}} + \xi_{\mathbf{k}})}{\left(i\omega_m + E_{\mathbf{q}+\mathbf{k}} - E_{\mathbf{k}}\right)\left(i\omega_m - E_{\mathbf{q}+\mathbf{k}} - E_{\mathbf{k}}\right)} \right.$$

$$\left. - \frac{\left(i\omega_m + E_{\mathbf{k}} + \xi_{\mathbf{q}+\mathbf{k}}\right)(E_{\mathbf{k}} - \xi_{\mathbf{k}})}{\left(i\omega_m + E_{\mathbf{k}} + E_{\mathbf{q}+\mathbf{k}}\right)\left(i\omega_m + E_{\mathbf{k}} - E_{\mathbf{q}+\mathbf{k}}\right)} \right). \tag{163}$$

The fermionic Matsubara summation in M_{12} results in

$$M_{12}(\mathbf{q}, i\omega_m) = -\frac{\Delta^2}{V}\sum_{\mathbf{k}}\frac{\sinh(\beta E_{\mathbf{k}})}{\cosh(\beta E_{\mathbf{k}}) + \cosh(\beta\zeta)}\frac{1}{2E_{\mathbf{k}}}$$

$$\times\left(\frac{1}{\left(i\omega_m + E_{\mathbf{q}+\mathbf{k}} - E_{\mathbf{k}}\right)\left(i\omega_m - E_{\mathbf{q}+\mathbf{k}} - E_{\mathbf{k}}\right)}\right.$$

$$\left.+\frac{1}{\left(i\omega_m + E_{\mathbf{q}+\mathbf{k}} + E_{\mathbf{k}}\right)\left(i\omega_m - E_{\mathbf{q}+\mathbf{k}} + E_{\mathbf{k}}\right)}\right). \tag{164}$$

The above equations provide the necessary ingredients to calculate the free energy, which in turn gives access to the thermodynamic variables of the system. Analyzing competing minima or evolving minima in the free energy also allows to derive phase diagrams for the system. In particular, from the saddle-point free energy (158) we derive the gap equation (the saddle-point condition):

$$\left.\frac{\partial\Omega_{sp}(T,\mu,\zeta;\Delta)}{\partial\Delta}\right|_{T,\mu,\zeta} = 0 \tag{165}$$

from which we can extract $\Delta(T,\mu,\zeta)$, and the number equations

$$n = -\left.\frac{\partial\Omega\left[T,\mu,\zeta;\Delta\right]}{\partial\mu}\right|_{T,\zeta,\Delta} - \left.\frac{\partial\Omega_{fl}\left[T,\mu,\zeta;\Delta\right]}{\partial\Delta}\right|_{T,\zeta,\mu}\left.\frac{\partial\Delta(T,\mu,\zeta)}{\partial\mu}\right|_{T,\zeta} \tag{166}$$

$$\delta n = -\left.\frac{\partial\Omega\left[T,\mu,\zeta;\Delta\right]}{\partial\zeta}\right|_{T,\mu,\Delta} - \left.\frac{\partial\Omega_{fl}\left[T,\mu,\zeta;\Delta\right]}{\partial\Delta}\right|_{T,\zeta,\mu}\left.\frac{\partial\Delta(T,\mu,\zeta)}{\partial\zeta}\right|_{T,\mu}. \tag{167}$$

The gap and two number equations have to be fulfilled simultaneously, and need to be solved together to determine Δ, μ, ζ from $T, n, \delta n$.

The different approaches to fermionic superfluidity can be catalogued through the gap and number equations that they consider. The full set, shown here, corresponds to the so-called "Gaussian pair fluctuation" (GPF) approach advocated by Hu, Liu and Drummond [14]. If the last terms in the number equations are dropped, we obtain the famous Nozières and Schmitt-Rink results [16], ported to the path-integral formalism by Sá de Melo, Randeria and Engelbrecht [17]. Finally, if also the fluctuation part Ω_{fl} is disregarded, we simply have the mean-field or saddle-point results discussed earlier [18]. The results that we summarized here form the starting point of many on-going investigations: the effects of imbalance, the effects of reducing the dimensionality (to study the Berezinskii-Kosterlitz-Thouless transitions [15]) and of optical potentials, the search for the Fulde-Ferrell-Larkin-Ovchinnikov state [19] and other exotic pairing states,... We hope that the above derivation of the basic functional integral formalism will be of use to the reader, not only to gain a deeper insight in these applications of the formalism, but also to extend and modify it.

Acknowledgments

Extensive discussions with S.N. Klimin and E. Vermeyen are gratefully acknowledged, as well as discussions with H. Kleinert, C. Sá de Melo, E. Zaremba and L. Salasnich. Financial support was provided by the Fund for Scientific Research Flanders, through FWO projects G.0180.09.N, G.0119.12N, and G.0115.12N. One of us (J.P.A.D.) acknowledges a Ph. D. fellowship of the Research Foundation - Flanders (FWO).

Author details

Jacques Tempere and Jeroen P.A. Devreese
TQC, Universiteit Antwerpen, Universiteitsplein 1, B-2610 Antwerpen, Belgium

13. References

[1] R.P. Feynman, *Quantum Mechanics and Path Integrals* (McGraw-Hill, 1965).

[2] H. Kleinert, *Path Integrals in Quantum Mechanics, Statistics, Polymer Physics, and Financial Markets* (World Scientific, Singapore, 2009).

[3] L.S. Schulman, *Techniques and Applications of Path Integration* (Wiley–Interscience, New York, 1996).

[4] G. Roepstorff, *Path Integral Approach to Quantum Physics: An Introduction* (Springer, 1996).

[5] J. Zinn-Justin, *Path Integrals in Quantum Mechanics* (Oxford University Press, Oxford, UK, 2010).

[6] J.L. Martin, *The Feynman principle for a Fermi System*, Proc. Roy. Soc. A 251, 542 (1959).

[7] N. Nagaosa, *Quantum Field Theory in Condensed Matter Physics* (Springer, 1999).

[8] F.A. Berezin, *The Method of Second Quantization* (Academic Press, 1966).

[9] A. Zee, *Quantum Field Theory in a Nutshell* (Princeton University Press, 2003).

[10] M. Inguscio, W. Ketterle, and C. Salomon, *Ultra-cold Fermi gases* (IOS press, Amsterdam, the Netherlands, 2007).

[11] I. Bloch, J. Dalibard, and W. Zwerger, *Many-body physics with ultracold gases*, Rev. Mod. Phys. 80, 885 (2008).

[12] C.J. Pethick and H. Smith, *Bose-Einstein Condensation in Dilute Gases* (Cambridge University Press, Cambridge UK, 2008).

[13] H.T.C. Stoof, K.B. Gubbels, and D.B.M. Dickerscheid, *Ultracold Quantum Fields* (Springer, 2009).

[14] H. Hu, X.-J. Liu, and P.D. Drummond, Europhys. Lett. 74, 574 (2006); New J. Phys. 12, 063038 (2010).

[15] J. Tempere, S. Klimin, J.T. Devreese, Phys. Rev. A 79, 053637 (2009); S.N. Klimin, J. Tempere, Jeroen P.A. Devreese, J. Low Temp. Phys. 165, 261 (2011).

[16] P. Nozières and S. Schmitt-Rink, J. Low Temp. Phys. 59, 195 (1985).

[17] C.A.R. Sá de Melo, M. Randeria, and J. R. Engelbrecht, Phys. Rev. Lett. 71, 3202 (1993).

[18] A. J. Leggett, in *Modern Trends in the Theory of Condensed Matter* (eds A. Pekalski and R. Przystawa, Springer, 1980).

[19] Jeroen P.A. Devreese, M. Wouters, J. Tempere, Phys. Rev. A 84, 043623 (2011).

Permissions

The contributors of this book come from diverse backgrounds, making this book a truly international effort. This book will bring forth new frontiers with its revolutionizing research information and detailed analysis of the nascent developments around the world.

We would like to thank Alex Gabovich, for lending his expertise to make the book truly unique. He has played a crucial role in the development of this book. Without his invaluable contribution this book wouldn't have been possible. He has made vital efforts to compile up to date information on the varied aspects of this subject to make this book a valuable addition to the collection of many professionals and students.

This book was conceptualized with the vision of imparting up-to-date information and advanced data in this field. To ensure the same, a matchless editorial board was set up. Every individual on the board went through rigorous rounds of assessment to prove their worth. After which they invested a large part of their time researching and compiling the most relevant data for our readers. Conferences and sessions were held from time to time between the editorial board and the contributing authors to present the data in the most comprehensible form. The editorial team has worked tirelessly to provide valuable and valid information to help people across the globe.

Every chapter published in this book has been scrutinized by our experts. Their significance has been extensively debated. The topics covered herein carry significant findings which will fuel the growth of the discipline. They may even be implemented as practical applications or may be referred to as a beginning point for another development. Chapters in this book were first published by InTech; hereby published with permission under the Creative Commons Attribution License or equivalent.

The editorial board has been involved in producing this book since its inception. They have spent rigorous hours researching and exploring the diverse topics which have resulted in the successful publishing of this book. They have passed on their knowledge of decades through this book. To expedite this challenging task, the publisher supported the team at every step. A small team of assistant editors was also appointed to further simplify the editing procedure and attain best results for the readers.

Our editorial team has been hand-picked from every corner of the world. Their multi-ethnicity adds dynamic inputs to the discussions which result in innovative outcomes. These outcomes are then further discussed with the researchers and contributors who give their valuable feedback and opinion regarding the same. The feedback is then collaborated with the researches and they are edited in a comprehensive manner to aid the understanding of the subject.

Apart from the editorial board, the designing team has also invested a significant amount of their time in understanding the subject and creating the most relevant covers. They scrutinized every image to scout for the most suitable representation of the subject and create an appropriate cover for the book.

The publishing team has been involved in this book since its early stages. They were actively engaged in every process, be it collecting the data, connecting with the contributors or procuring relevant information. The team has been an ardent support to the editorial, designing and production team. Their endless efforts to recruit the best for this project, has resulted in the accomplishment of this book. They are a veteran in the field of academics and their pool of knowledge is as vast as their experience in printing. Their expertise and guidance has proved useful at every step. Their uncompromising quality standards have made this book an exceptional effort. Their encouragement from time to time has been an inspiration for everyone.

The publisher and the editorial board hope that this book will prove to be a valuable piece of knowledge for researchers, students, practitioners and scholars across the globe.

List of Contributors

I. Zakharchuk, P. Belova, K. B. Traito and E. Lähderanta
Lappeenranta University of Technology, Finland

Valerij A. Shklovskij
Institute of Theoretical Physics, NSC-KIPT, 61108 Kharkiv, Ukraine
Physical Department, Kharkiv National University, 61077 Kharkiv, Ukraine

Oleksandr V. Dobrovolskiy
Physikalisches Institut, Goethe-University, 60438 Frankfurt am Main, Germany

R. De Luca
Dipartimento di Fisica "E. R. Caianiello", Università degli Studi di Salerno, Italy

Gabovich Alexander M. and Voitenko Alexander I.
Institute of Physics, National Academy of Sciences of Ukraine, 46, Nauka Ave., Kyiv 03028, Ukraine

Mai Suan Li and Szymczak Henryk
Institute of Physics, Polish Academy of Sciences, 32/46, Al. Lotników, PL-02-668 Warsaw, Poland

Loder Florian
Experimental Physics VI & Theoretical Physics III, Center for Electronic Correlations and Magnetism,
Institute of Physics, University of Augsburg, 86135 Augsburg, Germany

Kampf Arno P.
Theoretical Physics III, Center for Electronic Correlations and Magnetism, Institute of Physics,
University of Augsburg, 86135 Augsburg, Germany

Kopp Thilo
Experimental Physics VI, Center for Electronic Correlations and Magnetism, Institute of Physics,
University of Augsburg, 86135 Augsburg, Germany

Masaru Kato
Department of Mathematical Sciences, Osaka Prefecture University1-1, Gakuencho, Sakai, Osaka,
Japan

Takekazu Ishida
Department of Physics and Electronics, Osaka Prefecture University, 1-1, Gakuencho, Sakai, Osaka,
Japan

Tomio Koyama
Institute of Material Sciences, Tohoku University, Sendai, Japan

Masahiko Machida
CCSE, Japan Atomic Energy Agency, Tokyo, Japan

C. A. C. Passos, M. S. Bolzan, M. T. D. Orlando, H. Belich Jr, J. L. Passamai Jr. and J. A. Ferreira
Physics Department, University Federal of Espirito Santo - Brazil

E. V. L. de Mello
Physics Department, University Federal Fluminense – Brazil

Dimo I. Uzunov
Collective Phenomena Laboratory, G. Nadjakov Institute of Solid State Physics, Bulgarian Academy of Sciences, BG-1784 Sofia, Bulgaria

Jacques Tempere and Jeroen P.A. Devreese
TQC, Universiteit Antwerpen, Universiteitsplein 1, B-2610 Antwerpen, Belgium

Printed in the USA
CPSIA information can be obtained
at www.ICGtesting.com
JSHW011437221024
72173JS00004B/835